A Handbook of Organic Analysis
Qualitative and Quantitative

FIFTH EDITION

H. T. Clarke, D.Sc. (Lond.), F.R.I.C.
Formerly Professor of Biochemistry in the College of
Physicians and Surgeons of Columbia University

Revised by

B. Haynes, B.Sc., Dip. Ed., F.R.I.C.
Principal, Ewell County Technical College
with
E. C. Brick, M.Sc., F.R.I.C. and **G. G. Shone,** M.Sc., F.R.I.C.

Edward Arnold

© H. T. Clarke and B. Haynes 1975

First published 1911
by Edward Arnold (Publishers) Limited
25 Hill Street, London W1X 8LL

Second Edition 1916, *Reprinted* 1919
Third Edition 1920, *Reprinted* 1921, 1922, 1923
Fourth Edition 1926, *Reprinted* 1927, 1931, 1934, 1937, 1939, 1940, 1942, 1943, 1946, 1947, 1949, 1952, 1956, 1957, 1960, 1963, 1967
Fifth Edition 1975

ISBN: 0 7131 2460 1

Printed in Northern Ireland at The Universities Press, Belfast

Preface to the Fifth Edition

The long period of success of Professor H. T. Clarke's *Handbook of Organic Analysis*, which was first published in 1911, is due in large measure to the initial uniqueness of the treatment of the subject matter and in particular the collection of data relating to specific compounds into his 'Tables of Organic Compounds'. Over the years, there have been many developments in the methods available for the analysis of organic substances, but despite these factors, the general plan laid down by Professor Clarke in the first edition still has merit sufficient to justify its retention. However, major changes have been necessary in order to introduce modern methods of quantitative analysis, and separate chapters on the preparation of derivatives and the use of some spectroscopic methods have been included. A considerable volume of checking and inclusion of new data has been carried out in assembling the new Tables of Organic Compounds, from which the rather specialised sections on dyestuffs and alkaloids have now been omitted.

The *Handbook of Organic Analysis* has for long been regarded as one of the standard student textbooks, and it is hoped that in its revised form it will continue to provide for the needs of the student of organic analysis, as well as for those requiring a compact book of easy reference.

Ewell
June 1974   B. HAYNES

Contents

1 Preliminary Investigation

Purity

In undertaking the analysis of an organic substance the first consideration is that of purity, for without the certainty of this, all experimental examination is likely to lead to unreliable results.

(a) Liquids

When the substance is a liquid, the initial procedure is to heat a small quantity in a test-tube, in order to ascertain whether it will boil without decomposition.* If this is found to be the case, the entire quantity is submitted to fractional distillation, the liquid being boiled in a suitable distilling flask provided with a long, water-jacketed side-arm (Fig. 1). The temperature of the evolved vapour is continually

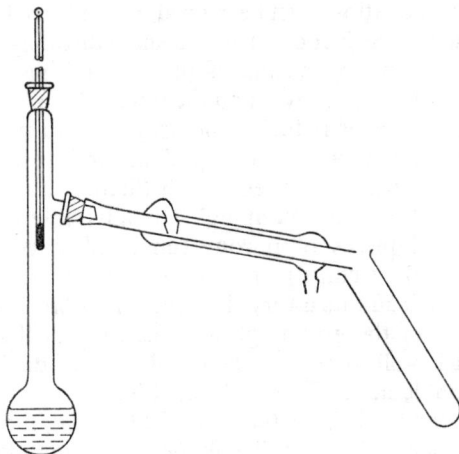

Fig. 1

observed, and the receiver is changed as soon as the temperature remains constant. A pure liquid should pass over within a range of not more than two degrees.† When no constant boiling point is observable (a sure indication of the presence of more than one compound)

---

* The boiling point of a pure liquid can be estimated with a fair degree of accuracy merely by holding a thermometer in the vapour of the liquid boiling in the test-tube. The reading is generally a few degrees too low.
† Liquid mixtures are known which distil unchanged at constant temperature ('constant boiling mixtures'); however, they are unlikely to be encountered in simple organic analysis.

the liquid must be systematically fractionated, preferably with the use of a distilling column, care being taken to carry out the distillation as slowly as possible at all times. By such treatment the liquid will tend to accumulate in two or more fractions, the boiling ranges of which include the boiling points of the pure components under atmospheric pressure.

For temperatures materially above 100°C a stem correction must be applied unless the entire thread of the thermometer is surrounded by the vapour. A formula for this purpose has been worked out, but as this involves a knowledge of the average temperature of the exposed portion of the thread (a value difficult to ascertain, depending upon the thermal conductivities of the mercury thread and the glass wall of the thermometer), it is advisable to calibrate the thermometer by direct comparison with short stem standard instruments.

When the liquid appears to be in any degree decomposed at its boiling point, fractionation must be carried out under reduced pressure. For this purpose the receiver consists of a small distilling flask attached to a water filter-pump by means of pressure tubing, a manometer being placed between the receiver and the pump. An oil-pump provides a more reliable device for reducing the pressure, but if used, must be protected from the vapour of the liquid under investigation by the insertion of a cold trap* between the distillation apparatus and the pump. In order to obviate 'bumping' it is advisable to place in the liquid a few small pieces of porous (unglazed) earthenware or of carborundum ('boiling chips') or else to lead into the liquid a fine capillary tube which admits a very slow current of air. If the apparatus being used is not of the ground-glass jointed type, all stoppers must be of rubber. It is well to replace the usual form of distilling flask by one having a double neck (Claisen flask), Fig. 2.

Siwoloboff's method. This method for determining the boiling point is very useful when only a small volume of the sample (<1 cm³) is available. About 0.5 cm³ of the liquid is introduced into a test-tube (100 mm × 5 mm) and a capillary melting point tube (*ca* 70 mm long) sealed at one end is placed inside the test-tube with its open end immersed in the liquid. The test-tube is attached to a thermometer stem by means of a rubber band and the thermometer put into a heating bath of medicinal mineral oil or concentrated sulphuric acid so that the stem is immersed to a depth of about 40 mm. The organic liquid sample must be adjacent to the bulb of the thermometer and the mouth of the test-tube well clear of the heating bath liquid surface.

The bath is heated slowly with continual stirring; at first air bubbles slowly escape from the open end of the capillary tube, but as the boiling

* A wide empty U-tube immersed in a solid carbon dioxide/acetone mixture contained in a Dewar vessel.

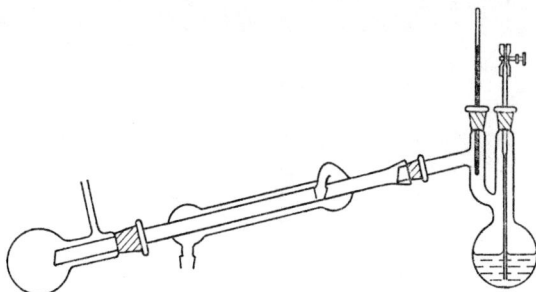

Fig. 2

point is reached a rapid and continuous escape of bubbles is observed. A more accurate result may be obtained by allowing the heating bath to cool and noting the temperature at which the last bubble appears at the end of the capillary tube and liquid begins to suck back into this tube.

(b) Solids

In the case of solids, a little of the dry original substance is finely powdered and submitted to melting point determination. A small quantity of the sample is forced into the open end of a thin-walled glass capillary tube 70 mm long and sealed at one end; it is then shaken down to the closed end by repeatedly tapping the capillary tube on a hard surface. The fine tube is then attached (either by moistening it with the liquid of the bath or by the use of a small rubber band) to a calibrated thermometer so that the enclosed sample is as near as possible to the middle of the thermometer bulb. The thermometer is now suspended in a bath of concentrated sulphuric acid, or medicinal mineral oil, with the length of capillary tube immersed in the bath liquid approximately equal to the length protruding above the surface of the bath. A wide test-tube forms a suitable vessel for the bath, which should consist of the acid for temperatures below about 150°C and of the oil for higher temperatures. The bath (Fig. 3) is heated steadily by a small flame, with continual stirring, and the temperature at which the sample melts is noted. A stem correction (see p. 2) must be applied for temperatures above 100°C. It is advisable to repeat the determination using a fresh sample, heating the bath rapidly to within some ten degrees of the melting point and thereafter in such a way that the temperature rises about two degrees per minute. In this way the range over which the substance melts, that is to say the temperature interval

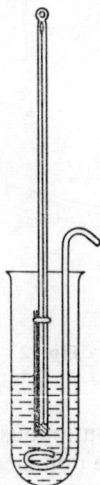

Fig. 3

through which softening begins and finally a clear melt is formed, can be accurately observed.

For solids of low melting point it is often more convenient to record the setting point: a sample of the substance, sufficient in quantity to cover the bulb of the thermometer, is melted in a test-tube and then slowly cooled, stirring continually with the thermometer. The temperature at which crystals first appear and that at which the substance becomes too solid to stir constitute the setting range. For a relatively pure compound this range extends over only one or two degrees, since the latent heat of solidification checks the fall in temperature; a wide setting range, like a wide melting range in the capillary tube method, indicates the presence of impurity.

It should be noted that although a sharp melting point is usually indicative of the high purity of a substance, there are exceptions, such as eutectic mixtures which may have sharp melting points; in these cases however the melting point would be changed by recrystallisation from a suitable solvent. In practice, eutectic mixtures are not commonly encountered.

In recent years electromechanical devices for determining melting points have been developed. The more complex of these instruments function rapidly and automatically, producing a recording of the accurate melting points of several different samples determined simultaneously. The automatic type of apparatus is fairly expensive but is very efficient in terms of time saved and allows melting points to be determined under reproducible conditions free from operator error.

If the melting point is not sharp (i.e. the range extends over appreciably more than one degree), the original sample must be recrystallised from some suitable solvent. Solvents may be tried in the following order: alcohol, water, ligroin, acetone, benzene, acetic acid, chloroform, ether. The recrystallisation is effected by maintaining a certain quantity of the solvent at its boiling point in a conical flask—under reflux if the solvent is highly volatile—and gradually adding small quantities of the original substance until no more is taken up into solution. A small volume of the solvent is then added, the hot solution rapidly filtered through a fluted filter-paper on a funnel without a stem, and the filtrate allowed to cool, the vessel being occasionally scratched on the inside with a glass rod, to induce crystallisation. When cold, the solid which has separated is filtered off by suction, pressed upon a porous plate, dried in a vacuum desiccator, and the melting point determined. This process must be repeated until the melting point is sharp and shows no change on further recrystallisation; it may then be regarded as pure. If, however, the melting point after one recrystallisation is identical with that of the original substance, no further purification is necessary.

Should the substance appear to contain tarry or coloured impurities, it is advisable to add some decolorising carbon to the hot solution before filtering.

Mixed melting point. Usually, the melting point of a compound is lowered and its melting range increased when it is mixed with another substance. This fact can be utilised in the identification of an unknown compound. An intimate mixture is prepared from powdered samples of the unknown substance and the known compound with which it is suspected to be identical, and the melting point of the mixture determined. If the mixture melts at the same temperature as its components it is very likely that the two compounds are identical. There are exceptions to the mixed melting point rule, such as the (+) and (−) forms of optically active compounds, and a small number of pairs of compounds e.g. naphthalene picrate and thionaphthalene picrate, which show no depression of the melting point upon admixture, but it is of wide applicability and is frequently of value.

It is well to ascertain whether a solid can be distilled or sublimed. Not only is the boiling point of a substance solid at ordinary temperatures a valuable additional characteristic to be taken into consideration, but solids can often be obtained in a higher state of purity by distillation or sublimation than by crystallisation. The precaution should of course be taken of heating a small quantity, in order to make sure that it can be distilled without decomposition. When this is possible, a distilling flask with a wide side-tube should be employed, without a condenser, and any distillate solidifying at the outset melted by gently warming the tube. This distillation of solids may often advantageously be carried out under reduced pressure; in such a case a distilling flask with a wide (10–15 mm) side-arm should be employed, in order to avoid stoppage by solidified distillate.

General and physical characteristics

It is of great importance to investigate the *solubility* of the substance in various solvents, for information as to the chemical nature of the compound can frequently be obtained from consideration of this characteristic property.

As a general rule, compounds are dissolved by liquids containing similar groups of atoms; thus polar compounds are usually more readily soluble in polar solvents than are non-polar compounds such as hydrocarbons. The majority of organic compounds consist of polar and non-polar parts, and their solubilities in different types of solvent will be affected by the relative influences of the polar and non-polar parts of each molecule. For simple monofunctional compounds such as alcohols, acids, esters, aldehydes, ketones, amides, amines, water-solubility is low if the substance contains more than about four carbon atoms per molecule. Simple monofunctional compounds usually dissolve in ether unless a very polar group is present (e.g. sulphonic acids) or they are highly associated. Substances containing two or more polar groups can be expected to have a lowered solubility in ether but an increased solubility in water, e.g. ethylene glycol, succinic acid, sugars. In general, salts are more or less soluble in water but insoluble in ether, acids are often soluble in hot water and less so in cold water while being as a rule soluble in ether or alcohol, hydrocarbons are all insoluble in water but are soluble in ether, and so on.

The boiling and melting points should always be kept in mind, so as to exclude erroneous conclusions which might otherwise be deduced from considerations of other properties and reactions.

Whether the substance is soluble in water or not, its reaction towards *litmus paper* should be examined, since this at once affords a clue for the allotment of the compound into one of three major classes. It frequently happens that substances, such as certain acids, though but sparingly soluble in water are sufficiently so to have an appreciable effect upon litmus paper.

The *odour* of a substance may sometimes give an indication of the class to which it belongs, though this is not to be relied upon with too great assurance, as the sense of smell varies greatly with individuals; moreover, substances of different constitution may have similar odours.

Similarly the *colour* of the substance may afford an indication of the class to which the compound is likely or unlikely to belong. However, the removal of the last traces of coloured impurities from colourless substances is occasionally a matter of extreme difficulty, and unrecognised failure in this respect may lead to faulty conclusions.

It is often advantageous, in the case of a liquid, to know its density as definite generalisations can be drawn from this property, which also aids in the final identification of liquids. The details for procedure will be found on pages 262–264.

Much information is afforded in doubtful cases by a knowledge of the approximate molecular weight, based upon cryoscopic or vapour pressure methods. The methods are described on pages 261 and 259. Rapid quantitative determinations, such as the titration of known weights of organic acids with standard alkali, the volumetric estimation of ionisable halogen in salts, or the quantitative saponification of esters, are of value, and should be carried out whenever possible.

Many compounds occurring in nature, such as sugars, glycosides, hydroxy and amino acids, exist in optically active forms. When the presence of such a substance is suspected, it is well to determine the rotatory power according to the method given on page 266.

Examination for constituent elements

(a) Ignition

A small quantity of the substance should at first be heated upon a nickel spatula until completely ignited, in order to ascertain the presence of any non-volatile residue. Should a residue be found which appears to consist of a heavy metal or its oxide, it will be necessary to ignite another small portion in a porcelain crucible. Substances leaving a residue consisting of an alkali carbonate or an oxide of a metal of the calcium group may preferably be heated on a piece of platinum foil. All such residues must be subjected to a complete qualitative analysis. The nature of the flame formed, as well as the odour of the vapours evolved, should be observed and recorded.

(b) Lassaigne's Test

The presence of carbon and hydrogen in an organic compound is usually indicated by its behaviour on ignition and it is not usual to perform specific tests for these elements. To determine the presence of nitrogen, sulphur, halogens, phosphorus, it is necessary to fuse the organic substance with metallic sodium in *Lassaigne's Test*. About 0.04 g of sodium (*ca.* cube 4 mm side) is heated gently in a soft glass ignition tube (*ca.* 50 × 10 mm) until it melts. The substance (0.04 g), powdered if a solid, is added portionwise directly onto the molten sodium (CAUTION). In dealing with volatile liquids, it may be found preferable to suspend the tube vertically from its rim by passing it through a hole in a piece of asbestos board. The tube is then heated strongly until the entire end of the tube is red hot, and it is maintained in this condition for about two minutes, after which it is dropped into a large test-tube containing about 10 cm³ of distilled water and the mouth of the test-tube covered *immediately* with a clean wire gauze. Caution is necessary in this operation, as the ignition tube will shatter and the sodium will react with the water. The test-tube contents are heated to boiling, and filtered. The filtrate should be water-clear and alkaline. If it is dark coloured the whole fusion operation should be repeated.

Nitrogen. To a portion of the filtrate 0.2 g of powdered ferrous sulphate crystals is added, the mixture is heated gently until it boils and sufficient dilute sulphuric acid added to the hot mixture to dissolve the iron hydroxides which have been precipitated and to give the solution an acid reaction. A Prussian blue precipitate or colouration indicates the presence of nitrogen in the original substance. (The addition of a little dilute potassium fluoride solution may lead to a purer Prussian Blue.) The action of sodium will have converted the carbon and nitrogen of the substance into sodium cyanide which with ferrous sulphate produces sodium ferrocyanide.

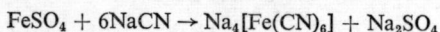

$$FeSO_4 + 6NaCN \rightarrow Na_4[Fe(CN)_6] + Na_2SO_4$$

Alkaline solutions containing ferrous ions are subject to ready aerial oxidation, especially when hot, so that under the conditions of this test, some ferric ions are inevitably produced. They react with the ferro-cyanide to give a precipitate of ferric ferrocyanide (Prussian blue).

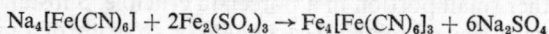

$$Na_4[Fe(CN)_6] + 2Fe_2(SO_4)_3 \rightarrow Fe_4[Fe(CN)_6]_3 + 6Na_2SO_4$$

Occasionally, the only result of this test is an indefinite colouration of greenish-blue character; in such cases the fusion test should be repeated using a mixture of the unknown substance and pure sucrose.

If sulphur is present in the original compound, a black precipitate of ferrous sulphide will be produced when the ferrous sulphate crystals dissolve. It is not usually necessary to remove this precipitate by filtering before proceeding with the test for nitrogen.

Sulphur. Organic compounds yield sodium sulphide by the action of sodium at high temperatures. Sulphide ions may be readily detected in a sample of the filtrate of the sodium fusion by treating the latter with a few drops of a freshly prepared dilute solution of sodium nitroprusside $Na_2[Fe(CN)_5NO]$. An intense pink to purple colouration indicates sulphur.

Halogens. (i) *Nitrogen and sulphur absent.* A portion of the filtrate should be rendered acid with nitric acid and aqueous silver nitrate added. A precipitate of silver halide indicates the presence of chlorine, bromine or iodine in the original substance. *If only one halogen is present*, it may be identified by decanting the mother liquor and treating the precipitate with dilute aqueous ammonia. If the precipitate is white and readily soluble in the ammonia solution, chlorine is present; if the precipitate is pale yellow and soluble only with difficulty, bromine is present; if it is yellow and insoluble, iodine is present. *If more than one halogen may be present* the following tests should be performed on portions of the sodium fusion filtrate. About 2 cm³ of the filtrate is treated with an excess of glacial acetic acid and 1 cm³ of carbon tetrachloride added followed by drops of 20% sodium nitrite solution, with continual shaking. A purple colour in the organic layer indicates iodine.

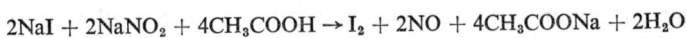

$$2NaI + 2NaNO_2 + 4CH_3COOH \rightarrow I_2 + 2NO + 4CH_3COONa + 2H_2O$$

If iodine has been found, further small portions of sodium nitrite solution should be added, the mixture gently warmed and extracted with small portions of carbon tetrachloride until all iodine has been removed as shown by a colourless organic layer. The solution should then be boiled until no more nitrous fumes are evolved, and cooled. This solution may now be used to test for the presence of bromine. (If iodine is absent, 1 cm³ of the fusion filtrate which has been acidified with glacial acetic acid may be used.) To it is added a small amount of lead dioxide. A piece of filter paper moistened with an ethanolic solution of fluorescein is placed across the mouth of the tube and the contents of the tube are heated to boiling. If bromide is present in the solution, the vapours of bromine cause the yellow fluorescein to turn pink owing to the formation of eosin.

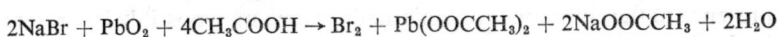

$$2NaBr + PbO_2 + 4CH_3COOH \rightarrow Br_2 + Pb(OOCCH_3)_2 + 2NaOOCCH_3 + 2H_2O$$

To identify chlorine in the presence of bromine and iodine, a portion of the fusion filtrate should be boiled with glacial acetic acid and a slight excess of lead dioxide until all iodine and bromine are liberated. The mixture is then diluted, filtered and the filtrate tested for chloride with nitric acid and silver nitrate.

(ii) *Nitrogen and/or sulphur present.* A portion of the filtrate is rendered acid with dilute nitric acid and the solution evaporated to half its original volume in order to expel hydrogen cyanide or hydrogen sulphide. The residual solution is then diluted with an equal volume of distilled water and used for chlorine, bromine, iodine tests as in (i) above.

(iii) *Fluorine.* A portion of the fusion filtrate is strongly acidified with glacial acetic acid and evaporated to about half volume. One drop of the cooled residual solution is placed on zirconium–alizarin red S paper. A yellow colour on the red paper indicates the presence of fluoride. The test paper may be prepared by dipping a strip of filter-paper into a mixture of 3 cm³ of 1% ethanolic alizarin solution and 2 cm³ of a 0.4% solution of zirconium chloride. The dried strip is treated with one drop of 50% acetic acid immediately before use.

Phosphorus. Phosphorus may be detected by boiling a portion of the sodium fusion filtrate with a slight excess of concentrated nitric acid for one minute, cooling and adding an equal volume of ammonium molybdate solution. A yellow precipitate of ammonium phosphomolybdate indicates the presence of phosphorus in the original substance. It is however preferable to fuse a mixture of 0.05 g of the original substance with 3 g of sodium peroxide and 2 g of anhydrous sodium carbonate in a nickel crucible, heating the mixture gently at first. After about ten minutes in the molten state, the mixture is allowed to cool and extracted with water. The extract is treated with concentrated nitric acid and ammonium molybdate solution and yields a yellow precipitate if phosphorus is present.

Approximate constitution

If the substance is soluble in water, ionised radicals must be at once tested for in the solution. When halogens, sulphur or phosphorus have been detected in the substance, the usual routine tests for inorganic acid radicals should be applied, in order to determine whether these elements exist wholly or in part as ionisable radicals. When nitrogen has been shown to be present, the solution must be examined for nitrate or nitrite ions. The metal in a water-soluble metallic salt will generally exist in the ionised condition. It is also well to test for the presence of the commoner organic acids, which may possibly be present in the form of salts of organic bases. The tests for such acids are given in detail in the appended tables.

In all cases where the presence of ionised acid radicals has been shown, a cold solution of caustic soda must be added, in order to liberate the free base. This may then be filtered off, distilled out, or extracted with ether, and examined independently. The presence of a salt of ammonia or of a volatile amine, such as methylamine, will be indicated by the evolution of ammoniacal odours in the cold. Similarly, from a substance ascertained to be a salt of an organic acid, the organic acid should be liberated by the action of dilute sulphuric acid, and prepared in a pure state for examination.

It should be borne in mind when testing for ionised halogen that acid halides, on treatment with water, yield a mixture of the corresponding acid and halogen hydracid, and in such a case it will of course be futile to attempt to isolate any base. In many instances, however, an acid halide will be detected by the action of a drop of water on a small quantity of the substance, fumes of the hydracid being thereby liberated.

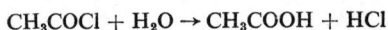

$$CH_3COCl + H_2O \rightarrow CH_3COOH + HCl$$

In addition, most *a*-chloro ethers evolve hydrochloric acid on treatment with water.

The nature of the vapours evolved on strongly heating the substance, as well as the character of the flame formed on ignition, constitute fairly definite indications of the general class to which it belongs. Thus the odour of phenols, aromatic nitro-compounds and aldehydes, and amines, as well as the decomposition products of carbohydrates and certain hydroxy acids, are all more or less characteristic; while the rule that unsaturated substances, and saturated compounds containing more than four or five carbon atoms in the molecule, burn with a smoky flame is generally applicable.

The action of hot and cold concentrated sulphuric acid should be tried. Although it is difficult to draw up a complete table of inferences, it should be observed whether the substance dissolves and whether change of colour, charring, or effervescence is produced. The nature of any gases evolved should also be examined.

The effect of heating in a tube with soda-lime and with zinc dust should also be observed, any well defined distillate or sublimate being isolated and examined.

Many phenols and enols give colours when treated with ferric chloride solution. The colours produced range from green to red or blue and are often transient. For compounds that fail to produce a colour with aqueous ferric chloride, an alternative procedure is as follows: to a dilute solution or suspension of the compound in chloroform is added a solution of anhydrous ferric chloride in chloroform and a drop of pyridine. It should be noted that a few other classes of compounds, e.g. oximes, also respond positively to the test.

Unsaturated linkages may be detected by the action of bromine water and of neutral permanganate solution. Unsaturated compounds decolorise these reagents, while in general saturated compounds do not. These reactions, however, do not afford a very certain test, as most polyhydroxylic compounds, many aldehydes, certain acids and esters, most phenols and ketones, and several other types of compound are thus attacked.

Unsaturated compounds in which an ethenoid linkage is conjugated with an aromatic nucleus may absorb bromine but slowly. Hence this test must not be considered to have failed until, after gentle warming, the mixture has stood for at least five minutes without appreciable diminution in the intensity of the colour of the bromine. Thus cinnamic acid scarcely decolorises bromine water until the solution is warmed. Phenolic and certain other classes of compounds yield in general precipitates of bromo-derivatives when treated with bromine water. Such precipitates should be collected and purified for examination.

A solution of bromine in carbon tetrachloride, chloroform, or glacial acetic acid is similarly decolorised by substances containing unsaturated linkages. In all cases such bromine addition products should be isolated for examination. Hydrobromic acid is evolved on warming most aromatic compounds with bromine in carbon tetrachloride, owing to the facility with which the majority of substituted benzene derivatives are brominated. Bromine, in non-aqueous as in aqueous solution, is absorbed by many phenolic, ketonic and by certain other types of compound. In carbon tetrachloride or chloroform, however, evolution of hydrobromic acid will take place on warming, owing to its slight solubility in these non-hydroxylic solvents. Amines also absorb bromine with formation of addition or substitution products, liberation of hydrobromic acid being, however, not necessarily a concomitant.

Unsaturated substances in general are also attacked by either fuming sulphuric acid or fuming nitric acid, in the latter case decomposition often taking place with violence.

Many compounds containing a triple linkage form insoluble metallic derivatives when treated with an ammoniacal solution of silver nitrate or cuprous chloride, and from these derivatives the free acetylenic compounds can be regenerated by warming with very dilute mineral acid. Acetylenic copper and silver compounds are not formed by disubstituted acetylenes of the general formula $RC{\equiv}CR$, owing to the absence of a replaceable hydrogen atom.

A rough tabular summary showing the effects produced by the foregoing tests on different classes of compound is given on the following pages. By its aid, some indication of the type of compound to be especially tested for, by the specific methods later enumerated, will be obtained.

Tabular summary of preliminary tests

CARBON and HYDROGEN detected

I. **Treat with cold and hot water, and test solution or mixture with litmus.**
A. Soluble in the cold.
(i) Strongly acid: simple aliphatic carboxylic acids of low mol. wt.; most aliphatic hydroxy acids; most polyhydroxylic phenolic acids; a few simple esters of very low mol. wt.
(ii) Faintly acid: some simple phenols; most polyhydroxylic phenols.

     (iii) Neutral: alcohols of low mol. wt.; most polyhydroxylic alcohols; aldehydes and ketones of low mol. wt.; sugars; most glycosides.

B. Sparingly soluble in the cold, more soluble on warming.
     (i) Strongly acid: simple carboxylic acids of fairly high mol. wt. (including some aromatic acids); many aromatic hydroxy acids and their acyl derivatives.
     (ii) Faintly acid: most monohydroxylic phenols.
     (iii) Neutral: a few carbohydrates and glycosides; simple quinones.

C. Insoluble.
     (i) Strongly acid: some acids of very high mol. wt.; acid anhydrides (slowly decomposed on warming with water).
     (ii) Faintly acid: some phenols of high mol. wt.; keto-enolic esters.
     (iii) Neutral: hydrocarbons; simple ethers; alcohols, aldehydes, ketones, and quinones of very high mol. wt.; almost all esters; a few fatty acids of very high mol. wt.

II. **Treat with $NaHCO_3$ solution all substances that show an acid reaction.**
  A. CO_2 evolved: all carboxylic acids.
  B. No CO_2 evolved: phenols; keto-enolic esters, etc.

III. **Treat with cold and hot concentrated NaOH solution.**
  A. Substances which are insoluble or sparingly soluble in cold water.
     (i) Soluble in the cold: all carboxylic acids; all phenols; keto-enolic esters and similar compounds.
     (ii) Gradually dissolve on warming: a few esters and lactones; acid anhydrides.
     (iii) Insoluble: hydrocarbons; ethers; alcohols and ketones of very high mol. wt.; many esters (in general only very slowly decomposed).
  B. The following classes of compounds undergo pronounced decomposition on warming with aqueous alkali: most acid anhydrides; a few esters and lactones; aldehydes; sugars; glycosides.

IV. **Treat with concentrated H_2SO_4.**
  A. Cold.
     (i) Soluble:
        (*a*) without decomposition: some aromatic hydrocarbons; most dialkyl and arylalkyl ethers; most

alcohols; most phenols; some ketones; simple carboxylic acids; most aromatic hydroxy acids; a few esters.

(b) with decomposition: almost all unsaturated compounds; some aromatic hydrocarbons; most aliphatic hydroxy acids; most esters; sugars (brownish colours); some glycosides (red or other pronounced colours).

(ii) Insoluble: saturated hydrocarbons; some aromatic hydrocarbons.

B. Hot
(i) Gases evolved:
(a) with charring: aldehydes; ketones; acetals; carbohydrates; glycosides.
(b) without charring: simple alcohols of low mol. wt. (evolve gaseous unsaturated hydrocarbons); formic and oxalic acids and their derivatives (yield CO).
(ii) Pungent vapours evolved, without charring: simple phenols; certain simple carboxylic acids; many esters.
(iii) No gases evolved, with charring: most polyhydroxylic phenols; many aromatic hydroxy acids and certain of their derivatives.
(iv) Soluble unchanged: some carboxylic acids; some aromatic ketones of high mol. wt.

V. **Dissolve in water or alcohol and treat with one drop of FeCl₃ solution.**
A. Reddish colouration or precipitate: almost all simple carboxylic acids.
B. Intense yellow colouration: aliphatic a-hydroxy acids.
C. Green, blue, or violet colourations: most phenols and phenolic compounds (some in alcoholic solution only); keto-enolic esters and similar compounds.

VI. **Treat with a solution of KMnO₄ in dilute H₂SO₄.**
Decolourisation by:
(a) almost all unsaturated compounds;
(b) certain easily oxidisable substances such as formic acid and malonic acid and their esters; many aldehydes; simple quinones; some aliphatic hydroxy acids; many polyhydric alcohols and phenols; certain sugars.

VII. **Treat with bromine water in the cold or warm.**
A. Decolourisation without formation of much acid: almost all unsaturated compounds.

B. Decolourisation with formation of much acid: many aldehydes and ketones; other compounds readily brominated,
   such as phenols and their derivatives.

VIII. Treat with a solution of bromine in CCl_4, $CHCl_3$, or CS_2.

A. Instant decolourisation in the cold, without evolution of
   HBr: almost all unsaturated compounds.

B. Decolourisation with evolution of HBr on warming; substances which are readily brominated, such as many aldehydes and ketones; most phenols and phenolic compounds;
   certain unstable hydrocarbons (such as terpenes).

C. Rapid decolourisation only on warming, without evolution
   of HBr: unsaturated compounds in which the unsaturated
   linkages are either conjugated with aromatic and similar
   residues or largely surrounded by substituents (e.g. (i)
   cinnamic acid; (ii) tetrasubstituted ethylenes).

IX. Ignite with dry soda-lime.

A. Hydrogen or hydrocarbons evolved: simple aliphatic and
   aromatic carboxylic acids.

B. Phenols evolved: aromatic hydroxy acids.

C. Odour of 'burnt sugar': most aliphatic hydroxy acids;
   sugars; glycosides.

X. Ignite with zinc dust.

Hydrocarbons are produced from many phenols, quinones, and
aromatic ketones of high mol. wt. This treatment may also result in the
disruption of some carboxylic acids, with formation of hydrocarbons
and other compounds.

CARBON, HYDROGEN, and NITROGEN detected

I. Treat with cold and hot water, and test solution or mixture with litmus.

A. Soluble in the cold.

   (i) Acid or faintly acid: a few aromatic amino acids; some
       aliphatic simple amides of low mol. wt.; a few simple
       urethanes; a few oximes of low mol. wt.; some nitro-
       phenols; nitrates of weak organic bases.

   (ii) Neutral: aliphatic amino carboxylic acids; a few
        aliphatic substituted amides; some purines; a few
        aromatic nitroamines; salts of organic acids with
        nitrogenous bases or ammonia; nitrates of strong
        organic bases.

   (iii) Alkaline or faintly alkaline: aliphatic primary, secon-
         dary, and tertiary amines of fairly low mol. wt.;

guanidine and its alkyl derivatives; some aromatic diamines.

B. Sparingly soluble in the cold, more so on warming.

(i) Acid or faintly acid: some simple amides; some nitrophenols; some nitrocarboxylic acids; formanilide.

(ii) Neutral: some aliphatic and aromatic substituted amides; a few purines; some aromatic nitroamines.

(iii) Alkaline or faintly alkaline: some aromatic diamines and aminophenols.

C. Insoluble.

(i) Acid or faintly acid: a few purines; alkyl nitrates and nitrites.

(ii) Neutral: some aromatic amines of high mol. wt.; simple and substituted amides of very high mol. wt.; simple nitriles; isocyanides; most oximes; hydrazones; most substituted urethanes; nitrohydrocarbons; nitro ethers; nitroso, azoxy, azo, and hydrazo compounds.

(iii) Alkaline or faintly alkaline: most simple aromatic amines; most substituted hydrazines.

II. Treat substances insoluble or sparingly soluble in cold water with dilute acid and with dilute alkali.

A. Soluble in dilute acid: all primary amines; all aliphatic and most aromatic secondary and tertiary amines; many substituted hydrazines; some simple and substituted amides; some oximes; some purines.

B. Soluble in dilute alkali: many simple amides and imides; a few primary substituted amides; aminocarboxylic acids; nitrocarboxylic acids; oximes; nitrophenols; some purines.

III. Treat with cold and hot concentrated alkali.

A. Ammonia or ammoniacal vapours evolved in the cold: ammonium salts of organic acids; salts of simple aliphatic amines.

B. Ammonia or ammoniacal vapours evolved only on heating: simple amides and imides; urea and monosubstituted ureas; urethanes; nitriles (slowly); acyl derivatives of simple aliphatic primary and secondary amines; some aromatic nitroamines; most polynitroaromatic compounds; guanidine and its alkyl derivatives.

C. Separation of an insoluble compound in the cold: salts of insoluble bases.

D. Separation of an insoluble compound on heating: acyl derivatives of insoluble primary and secondary amines; many substituted urethanes of high mol. wt.

IV. Boil with Sn and concentrated HCl, then add alkali in excess.

A. Ammonia or ammoniacal vapours produced: simple amides and imides; acyl derivatives of amines of low mol. wt.; nitriles; isocyanates of aliphatic radicals of low mol. wt.; isocyanides; aliphatic oximes; ammonium salts.

B. Liquid or solid bases produced: nitro, nitroso, azoxy, azo, and hydrazo compounds; acyl derivatives of amines of high mol. wt.; nitriles, isocyanates, and isocyanides of high mol. wt.; hydrazones.

V. Heat with soda-lime.

A. Ammonia or ammoniacal vapours produced: simple amides and imides; nitriles; purines; many substituted hydrazines; many aminocarboxylic acids of low mol. wt.; simple urethanes; many aromatic nitroamines.

B. Liquid or solid bases produced: acyl derivatives of primary and secondary amines; aminocarboxylic acids of high mol. wt.; many hydrazine derivatives; substituted urethanes of high mol. wt.

CARBON, HYDROGEN, and SULPHUR detected

I. Treat with cold and hot water, and test solution or mixture with litmus.

A. Soluble in the cold.

    (i) Acid: most sulphonic acids; a few thiocarboxylic acids; some sulphinic acids.

    (ii) Neutral or faintly alkaline: aliphatic sulphoxides of low mol. wt.

B. Sparingly soluble in the cold, more soluble on warming.

    (i) Acid: some sulphonic acids; many sulphinic acids.

    (ii) Neutral: a few sulphones.

C. Insoluble.

    (i) Acid or faintly acid: some hydroxysulphones; alkyl sulphates and esters of sulphonic acids (slowly decomposed).

    (ii) Neutral: mercaptans; sulphides; disulphides; sulphoxides and sulphones of high mol. wt.; aryl esters of sulphonic acids.

II. Treat substances insoluble or sparingly soluble in cold water, with warm dilute alkali.

A. Soluble without pronounced decomposition: all sulphonic acids; sulphinic acids; sulphur compounds containing carboxyl or phenolic hydroxyl groups; mercaptans.

B. Soluble with pronounced decomposition: alkyl sulphates and esters of sulphonic acids; thiocarboxylic acids and esters.

III. **Shake with water and HgCl₂ solution.**
Precipitates formed with: mercaptans, sulphides, and some disulphides.

IV. **Ignite with soda-lime.**
Phenols produced from most sulphonic acids; hydrocarbons from sulphinic acids and a few sulphonic acids.

CARBON, HYDROGEN, and HALOGEN detected

I. **Treat with cold and hot water, and test solution or mixture with litmus.**
    A. Soluble in the cold.
        (i) Acid or faintly acid: aliphatic halogen substituted carboxylic acids.
        (ii) Neutral: halogen substituted alcohols and aldehydes; some halogen substituted phenols.
    B. Sparingly soluble in the cold, more soluble on warming.
        (i) Strongly acid: simple aromatic halogen substituted carboxylic acids.
        (ii) Faintly acid: simple halogen substituted phenols.
    C. Insoluble.
        (i) Faintly acid: some halogen substituted esters; poly-halogen substituted phenols.
        (ii) Neutral: halogen substituted hydrocarbons; esters of halogen substituted acids; halogen substituted ketones; halogen substituted aromatic ethers.
    D. Decomposed with liberation of halogen hydracid.
        (i) Rapidly: aliphatic carboxylic halides; aliphatic *a*-halogen substituted ethers.
        (ii) Slowly: aromatic carboxylic halides.

II. **Boil under reflux with alcoholic AgNO₃.**
    A. Silver halide rapidly produced: aliphatic iodo compounds; carboxylic halides; aliphatic *a*-halogen substituted ethers.
    B. Silver halide slowly produced: aliphatic *a*-halogen substituted acids, esters, aldehydes, and ketones; some unsaturated aliphatic and some aromatic *a*-halogen substituted hydrocarbons (such as allyl bromide and benzyl chloride).
    C. Silver halide produced very slowly or not at all: saturated aliphatic chloro and bromo substituted compounds in general; most aromatic chloro, bromo, and iodo compounds in which halogen is attached to aromatic nucleus.

III. Boil under reflux with alcoholic KOH.

A. Potassium halide precipitated: aliphatic chloro and bromo compounds; aromatic chloro and bromo compounds in which halogen is not attached to aromatic nucleus.

B. No precipitate of potassium halide: iodo compounds (KI is soluble in alcohol); aryl halides in general.

CARBON, HYDROGEN, NITROGEN and SULPHUR detected

I. Treat with cold and hot water, and test solution or mixture with litmus.

A. Soluble in the cold: thiocarbamide; some nitrosulphonic acids; salts of thiocyanic acid.

B. Sparingly soluble in the cold, more soluble on warming: many substituted thiocarbamides; many aminosulphonic acids; some nitrosulphonic acids; some simple and substituted sulphonamides.

C. Insoluble: alkyl thiocyanates; alkyl isothiocyanates; many aromatic aminosulphonic acids; many simple and substituted sulphonamides.

II. Treat with cold and hot NaOH solution.

A. Cold.
  (i) Soluble: simple and primary substituted sulphonamides; carbosulphonimides; amino, nitro, azo, etc. sulphonic acids.
  (ii) Insoluble: sulphonyl derivatives of secondary amines; alkyl thiocyanates and isothiocyanates.

B. Hot.
  (i) Ammonia evolved: thiocarbamide and monosubstituted thiocarbamides; simple sulphonamides (very slowly).
  (ii) Bases produced: substituted thiocarbamides; sulphonyl derivatives of primary and secondary amines (very slowly).

III. Boil under reflux with concentrated HCl.

Alkyl thiocyanates yield alkyl sulphides, CO_2, and NH_4Cl.
Alkyl isothiocyanates yield H_2S, CO_2, and primary amines.
Sulphonamides yield sulphonic acids and NH_4Cl (slowly).
Substituted sulphonamides yield primary or secondary amines and sulphonic acids.
Thiocarbamide and substituted thiocarbamides yield H_2S and guanidine or substituted guanidines.
Aminosulphonic acids form hydrochlorides unchanged.

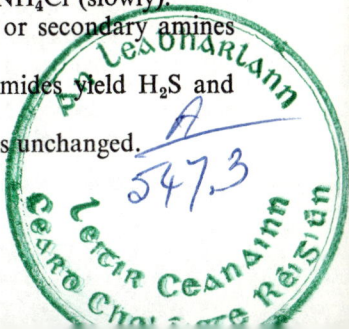

IV. Boil with zinc dust and dilute HCl.

Amino compounds produced from: nitro and azo sulphonic acids; substituted thiocarbamides; alkyl isothiocyanates (with evolution of H$_2$S).

CARBON, HYDROGEN, SULPHUR and HALOGEN detected

I. Heat with water and cool. Add dilute HNO$_3$ and aqueous silver nitrate.

White to yellow precipitate: sulphonyl halides.

II. Boil with dilute hydrochloric acid and granulated tin.

Unpleasant penetrating odour of thiol: sulphonyl halides.

METALLIC RESIDUE LEFT AFTER IGNITION

Treat with dilute HCl

Carbon and **Hydrogen** detected

Salts of carboxylic acids: free acids liberated.

Carbon, Hydrogen, and **Nitrogen** detected

Salts of nitrocarboxylic acids: free acids liberated.
Salts of azocarboxylic acids: free acids liberated.
Salts of aminocarboxylic acids: soluble hydrochlorides produced.
Salts of nitrophenols: free nitrophenols liberated.
Metallic derivatives of imides of dicarboxylic acids: free imides produced.

Carbon, Hydrogen, and **Halogen** detected

Salts of halogen substituted carboxylic acids: free acids liberated.

Carbon, Hydrogen, and **Sulphur** detected

Boil with solution of BaCl$_2$ and concentrated HCl.
- (i) SO$_2$ evolved: bisulphite compounds of aldehydes or ketones.
- (ii) BaSO$_4$ slowly produced: salts of alkylsulphuric acids.
- (iii) No BaSO$_4$ produced: salts of sulphonic acids, or carbosulphonimides.

Carbon, Hydrogen, Nitrogen, and **Sulphur** detected

Acidify strongly with concentrated HCl: precipitates of free acids produced from salts of many amino and azosulphonic acids.

Carbon, Hydrogen, Nitrogen, Sulphur and **Halogen** detected

Salts of halogen sulphonamides.

2 Examination for Radicals

This chapter is intended to serve as a guide for the detection of the various salient groups which may be present in the molecule. When one radical has been found, it will, of course, still be necessary to search for others which may also exist in the compound under examination; and since the great majority of organic compounds contain more than one characteristic group, the manner in which these different groups may affect the general properties of the substance must be the subject of careful consideration.

When an element other than carbon and hydrogen has been detected, the first important point to be established is the form in which this element exists. To take the case of a substance in which carbon, hydrogen, and nitrogen have been shown to exist, it must first be established whether it is a primary, secondary, or tertiary amine, an amide, a hydrazine or hydroxylamine derivative, an azo, azoxy, nitroso, nitro or other type of nitrogen compound. In short, the first problem is to determine the class of the body in reference to the most uncommon element occurring therein, and not until this point is settled will it be profitable to examine the substance with a view to ascertaining the constitution of the remaining portion of the molecule.

For this reason the section dealing with substances in which only carbon and hydrogen have been detected has been reserved until the end of this chapter, so that after the 'special elements' have been allotted to their proper classes, the substance may be examined according to the guide given for compounds in which carbon and hydrogen have been found, due allowances being made for the presence of the known 'special' groups.

When all the salient radicals have been definitely determined, reference should be made to the appended tables for the list of compounds, tabulated mainly according to their class, and usually arranged in the order of boiling and melting points; the characteristic tests and properties of derivatives of each substance being, in so far as possible, enumerated.

Most important points, not to be neglected in the examination for radicals, are quantitative estimation of groups and preparation of derivatives. The importance of both these operations cannot be too strongly emphasised. Unfortunately the quantitative work frequently requires a comparatively long time; but many simple estimations, such as the titration of an acid or the salt of a base, or a volumetric estimation of ionised halogen, can be expeditiously carried out, and may contribute much to the certainty of an identification. In fact, owing to the absence of any definite system for the identification of organic compounds, after the establishment of the presence of the various characteristic groups in the substance the utlimate identification when spectroscopic methods are unavailable must occasionally depend upon such quantitative experiments. The preparation of derivatives is even more important, and no excuse is valid for failure to prepare and examine at least one pure derivative in addition to the determination of a mixed melting point if this is feasible.

The preliminary tests described and tabulated in the previous chapter should have led to some clue as to the nature of the group or groups present in the molecule, and the properties of the various types are therefore discussed without further reference to the general methods of distinction. The equations given in the text refer to simple typical examples of the reactions under discussion. The page references refer to the tables of organic compounds given in Chapter 4.

Types of Radical Involving the Detection of
CARBON, HYDROGEN and NITROGEN

On treatment with cold aqueous alkali, ammonia is liberated from *ammonium salts* with formation of products which may more profitably be investigated independently, with a view to identification.

Aldehyde-ammonias also yield ammonia on gently warming with aqueous alkali. For preparation of the free aldehyde from an aldehyde-ammonia, however, decomposition by distillation with dilute sulphuric acid is advisable. Salts of organic acids may also be decomposed by means of dilute mineral acid, subsequently distilling the mixture when the acid is volatile with steam, filtering when it is insoluble or sparingly soluble in cold water, or extracting with ether. Acids which cannot be isolated by any of these methods must be examined in solution before proceeding with their isolation by some special method.

By the action of hot 30% aqueous alkali, ammonia is liberated comparatively rapidly from *simple amides* (pp. 71, 176) and slowly from most *nitriles* (pp. 85, 192).

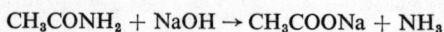

$$CH_3CONH_2 + NaOH \rightarrow CH_3COONa + NH_3$$

$$C_6H_5CN + NaOH + H_2O \rightarrow C_6H_5COONa + NH_3$$

These types of substance yield no ammonia by the action of alkali in the cold. In most cases it will be found expedient to carry out the hydrolysis by boiling 1 g of substance under reflux with 10 cm^3 of 70% sulphuric acid or phosphoric acid, for about 15 minutes.

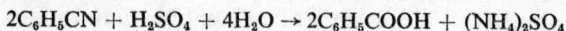

$$2C_6H_5CN + H_2SO_4 + 4H_2O \rightarrow 2C_6H_5COOH + (NH_4)_2SO_4$$

The acid formed should in every case be examined, after isolation by distillation with steam, filtration, or extraction with ether, according to the nature of the acid.

Phosphorus pentoxide reacts with amides, nitriles being produced on heating.

$$CH_3CONH_2 \rightarrow CH_3CN + H_2O$$

These can best be isolated by distilling the nitrile from the resulting mixture, or failing this, by addition of cold water and extraction with ether.

Amides may be transformed into anilides by refluxing with aniline for 1–2 hr., ammonia being evolved

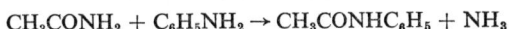

$$CH_3CONH_2 + C_6H_5NH_2 \rightarrow CH_3CONHC_6H_5 + NH_3$$

Aromatic nitriles when treated with a warm alkaline solution of 20 volume hydrogen peroxide are converted into the corresponding amides

$$2C_6H_5CN + 2H_2O_2 \rightarrow 2C_6H_5CONH_2 + O_2$$

On reduction with tin and hydrochloric acid or by adding sodium to a boiling ethanolic solution of the substance, nitriles are converted into the related primary amines.

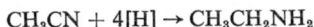

$$CH_3CN + 4[H] \rightarrow CH_3CH_2NH_2$$

Imides of dicarboxylic acids present properties similar to those of simple amides, except that on treatment of a saturated methanolic solution with saturated methanolic potash a precipitate of the potassium derivative is formed.

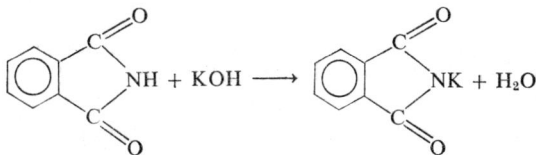

On boiling with an alkaline solution of sodium hypobromite an amino acid is produced.

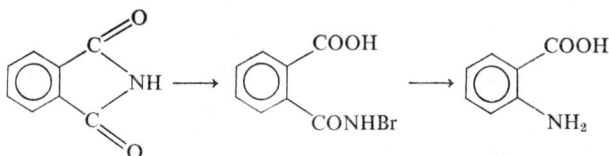

Closely related to the amides are **urea** and its *monosubstitution products* (pp. 71, 176, 181), which behave as amides derived from carbonic acid. On acid hydrolysis they yield carbon dioxide, an ammonium salt, and, in the case of substituted ureas, a salt of an amine. On alkaline hydrolysis ammonia is evolved, a free amine is liberated, and alkali carbonate is formed. On heating with aromatic amines, disubstituted ureas are produced, with evolution of ammonia. Thus both urea and phenylurea on heating with aniline yield carbanilide and ammonia.

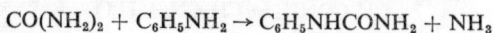

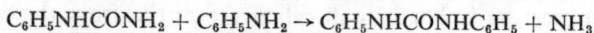

$$CO(NH_2)_2 + C_6H_5NH_2 \rightarrow C_6H_5NHCONH_2 + NH_3$$

$$C_6H_5NHCONH_2 + C_6H_5NH_2 \rightarrow C_6H_5NHCONHC_6H_5 + NH_3$$

The same reactions may be brought about by warming the urea on the water bath with a solution of aniline hydrochloride; in the case of urea it is possible to isolate the mono and disubstituted derivatives by taking advantage of the solubility of the latter in hot water.

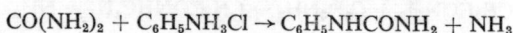

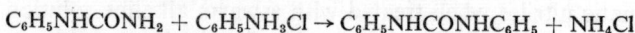

$$CO(NH_2)_2 + C_6H_5NH_3Cl \rightarrow C_6H_5NHCONH_2 + NH_3$$

$$C_6H_5NHCONH_2 + C_6H_5NH_3Cl \rightarrow C_6H_5NHCONHC_6H_5 + NH_4Cl$$

Urea and its monosubstitution products may be readily distinguished from true carboxylic amides by their ability to form, in extremely dilute solution, insoluble condensation products with xanthydrol.

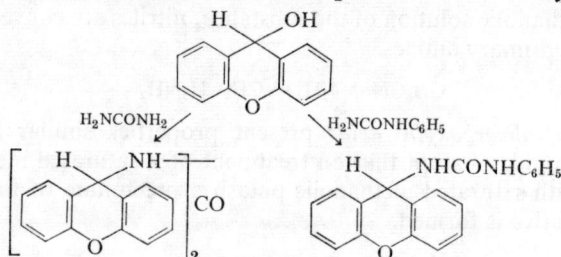

This reaction, which serves to detect minute quantities of urea in very dilute solution, is carried out by adding 1 cm³ of a 5% solution of xanthydrol in methyl alcohol to a cold solution of 0.001 g of the urea in 10 cm³ of 50% acetic acid, when the condensation product separates in quantitative yield in the form of colourless, microcrystalline flocks.

Urethanes, or *alkyl carbamates* (pp. 71, 186), behave in many ways like true amides. Thus on boiling with dilute alkali they yield ammonia, and on boiling with dilute acids they yield ammonium salts, a carbonate (or carbon dioxide) and an alcohol being produced.

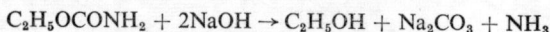

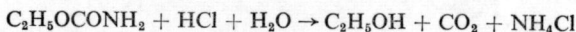

$$C_2H_5OCONH_2 + 2NaOH \rightarrow C_2H_5OH + Na_2CO_3 + NH_3$$

$$C_2H_5OCONH_2 + HCl + H_2O \rightarrow C_2H_5OH + CO_2 + NH_4Cl$$

On treatment with alcoholic potassium hydroxide, urethanes yield crystalline potassium cyanate, which may be identified by its reaction with aniline hydrochloride to form phenylurea.

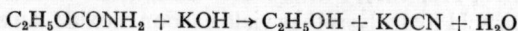

$$C_2H_5OCONH_2 + KOH \rightarrow C_2H_5OH + KOCN + H_2O$$

This reaction takes place slowly in the cold and rapidly on warming. On refluxing with aniline, urethanes yield the corresponding alcohol, ammonia, and carbanilide.

$$C_2H_5OCONH_2 + 2C_6H_5NH_2 \rightarrow C_2H_5OH + NH_3 + CO(NHC_6H_5)_2$$

Ammonia is also liberated from guanidine and its simple derivatives on boiling with alkali. These differ from amides and ureas in being strong bases, and are commonly met as salts of mineral acids. On boiling with barium hydroxide solution, guanidine yields urea and ammonia.

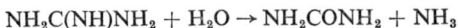

$$NH_2C(NH)NH_2 + H_2O \rightarrow NH_2CONH_2 + NH_3$$

The barium hydroxide has no action on the urea, which is hydrolysed further when sodium hydroxide is employed. On treatment with sodium hypobromite guanidine gives off two-thirds of its nitrogen in the elemental form. Guanidine and its alkyl derivatives form sparingly soluble addition products with picric acid; these crystallise well and possess characteristic melting points.

Cyanohydrins (p. 192) of aldehydes and ketones present properties differing somewhat from those of the simple nitriles. On treatment with hot concentrated hydrochloric acid normal hydrolysis takes place, the corresponding hydroxy acids being produced; but on treatment with alkaline reagents, hydrogen cyanide is eliminated with regeneration of the corresponding carbonyl compounds.

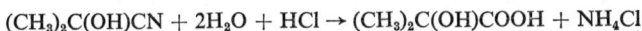

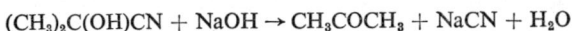

$$(CH_3)_2C(OH)CN + 2H_2O + HCl \rightarrow (CH_3)_2C(OH)COOH + NH_4Cl$$

$$(CH_3)_2C(OH)CN + NaOH \rightarrow CH_3COCH_3 + NaCN + H_2O$$

In some cases cyanohydrins are so unstable as to be decomposed merely on heating, with formation of hydrocyanic acid and aldehydes and ketones.

If, on treatment with cold alkali, an amine is produced, the substance is probably a *salt of an amine and an organic acid.* The alkaline mixture should be extracted with ether to remove the amine, and the aqueous residue acidified and worked up for the acid by steam distillation, filtration, or extraction with ether, as described for ammonium salts.

The majority of *substituted amides* (pp. 71, 181) yield the corresponding amines on treatment with hot aqueous or alcoholic alkali. *As in the case of simple amides, hydrolysis by means of acid reagents is more certain in its results*, since a considerable number of substituted amides, especially in the aromatic series, are decomposed only with the greatest difficulty by boiling under reflux even with alcoholic potash. The amine and the acid may be isolated by steam distillation, filtration, or extraction with ether, after rendering the mixture after hydrolysis alternately alkaline and acid, or vice versâ, according to whether hydrolysis has been effected by acid or by alkaline reagents.

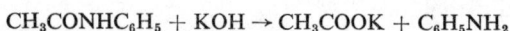

$$CH_3CONHC_6H_5 + KOH \rightarrow CH_3COOK + C_6H_5NH_2$$

$$CH_3CONHC_6H_5 + HCl \rightarrow CH_3COOH + C_6H_5NH_3Cl$$

After isolation, both acid and amine should be examined independently.

In the same way, substituted ureas are hydrolysed on prolonged boiling with mineral acids, the corresponding amines and carbon dioxide being formed.

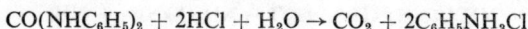

$$CO(NHC_6H_5)_2 + 2HCl + H_2O \rightarrow CO_2 + 2C_6H_5NH_3Cl$$

Substituted urethanes undergo an analogous reaction,

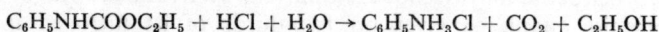

$$C_6H_5NHCOOC_2H_5 + HCl + H_2O \rightarrow C_6H_5NH_3Cl + CO_2 + C_2H_5OH$$

as do also the substituted guanidines.

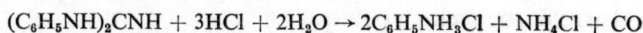

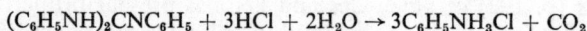

$$(C_6H_5NH)_2CNH + 3HCl + 2H_2O \rightarrow 2C_6H_5NH_3Cl + NH_4Cl + CO$$

$$(C_6H_5NH)_2CNC_6H_5 + 3HCl + 2H_2O \rightarrow 3C_6H_5NH_3Cl + CO_2$$

Oximes and *Hydrazones* are hydrolysed by boiling with concentrated hydrochloric acid or 30% sulphuric acid under reflux for 30 minutes.

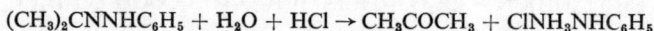

$$C_6H_5CHNOH + H_2O + HCl \rightarrow C_6H_5CHO + HONH_3Cl$$

$$(CH_3)_2CNNHC_6H_5 + H_2O + HCl \rightarrow CH_3COCH_3 + ClNH_3NHC_6H_5$$

After hydrolysis, the aldehyde or ketone may be isolated by distillation with steam, filtration, or extraction with ether, and the acid solution examined for hydroxylamine or a hydrazine. Many substituted hydrazines, especially in the aromatic series, are precipitated as oils or solids on the addition of alkali, hydroxylamine and unsubstituted hydrazine remaining, however, in aqueous solution. Hydroxylamine and hydrazines possess powerful reducing properties, which can be shown by testing with Fehling's solution or ammoniacal silver nitrate. Oximes dissolve in dilute caustic alkalis, from which solution they may be liberated by saturating with carbon dioxide.

Hydrazides, or *acyl derivatives of hydrazines,* undergo hydrolysis on boiling with concentrated hydrochloric acid or alcoholic potash under reflux, with formation of the corresponding acids and hydrazines.

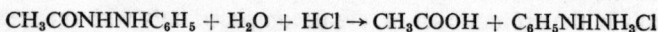

$$CH_3CONHNHC_6H_5 + H_2O + HCl \rightarrow CH_3COOH + C_6H_5NHNH_3Cl$$

The resulting acids and hydrazines should be isolated in the manner indicated for amides and hydrazones, and examined.

Isocyanides all have characteristic and unpleasant odours. On warming with hydrochloric acid they are hydrolysed, with formation of the hydrochlorides of primary amines and formic acid.

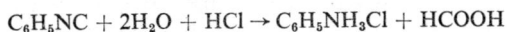

$$C_6H_5NC + 2H_2O + HCl \rightarrow C_6H_5NH_3Cl + HCOOH$$

Isocyanides are readily oxidised by mercuric oxide to the corresponding isocyanates, with production of metallic mercury.

$$C_6H_5NC + HgO \rightarrow C_6H_5NCO + Hg$$

Isocyanates (p. 195) possess pungent odours, the vapours irritating the mucous membranes. They are extremely reactive, reacting with water to form symmetrical disubstituted carbamides and carbon dioxide

$$2C_6H_5NCO + H_2O \rightarrow C_6H_5NHCONHC_6H_5 + CO_2$$

with alcohols and phenols, giving rise to substituted carbamic esters (substituted urethanes)

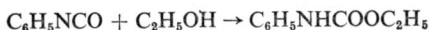

$$C_6H_5NCO + C_2H_5OH \rightarrow C_6H_5NHCOOC_2H_5$$

with ammonia and with primary or secondary amines, forming substituted carbamides

$$C_6H_5NCO + C_6H_5NH_2 \rightarrow C_6H_5NHCONHC_6H_5$$

and with carboxylic acids with production of substituted amides.

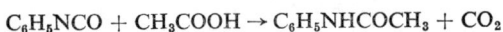

$$C_6H_5NCO + CH_3COOH \rightarrow C_6H_5NHCOCH_3 + CO_2$$

By the action of hot acid or alkaline hydrolytic reagents, primary amines are produced from isocyanates, owing to the initial formation of substituted carbamides or urethanes and subsequent hydrolysis of these with elimination of carbon dioxide.

Amines. If the substance should fail to undergo hydrolysis on treatment with acid or alkaline reagents, yet enter into solution in aqueous mineral acids, tests must be applied to determine whether it belongs to the family of amines (pp. 164 et seq.). These are almost without exception compounds possessing pronounced basic properties, and are frequently met in the form of a salt of an inorganic acid. In the free state they turn red litmus paper blue, or, when too feebly basic to show this colour change distinctly, they may be tested with Congo paper previously coloured blue by very dilute hydrochloric acid. Amines in which the nitrogen atom is directly attached to one or more aromatic nuclei are as a rule far weaker bases than aliphatic amines, and solutions of their salts give an acid reaction with litmus, though not (unless more than one aromatic radical is attached to the nitrogen atom) with Congo Red. Some aromatic amines have quite negligible salt-forming properties are are insoluble in hydrochloric acid or almost so, e.g.

triphenylamine, phenylpyrrole, trinitroaniline, while others are soluble in this acid but precipitate from concentrated solutions as their hydrochlorides, e.g. diphenylamine and the naphthylamines. Strong bases having a dissociation constant of 5×10^{-9} or greater are absorbed from dilute aqueous solutions of their salts by exchange resins, whereas weaker bases, such as aromatic amines, are not so taken up. The bases may be liberated from the resins by means of sodium hydroxide.

Salts of aliphatic amines are usually neutral or only faintly acid to litmus; for the separation of the free bases a concentrated solution of the salt is saturated with solid potassium hydroxide, since organic bases are only slightly soluble in concentrated alkali. Hydrochlorides of most secondary and tertiary amines are soluble in chloroform, while salts of primary amines are insoluble.

Amines of all classes form definite compounds with phosphotungstic acid, with platinic chloride, and with picric acid, the latter type of compound constituting a valuable aid to identification. The picrates are as a rule insoluble in water and sparingly soluble in cold alcohol; they are conveniently prepared by mixing concentrated alcoholic solutions of the free base and picric acid. Their melting points are generally sharp and characteristic.

Primary aliphatic amines and aromatic side-chain amines may be distinguished from ammonia by placing a drop of an aqueous solution of the base upon a test paper prepared by moistening filter paper with a saturated alcoholic solution of pure chloro-2,4-dinitrobenzene; an intense yellow colouration is produced by the amines, but not by ammonia. Many secondary amines also give this test (in particular dimethylamine, which responds as readily as methylamine), but the higher members, such as diethylamine, produce the colour more slowly and less intensely. Tertiary amines give only a faint reaction or none at all. With pyridine and aromatic amines the test fails entirely. This reaction may also be employed (by treating the base in alcoholic solution with chlorobenzene) for the preparation of characteristic derivatives.

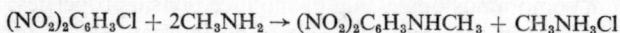

$$(NO_2)_2C_6H_3Cl + 2CH_3NH_2 \rightarrow (NO_2)_2C_6H_3NHCH_3 + CH_3NH_3Cl$$

The same condensation may be applied to aromatic primary amines, which form, however, the 2,4-dinitrophenyl derivatives much more slowly than the aliphatic primary amines.

To distinguish between primary, secondary and tertiary amines the action of *sodium nitrite* upon a solution of the substance in an excess of mineral acid should be investigated as follows: 0.5 g (or 0.5 cm³) of the substance is dissolved in a mixture of 3 cm³ of concentrated hydrochloric acid and 2 cm³ of water, and the solution cooled to about 5°C by immersion in an ice-bath. A similarly cooled solution of 0.4 g of crystalline sodium nitrite in 4 cm³ of water is added to the amine solution in small portions, stirring continually and keeping the temperature below 10°C during this operation, until one drop of the reaction mixture, withdrawn and diluted with 4 drops of water, gives an immediate positive test for free nitrous acid when spotted onto fresh potassium iodide–starch paper: an immediate blue colouration should be obtained. (If the amine is not appreciably soluble in dilute hydrochloric acid, it should be dissolved in concentrated hydrochloric acid or in a mixture of 3 cm³ of water and 4 cm³ of concentrated sulphuric acid.)

(i) *Primary aliphatic amines* give a copious evolution of nitrogen, especially when the mixture is warmed on a water bath.

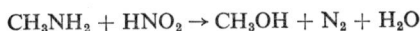

$$CH_3NH_2 + HNO_2 \rightarrow CH_3OH + N_2 + H_2O$$

The alcohol formed may in some cases be isolated and identified but it must be pointed out that the reaction is often less simple than the above equation indicates; thus n-propylamine yields n-propanol, isopropanol and propylene.

The presence of either an aliphatic or aromatic primary amine may be confirmed by warming a minute quantity with a mixture of alcoholic potash and chloroform, when the vile and characteristic odour of an isocyanide will be perceptible.

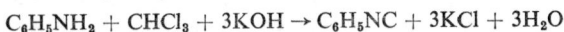

$$C_6H_5NH_2 + CHCl_3 + 3KOH \rightarrow C_6H_5NC + 3KCl + 3H_2O$$

Primary aliphatic amines and aromatic compounds in which the amino group is in a side-chain respond to Rimini's test; a violet-red colour develops within one minute when a solution or suspension of one drop of the substance in 3 cm³ of water is treated with 1 cm³ of pure acetone and one drop of a 1% aqueous solution of sodium nitroprusside.

(ii) *Primary aromatic amines* produce a clear solution of a diazonium compound in the nitrous acid test. The solution should be divided into two parts. One part should be poured slowly into a solution of 0.5 g of 2-naphthol in 5 cm³ of 5% sodium hydroxide solution; an orange-red azo dye is formed and may be removed by filtration for recrystallisation (ethanol) and identification.

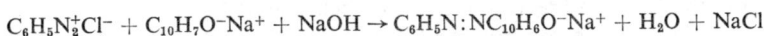

$$C_6H_5NH_2 + HCl + HNO_2 \rightarrow C_6H_5N_2^+Cl^- + 2H_2O$$

$$C_6H_5N_2^+Cl^- + C_{10}H_7O^-Na^+ + NaOH \rightarrow C_6H_5N:NC_{10}H_6O^-Na^+ + H_2O + NaCl$$

The other part should be warmed gently; nitrogen is evolved and a phenol produced, accompanied often by tarry matter.

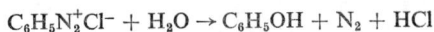

$$C_6H_5N_2^+Cl^- + H_2O \rightarrow C_6H_5OH + N_2 + HCl$$

Primary aromatic amines are readily oxidised by alkaline bleaching powder or sodium hypochlorite solution with formation of brightly coloured oxidation products (e.g. aniline gives a purple colour). They are also very readily brominated by the action of bromine water to give polybromo derivatives which after removal by filtration may be recrystallised from ethanol and may be useful in identification.

(iii) *Aliphatic or aromatic secondary amines* produce a turbid solution which may contain coloured oils or precipitates; no nitrogen is evolved.

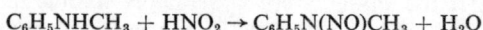

$$C_6H_5NHCH_3 + HNO_2 \rightarrow C_6H_5N(NO)CH_3 + H_2O$$

The resulting nitrosamines are often only sparingly soluble in water or dilute acid and are best extracted by means of ether. The ethereal solution should be washed with dilute alkali and then with water before the ether is evaporated. To confirm the presence of a nitrosamine, *Liebermann's test* should be performed on the resulting oil or solid. To two drops (or 0.02 g) of the oil or solid in a dry test-tube is added 0.05 g phenol and the mixture warmed for 20 seconds. The test-tube is cooled and the contents treated with 1 cm³ of concentrated sulphuric acid; an intense greenish-blue colour is developed which changes to pale red on pouring into 40 cm³ of water and to blue when the aqueous mixture is made alkaline with sodium hydroxide solution.

(iv) *Tertiary amines of the aliphatic series* (*together with pyridine, quinoline and their derivatives*) do not undergo any deep seated reaction with nitrous acid. Occasionally the bases may precipitate from the reaction mixture as their salts (e.g. hydrochlorides, nitrites) from which they may be regenerated by the action of base. *Mixed aliphatic-aromatic tertiary amines* such as dimethylaniline which have the nuclear position *para* to the dimethylamino group unsubstituted, react with nitrous acid to form *p*-nitroso compounds, which may be present as the yellow or brown insoluble salt (e.g. hydrochloride) in the reaction mixture. In such a case, the reaction liquor should be made alkaline in order to release the *p*-nitroso compound which may precipitate as a green or greenish-blue substance.

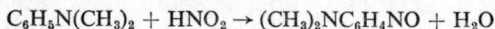

$$C_6H_5N(CH_3)_2 + HNO_2 \rightarrow (CH_3)_2NC_6H_4NO + H_2O$$

The *p*-nitrosodialkylarylamine may be extracted with ether, the extract dried with sodium sulphate, and after removal of the ether, the nitroso compound recrystallised from benzene or ligroin (b.p. 60–80°C).

Tertiary amines do not react with acylating agents such as benzoyl chloride, acetic anhydride and *p*-toluenesulphonyl chloride, which are so useful for characterising primary and secondary amines.

Hinsberg's method for distinguishing between primary, secondary and tertiary amines depends upon the unreactive nature of tertiary amines towards acylating reagents and the fact that primary amines, on reaction with sulphonyl chlorides, generally yield sulphonamides which are soluble in alkali,* while secondary amines give products which, lacking a replaceable H on the nitrogen atom, are insoluble in alkali.

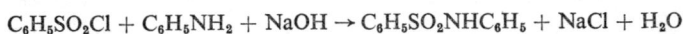

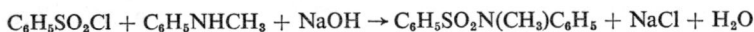

$$C_6H_5SO_2Cl + C_6H_5NH_2 + NaOH \rightarrow C_6H_5SO_2NHC_6H_5 + NaCl + H_2O$$

$$C_6H_5SO_2Cl + C_6H_5NHCH_3 + NaOH \rightarrow C_6H_5SO_2N(CH_3)C_6H_5 + NaCl + H_2O$$

p-Toluene sulphonyl chloride may replace the benzenesulphonyl chloride illustrated in the above equations.

To 0.3 cm³ or 0.3 g of the amine(s) is added 5 cm³ of 10% aqueous sodium hydroxide and 0.4 g of *p*-toluenesulphonyl chloride. The mixture is warmed on a water bath to complete the reaction, and the cooled liquor is made acid with hydrochloric acid, when sulphonamides from primary and secondary amines are precipitated. Filter and wash any solid with water: the tertiary amine will be present in the filtrate, from which it may be regenerated unchanged by addition of an excess of dilute alkali. Any precipitated sulphonamide should now be boiled under reflux with 0.4 g of sodium in 10 cm³ of absolute ethanol for 30 minutes in order to convert any disulphonamide (which may have been formed from a primary amine and which would mislead due to its insolubility in alkali) into the monosulphonamide. The mixture is now diluted with water, the alcohol removed by distillation, and the sulphonamide of a secondary amine removed by filtration. Upon acidification of the filtrate, the sulphonamide of a primary amine would precipitate.

It should be noted that original compounds which are amphoteric, such as amino acids and aminophenols, will yield sulphonamides which are soluble in alkali irrespective of whether the amino group is a primary or secondary one.

In *aliphatic amino acids* (pp. 73, 186) the functional characteristics of the amino group, while present, are to some extent masked by association of the carbonyl group. Thus the free compounds are practically neutral to litmus and the simpler members possess a sweet taste. The presence of one of these acids may be detected by treating an aqueous solution of the substance with a few drops of a 0.2% solution of ninhydrin reagent†, when a blue colour develops. Ferric chloride solution gives red colours when added to aqueous or alcoholic solutions of amino acids. Sörensen's reaction may also be useful in detecting their presence: about 0.2 g of the substance in water is neutralised with dilute sodium

---

* Some primary amines react to give disulphonamides which are not alkali-soluble and so simulate the behaviour of secondary amines.

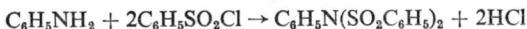

$$C_6H_5NH_2 + 2C_6H_5SO_2Cl \rightarrow C_6H_5N(SO_2C_6H_5)_2 + 2HCl$$

† The reagent consists of 0.2 g of triketohydrindene hydrate in 100 cm³ of 90% ethanol.

hydroxide to phenolphthalein. The addition of a few cm^3 of neutralised (phenolphthalein) formaldehyde causes the immediate disappearance of the pink colour, owing to the formation of a methylene derivative of an acidic nature. In a similar manner, acylating agents such as acetyl chloride, benzoyl chloride, p-toluenesulphonyl chloride yield derivatives which are strongly acidic.

Aromatic amino acids (pp. 73, 186) are more acidic than their aliphatic counterparts and cause a vigorous evolution of carbon dioxide from sodium bicarbonate. They usually give reddish colours with ferric chloride solutions. With nitrous acid (q.v.) they yield diazonium salts which couple normally with alkaline 2-naphthol solution to give azo dyes.

If the substance is hydrolysed by the action of alkaline reagents, yet yields no basic product, it may possibly be an ester of nitric acid or of nitrous acid.

Alkyl nitrates (p. 211), on hydrolysis with aqueous potash under reflux, yield potassium nitrate and the corresponding alcohol.

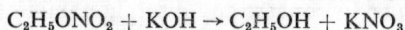

$$C_2H_5ONO_2 + KOH \rightarrow C_2H_5OH + KNO_3$$

This normal hydrolysis is nevertheless always accompanied to a greater or less degree by an abnormal reaction, whereby the alkyl group is partially oxidised at the expense of the nitrate, potassium nitrite being produced. Alkyl nitrates are reduced on treatment with tin and hydrochloric acid, with formation of hydroxylamine.

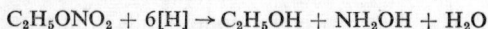

$$C_2H_5ONO_2 + 6[H] \rightarrow C_2H_5OH + NH_2OH + H_2O$$

Care must be exercised when purifying by distillation or determining the boiling point of an alkyl nitrate, as these esters are liable to undergo explosive decomposition on rapid heating, this property being forcibly exemplified in the case of nitroglycerol (glyceryl trinitrate) and nitrocellulose.

Alkyl nitrites (p. 211), on hydrolysis with alkaline reagents, yield the corresponding alcohols and a metallic nitrite.

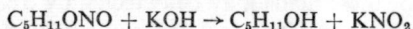

$$C_5H_{11}ONO + KOH \rightarrow C_5H_{11}OH + KNO_2$$

They are readily reduced, even by hydrogen sulphide, yielding the corresponding alcohol and ammonia,

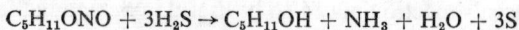

$$C_5H_{11}ONO + 3H_2S \rightarrow C_5H_{11}OH + NH_3 + H_2O + 3S$$

the same effect being also produced by more powerful reducing agents. On treatment of one molecular proportion of aniline dissolved in absolute alcohol in the presence of one-and-one-quarter molecular proportions of sulphuric acid with a slight excess of an alkyl nitrite, diazobenzene hydroxide is produced.

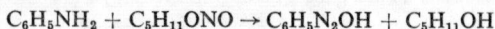

$$C_6H_5NH_2 + C_5H_{11}ONO \rightarrow C_6H_5N_2OH + C_5H_{11}OH$$

By the action of sodium ethoxide or gaseous hydrogen chloride, alkyl nitrites can be caused to react with ketones containing a methylene group adjacent to the carbonyl group, with formation of an isonitroso derivative.

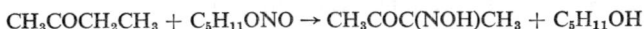

$$CH_3COCH_2CH_3 + C_5H_{11}ONO \rightarrow CH_3COC(NOH)CH_3 + C_5H_{11}OH$$

Care should be taken not to inhale the vapours of alkyl nitrites, since they distend the blood vessels and thus stimulate the action of the heart.

If the substance is insoluble in aqueous mineral acids, and remains unaffected by the action of hydrolytic reagents, it may be a nitro, nitroso, azo, azoxy, or hydrazo compound.

Nitro, azo, hydrazo, azoxy and nitroso compounds are all coloured substances as normally encountered, ranging from yellow to red. Most of them are insoluble in water and are usually insoluble in alkali (except for primary and secondary nitroalkanes) unless other groups such as —OH or —COOH are present. They are generally insoluble in dilute mineral acids. Hydrazo compounds reduce ammoniacal silver nitrate solution and Fehling's solution.

Compounds of these types may be detected by the following tests

(i) 0.1 g of the substance, dissolved in a mixture of 1 cm³ of ethanol and 1 cm³ of water by warming, is treated with 1 cm³ of ammonium chloride solution and 0.1 g of zinc dust. The mixture is briefly boiled, filtered and the filtrate tested immediately with ammoniacal silver nitrate solution.* Nitro and nitroso compounds produce a black or grey precipitate of silver owing to their reduction to hydroxylamines, e.g.

$$C_6H_5NO_2 + 4[H] \rightarrow C_6H_5NHOH + H_2O$$

$$C_6H_5NHOH + 2[Ag(NH_3)_2]OH \rightarrow C_6H_5NO + 2Ag + NH_3 + 2H_2O$$

The same result is produced by azoxy and azo compounds owing to their reduction to hydrazo and hydrazine compounds respectively. If the original compound is a hydrazo compound, it will reduce the silver nitrate reagent by itself and this should be investigated.

(ii) About 0.5 g of the substance is treated with 5 cm³ of concentrated hydrochloric acid and 2 cm³ of alcohol. A few small pieces of granulated tin are added and the ensuing reaction moderated by cooling as necessary. The mixture is then refluxed on a water bath until all of the original compound has passed into solution. The liquor is decanted from residual tin and a sample of 1 cm³ treated with 2 cm³ of water. The solution is chilled to 5°C in an ice-bath and very slowly, with

* The reagent is prepared by adding dilute ammonia solution to a solution of silver nitrate until the first formed precipitate just dissolves. The reagent is potentially explosive and should not be heated alone or stored. After use, all traces of reagent should be washed away and vessels rinsed with dilute nitric acid.

stirring, is added 1 cm³ of a 10% solution of sodium nitrite. A few drops of the resulting solution (which should contain a diazonium salt) are added to 5 cm³ of a solution of 2-naphthol in sodium hydroxide. All compounds of the types under discussion produce orange or red-brown azo dyes, except aliphatic nitro compounds. e.g.

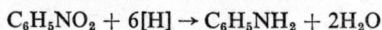

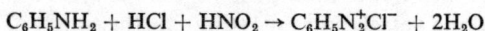

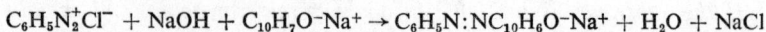

$$C_6H_5NO_2 + 6[H] \rightarrow C_6H_5NH_2 + 2H_2O$$

$$C_6H_5NH_2 + HCl + HNO_2 \rightarrow C_6H_5N_2^+Cl^- + 2H_2O$$

$$C_6H_5N_2^+Cl^- + NaOH + C_{10}H_7O^-Na^+ \rightarrow C_6H_5N:NC_{10}H_6O^-Na^+ + H_2O + NaCl$$

Oxidation with alkaline hydrogen peroxide, nitric acid or potassium permanganate converts C-nitroso compounds into the corresponding nitro compounds. Many aromatic nitroso compounds containing one or more hydroxyl groups behave in some reactions as quinone monoximes, with which they are tautomeric: e.g.

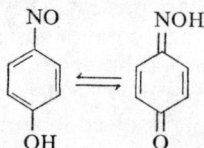

Aromatic hydrazo compounds may be acylated as for primary and secondary amines. They are oxidised to the corresponding azo compounds by the action of alkaline permanganate or ferricyanide.

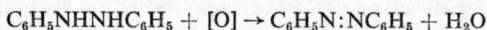

$$C_6H_5NHNHC_6H_5 + [O] \rightarrow C_6H_5N:NC_6H_5 + H_2O$$

They undergo intramolecular rearrangements when heated with concentrated acids, thus hydrazobenzene gives benzidine, which may

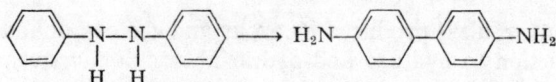

be isolated from the reaction mixture by addition of alkali.

Aromatic azoxy compounds are in general less intensely coloured than the corresponding azo compounds. They are reduced to azo compounds by alkaline reducing agents such as stannous chloride in alkali

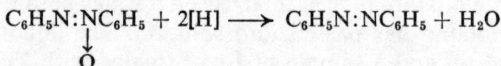

$$\underset{\downarrow}{C_6H_5N:NC_6H_5} + 2[H] \longrightarrow C_6H_5N:NC_6H_5 + H_2O$$
$$O$$

The action of zinc dust in glacial acetic acid leads to the corresponding hydrazo compounds.

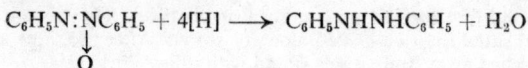

$$\underset{\downarrow}{C_6H_5N:NC_6H_5} + 4[H] \longrightarrow C_6H_5NHNHC_6H_5 + H_2O$$
$$O$$

As in test (ii) above, they are reduced by tin and hydrochloric acid to primary amines.

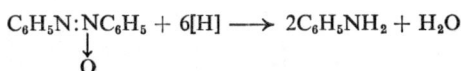

$$C_6H_5N:NC_6H_5 + 6[H] \longrightarrow 2C_6H_5NH_2 + H_2O$$
$$\downarrow$$
$$O$$

On warming with concentrated sulphuric acid they are converted into the isomeric hydroxyazo compounds which are usually highly coloured materials.

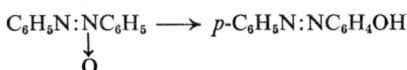

$$C_6H_5N:NC_6H_5 \longrightarrow p\text{-}C_6H_5N:NC_6H_4OH$$
$$\downarrow$$
$$O$$

Aromatic azo compounds are generally stable compounds which may be brominated, nitrated or sulphonated without fundamental change in the molecule. They may be reduced to amines by tin and hydrochloric acid as in test (ii) above;

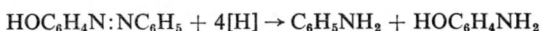

$$HOC_6H_4N:NC_6H_5 + 4[H] \rightarrow C_6H_5NH_2 + HOC_6H_4NH_2$$

while milder agents such as zinc powder in glacial acetic acid convert them into hydrazo compounds.

$$C_6H_5N:NC_6H_5 + 2[H] \rightarrow C_6H_5NHNHC_6H_5$$

Aliphatic nitro compounds, the majority of which are colourless liquids when pure, are reduced readily to primary amines.

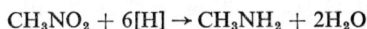

$$CH_3NO_2 + 6[H] \rightarrow CH_3NH_2 + 2H_2O$$

They dissolve in 20% sodium hydroxide solution to give yellow solutions from which they may be regenerated by addition of ice cold acetic acid. Primary nitroalkanes (RCH_2NO_2) when dissolved in concentrated sodium hydroxide solution and treated with an excess of sodium nitrite solution, give an intense red colour due to the salt of the nitrolic acid.

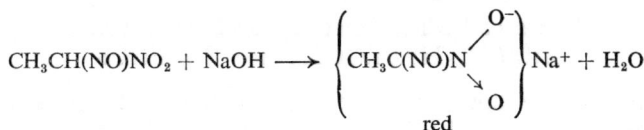

$$CH_3CH_2NO_2 + HNO_2 \longrightarrow CH_3CH(NO)NO_2 + H_2O$$

$$CH_3CH(NO)NO_2 + NaOH \longrightarrow \left\{ CH_3C(NO)N \begin{array}{c} O^- \\ \nearrow \\ \searrow \\ O \end{array} \right\} Na^+ + H_2O$$
$$\text{red}$$

Upon acidification of the reaction mixture with drops of sulphuric acid, the red colour is discharged.

Secondary nitroalkanes ($R'R''CHNO_2$), under the same conditions give dark blue or green colours due to nitroso derivatives which are soluble in chloroform but insoluble in water and alkali.

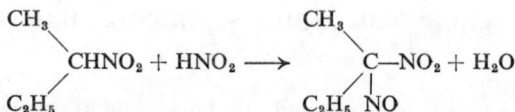

$$\begin{array}{c} CH_3 \\ \diagdown \\ CHNO_2 + HNO_2 \longrightarrow \\ \diagup \\ C_2H_5 \end{array} \qquad \begin{array}{c} CH_3 \\ \diagdown \\ C-NO_2 + H_2O \\ \mid \\ C_2H_5 \ NO \end{array}$$

Tertiary nitroalkanes give no colouration in this test.

Aromatic nitro compounds may be reduced by different reagents to amines, azo, hydrazo or azoxy compounds. On boiling with concentrated alkali, some aromatic nitro compounds undergo profound decomposition and this tendency is accentuated by the presence of more than one nitro group.

Polynitro aromatic compounds often give reddish colourations with aqueous alkali in the presence of acetone, more especially if the nitro groups are in positions *meta* to each other.

The majority of simple aromatic nitro compounds are either colourless or faintly yellow, but on introducing certain substituents into the molecule, such as hydroxyl or amino groups, a distinct colour is sometimes acquired. Thus all the nitroanilines possess brilliant yellow colours, while in the case of the nitrophenols, some are colourless and some distinctly yellow, but all form intensely coloured solutions in alkali. Thus *p*-nitrophenol is colourless, and forms a yellow sodium salt; *o*-nitrophenol is pale yellow, and forms a red sodium salt; while picric acid, which has a lemon yellow colour, yields an intensely yellow solution in alkali but colourless solutions in ligroin and in mineral acids. Nitrophenols also differ from simple phenols in that they are far more strongly acidic and are acylated with greater difficulty; this latter observation is applicable to aromatic nitroamines in contrast to other aromatic amino compounds.

Purines all respond to the murexide test: on evaporating a small quantity of the substance in a porcelain basin on the water-bath with chlorine water or nitric acid, and then adding a drop of ammonia or sodium hydroxide solution to the residue after cooling, a purple colour is produced.

Types of Radical Involving the Detection of
CARBON, HYDROGEN and SULPHUR

If the substance shows an acid reaction to litmus after warming with water, it may be a sulphonic acid, a sulphinic acid, a thiocarboxylic acid, an alkyl sulphate or sulphite, or an alkyl ester of a sulphonic acid. It is to be noted, however, that this evidence must not be relied upon with too much assurance, as carboxyl or phenolic hydroxyl groups may be present in the molecule.

When a metallic residue has also been detected in a sulphur containing substance which gives an acid reaction on boiling with water, it may possibly be a bisulphite compound of an aldehyde or a ketone, or a salt of an alkyl sulphuric acid.

Bisulphite compounds of aldehydes or ketones, on heating alone, yield water, sulphur dioxide, a metallic sulphite, and the free aldehyde or ketone.

$$2(CH_3)_2C(OH)SO_2ONa \rightarrow 2CH_3COCH_3 + Na_2SO_3 + SO_2 + H_2O$$

On warming with dilute aqueous acids or sodium carbonate they yield the free carbonyl compound, together with sulphurous acid or a sulphite. For the purpose of isolating the aldehyde or ketone it is advisable to add dilute sodium carbonate and to distil with or without injection of steam, or, after warming, to extract with ether, as may appear most convenient from the volatility or solubility of the desired compound.

Salts of alkyl sulphuric acids are slowly decomposed on boiling with dilute hydrochloric acid, yielding a sulphate and the corresponding alcohol.

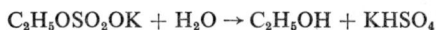

$$C_2H_5OSO_2OK + H_2O \rightarrow C_2H_5OH + KHSO_4$$

They are mild alkylating agents, reacting with aliphatic primary and secondary amines on long boiling in aqueous solution, with formation of sulphates of secondary or tertiary amines.

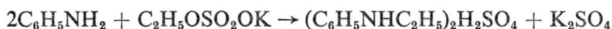

$$2C_6H_5NH_2 + C_2H_5OSO_2OK \rightarrow (C_6H_5NHC_2H_5)_2H_2SO_4 + K_2SO_4$$

They react readily on boiling with aqueous sodium sulphide to form the corresponding alkyl sulphides, together (especially in the cases of the higher alkyl derivatives) with some mercaptan.

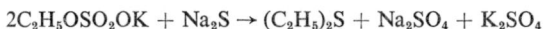

$$2C_2H_5OSO_2OK + Na_2S \rightarrow (C_2H_5)_2S + Na_2SO_4 + K_2SO_4$$

$$2C_2H_5OSO_2OK + 2Na_2S + 2H_2O \rightarrow 2C_2H_5SH + K_2SO_4 + Na_2SO_4 + 2NaOH$$

They are also hydrolysed on boiling with alkalis, with formation of the corresponding alcohol, but (with the exception of the case of the methyl derivative) this reaction is too slow for the purpose of identification.

Aromatic sulphonic acids are often crystalline solids readily soluble in water from which solutions they cannot be extracted with ether. The free acids are frequently hygroscopic and because of the difficulty of isolating them, they are usually encountered as their sodium, potassium, calcium or barium salts. They may be detected by *fusion with alkali* as follows: about 0.5 g of the substance, 3 g of potassium hydroxide and 5 drops of water are mixed in a nickel crucible which is carefully heated so that the mixture just melts and is held in the molten state, with occasional stirring, for 10 minutes. When the melt has cooled, about

5 cm³ of water are added and stirred to dissolve the mass, by warming if necessary. On acidification of the solution, sulphur dioxide is evolved and may be detected by means of a filter paper strip impregnated with acidified potassium dichromate solution.

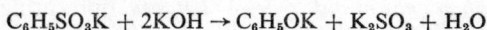

$$C_6H_5SO_3K + 2KOH \rightarrow C_6H_5OK + K_2SO_3 + H_2O$$

In some cases the phenol may be isolated by extracting the acidified solution with ether, but the yield is often low due to extensive decomposition during the fusion.

Aliphatic sulphonic acids are completely destroyed by the fusion process. [It should be noted that sulphoxides (RSOR'), sulphinic acids, and sulphones (RSO₂R') give an evolution of sulphur dioxide in the fusion test, while mercaptans (RSH), sulphides (RSR') and disulphides (RSSR') evolve hydrogen sulphide when the cooled melt is acidified.]

The action of phosphorus pentachloride upon the sodium salts of sulphonic acids or of thionyl chloride upon the free acids yields *sulphonyl chlorides* (RSO₂Cl). These may be purified by recrystallisation from benzene or ligroin or by distillation under reduced pressure. They are used as acylating agents, giving sulphonic esters with phenols:

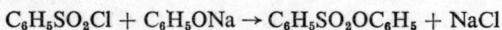

$$C_6H_5SO_2Cl + C_6H_5ONa \rightarrow C_6H_5SO_2OC_6H_5 + NaCl$$

and sulphonamides with ammonia or primary and secondary amines (see Hinsberg's method, p. 30).

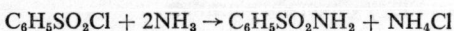

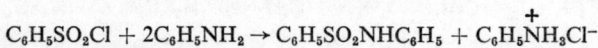

$$C_6H_5SO_2Cl + 2NH_3 \rightarrow C_6H_5SO_2NH_2 + NH_4Cl$$

$$C_6H_5SO_2Cl + 2C_6H_5NH_2 \rightarrow C_6H_5SO_2NHC_6H_5 + C_6H_5\overset{+}{N}H_3Cl^-$$

Sulphonyl chlorides are completely reduced to mercaptans by boiling with zinc and hydrochloric acid.

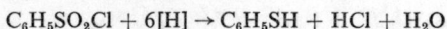

$$C_6H_5SO_2Cl + 6[H] \rightarrow C_6H_5SH + HCl + H_2O$$

With water, sulphonyl halides react but slowly unless heated, in which case the presence of halide ion is detectable by means of dilute nitric acid and silver nitrate solution.

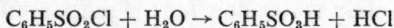

$$C_6H_5SO_2Cl + H_2O \rightarrow C_6H_5SO_3H + HCl$$

Aromatic sulphinic acids (p. 222) may be detected by Smiles' test: on dissolving in cold concentrated sulphuric acid and adding one drop of phenetole (or anisole) a blue colour is produced, due to the formation of a *para*-substituted aromatic sulphoxide by elimination of water:

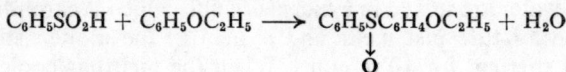

$$C_6H_5SO_2H + C_6H_5OC_2H_5 \longrightarrow C_6H_5SC_6H_4OC_2H_5 + H_2O$$
$$\underset{O}{\downarrow}$$

and on further addition of an excess of phenetole the blue colour is discharged, owing to the formation of the sulphate of a sulphonium base.

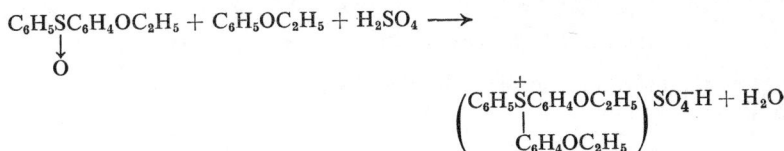

$$C_6H_5SC_6H_4OC_2H_5 + C_6H_5OC_2H_5 + H_2SO_4 \longrightarrow$$
$$\underset{O}{\downarrow}$$

$$\left(\underset{C_6H_4OC_2H_5}{\overset{+}{C_6H_5\underset{|}{S}C_6H_4OC_2H_5}} \right) SO_4^-H + H_2O$$

Aromatic sulphinic acids on fusion with potash yield hydrocarbons.

$$C_6H_5SO_2K + KOH \rightarrow C_6H_6 + K_2SO_3$$

They may be oxidised by permanganate to sulphonic acids, and reduced by zinc with hydrochloric acid to mercaptans.

Alkyl sulphates (p. 222) are in general water-insoluble liquids, which on boiling with water are converted into the corresponding alcohols and sulphuric acid, alkyl sulphuric acids being formed as intermediate products.

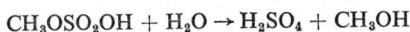

$$(CH_3O)_2SO_2 + H_2O \rightarrow CH_3OSO_2OH + CH_3OH$$
$$CH_3OSO_2OH + H_2O \rightarrow H_2SO_4 + CH_3OH$$

They are more powerful alkylating agents than salts of alkyl sulphuric acids, and can be employed, in the presence of alkali, for the alkylation of phenols, amines, and similar compounds.

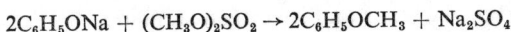

$$2C_6H_5ONa + (CH_3O)_2SO_2 \rightarrow 2C_6H_5OCH_3 + Na_2SO_4$$
$$2C_6H_5NH_2 + 2(CH_3O)_2SO_2$$
$$\rightarrow C_6H_5N(CH_3)_2CH_3OSO_2OH + C_6H_5NH_2CH_3OSO_2OH$$

Alkyl sulphites resemble carboxylic esters rather than alkyl sulphates in that they possess no alkylating properties. On hydrolysis with alkali they yield the corresponding alcohol and a sulphite.

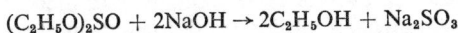

$$(C_2H_5O)_2SO + 2NaOH \rightarrow 2C_2H_5OH + Na_2SO_3$$

Alkyl esters of sulphonic acids (p. 221) exhibit properties similar to those of the alkyl sulphates, being hydrolysed to alcohols and sulphonates on boiling with aqueous alkalis,

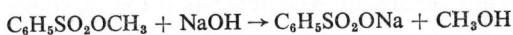

$$C_6H_5SO_2OCH_3 + NaOH \rightarrow C_6H_5SO_2ONa + CH_3OH$$

and acting as vigorous alkylating agents towards amines

$$2C_6H_5SO_2OCH_3 + 3C_6H_5NH_2 \rightarrow C_6H_5N(CH_3)_2 + 2C_6H_5SO_2OHNH_2C_6H_5$$

and phenols in alkaline solution.

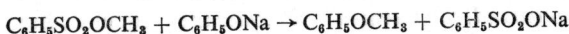

$$C_6H_5SO_2OCH_3 + C_6H_5ONa \rightarrow C_6H_5OCH_3 + C_6H_5SO_2ONa$$

Aryl esters of sulphonic acids (p. 221) are much more stable towards hydrolytic reagents than the foregoing, resisting the action of boiling 10% sodium hydroxide solution. They may, however, be broken up by heating with highly concentrated alkali to about 200°C.

$$C_6H_5SO_2OC_6H_5 + 2NaOH \rightarrow C_6H_5SO_2ONa + C_6H_5ONa + H_2O$$

On warming with alcoholic sodium ethoxide they yield a phenolic ether and a sulphonate.

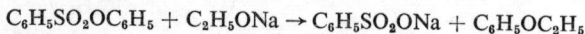

$$C_6H_5SO_2OC_6H_5 + C_2H_5ONa \rightarrow C_6H_5SO_2ONa + C_6H_5OC_2H_5$$

They do not react in an analogous manner with amines, and can be boiled with aniline without change. For their identification, they may be nitrated by the action of concentrated nitric acid upon a solution in concentrated sulphuric acid, when they yield dinitro derivatives which are readily broken up by alkalis into salts of nitrosulphonic acids and (generally *para*) nitrophenols.

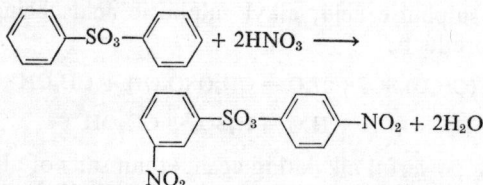

Thiocarboxylic acids (p. 222), on boiling with water, yield the corresponding carboxylic acids and hydrogen sulphide.

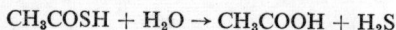

$$CH_3COSH + H_2O \rightarrow CH_3COOH + H_2S$$

They may be employed as acylating agents reacting with amines such as aniline with production of substituted amides and hydrogen sulphide.

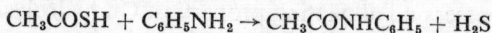

$$CH_3COSH + C_6H_5NH_2 \rightarrow CH_3CONHC_6H_5 + H_2S$$

They do not yield thio-esters on treatment with alcohols in the usual manner, but lose hydrogen sulphide with formation of carboxylic esters.

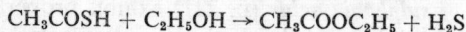

$$CH_3COSH + C_2H_5OH \rightarrow CH_3COOC_2H_5 + H_2S$$

On treatment with salts of heavy metals in aqueous solution they yield coloured precipitates which decompose into the sulphides on boiling.

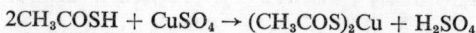

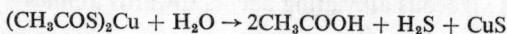

$$2CH_3COSH + CuSO_4 \rightarrow (CH_3COS)_2Cu + H_2SO_4$$
$$(CH_3COS)_2Cu + H_2O \rightarrow 2CH_3COOH + H_2S + CuS$$

If the substance is strongly alkaline, or is the salt of a strong base which is not liberated by aqueous alkali, it may possibly be a *sulphonium compound*. *Sulphonium hydroxides* are frequently more strongly basic than the hydroxides of the alkali metals; the free bases cannot for this reason be prepared by the action of caustic soda upon solutions of their salts. The process commonly employed for this purpose involves the action of silver oxide upon a solution of sulphonium halide.

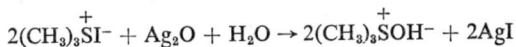

$$2(CH_3)_3\overset{+}{S}I^- + Ag_2O + H_2O \rightarrow 2(CH_3)_3\overset{+}{S}OH^- + 2AgI$$

Sulphonium salts in acetone solution dissolve mercuric iodide with formation of addition compounds, which crystallise out on evaporation of the solvent.

If the substance is practically neutral to litmus, it may be a sulphone, a sulphoxide, a sulphide or a disulphide, or a mercaptan. Due allowance must be made for the possible presence of carboxyl groups, and the fact is to be borne in mind that the presence in the molecule of a sulphone grouping considerably augments the acidic character of phenolic hydroxyl groups.

Sulphones (p. 218) exhibit no specific reactions, being extremely stable. They are unattacked by the action of the most powerful acids, alkalis, or oxidising or reducing agents. Aromatic sulphones may thus be boiled with fuming nitric acid with no further effect than nitration, while the simpler aliphatic members of this class are unaffected even by this violent treatment. They are, however, broken up into sulphinates and hydrocarbons on treatment with sodium in boiling toluene.

$$2C_6H_5SO_2C_6H_5 + 2Na \rightarrow 2C_6H_5SOONa + C_6H_5C_6H_5$$

The simpler sulphones may be distilled unchanged under atmospheric pressure; some disulphones such as sulphonal, on the other hand, break down on strongly heating, yielding carboxylic acids, mercaptans, sulphur dioxide, and other products.

Sulphones may be detected by the fusion test described under sulphonic acids; sulphur dioxide is evolved upon acidification of the cold melt.

Sulphoxides (p. 218) may exhibit faintly basic properties. Many give Smiles' test (*cf.* aromatic sulphinic acids),* and are far less stable than the sulphones. They may be oxidised by nitric acid, or better, by adding finely powdered potassium permanganate to a solution in glacial acetic acid, with formation of sulphones, and may be reduced to the corresponding sulphides by boiling with tin or zinc in hydrochloric acid.

Sulphides and mercaptans (thiols) are generally liquids with unpleasant odours. Upon fusion with potassium hydroxide and treatment with

---

* It is to be observed that aromatic sulphoxides which contain hydroxyl and similar groups dissolve in concentrated sulphuric acid with a more or less pronounced blue colour.

acid as described under sulphonic acids (*q.v.*), these compounds evolve hydrogen sulphide (detected by lead acetate paper).

Both mercaptans and sulphides react readily with mercuric chloride solution to give crystalline precipitates, mercaptans yielding mercaptides, which may be isolated and used for characterisation, e.g.

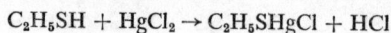

$$C_2H_5SH + HgCl_2 \rightarrow C_2H_5SHgCl + HCl$$

The liberation of hydrochloric acid in the reaction may easily be detected and serves to distinguish mercaptans from sulphides, which with mercuric chloride give precipitates that vary in composition according to the nature of the sulphide, some yielding crystalline adducts while others undergo fission to produce mercury mercaptides.

A distinction may be made between a primary thiol, a secondary thiol and a thiophenol by treating the substance dissolved in ethanol with solid sodium nitrite followed by dilute sulphuric acid. Primary and secondary mercaptans produce a red colour, tertiary mercaptans and thiophenols give a green colour which changes to red on standing. This test depends upon the formation of a nitrosyl-mercaptide (RSNO).

Sulphides, on oxidation by standing with the calculated amount of hydrogen peroxide in acetone solution, yield sulphoxides, while on treatment with stronger oxidising agents, such as nitric acid or potassium permanganate in glacial acetic acid, they are oxidised to sulphones.

Mercaptans, on oxidation with dilute nitric acid, are converted into the corresponding sulphonic acids,

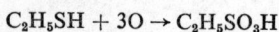

$$C_2H_5SH + 3O \rightarrow C_2H_5SO_3H$$

milder oxidising agents, such as bromine water, iodine, or ferric chloride, giving rise to disulphides.

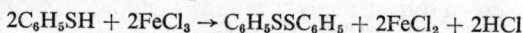

$$2C_6H_5SH + 2FeCl_3 \rightarrow C_6H_5SSC_6H_5 + 2FeCl_2 + 2HCl$$

On treatment with cold concentrated sulphuric acid they are oxidised with evolution of sulphur dioxide. Mercaptans exhibit their close relationship to hydrogen sulphide in their ability to form metallic derivatives, yielding (generally yellow) precipitates with dilute silver nitrate and dissolving in aqueous sodium hydroxide. With the exception of that of methyl mercaptan, the sodium salts of the simple aliphatic mercaptans are so readily hydrolysed that on boiling their alkaline solutions the free mercaptans are liberated and pass out with the steam. Like the corresponding sulphides, the alkali mercaptides of the aliphatic series yield intense purple colours on treatment with sodium nitroprusside solution; in the aromatic series these colours are extremely transitory, owing to the oxidation of the mercaptan to the disulphide.

Mercaptans may be acetylated by the action of acetyl chloride in dry chloroform or benzene solution, with formation of thioacetic esters.

$$C_2H_5SH + CH_3COCl \rightarrow CH_3COSC_2H_5 + HCl$$

Disulphides (p. 217) exhibit properties similar to those of the sulphides. They may, however, be reduced to the mercaptans by treatment with zinc dust in dilute mineral acids

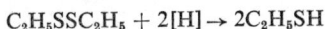

$$C_2H_5SSC_2H_5 + 2[H] \rightarrow 2C_2H_5SH$$

and yield sulphonic acids on oxidation with nitric acid. Solutions of aromatic disulphides in concentrated sulphuric acid, on treatment in the cold with phenetole, yield mixed *p*-phenetyl sulphides with evolution of sulphur dioxide.

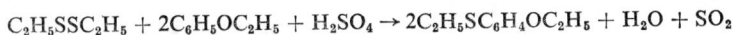

$$C_2H_5SSC_2H_5 + 2C_6H_5OC_2H_5 + H_2SO_4 \rightarrow 2C_2H_5SC_6H_4OC_2H_5 + H_2O + SO_2$$

Types of Radical Involving the Detection of CARBON, HYDROGEN and HALOGEN

Throughout the series of organic halogen compounds it may be taken that the fluorine compounds usually part less readily with their halogen than do the corresponding chlorine and bromine compounds, and that iodine compounds are the most reactive of their class. Thus alkyl iodides are rapidly decomposed on warming with alcoholic silver nitrate, the other alkyl halides being decomposed by this reagent far less readily, if at all. The only iodo compounds which are not decomposed in this way are those in which the iodine atom is directly attached to an aromatic nucleus, an example of this type being iodobenzene.

On boiling the substance with alcoholic potash, all types of halogen compounds, with the exception of the majority of those in which the halogen atom is united to an aromatic nucleus, yield potassium halide. This, if formed in considerable quantity, is precipitated from the alcoholic solution, or if formed in only small amount, can be detected by dilution, acidification with dilute nitric acid, removal of by-products by filtration or extraction with ether, and addition of silver nitrate to the clear aqueous solution. Aromatic halogen compounds in which a nitro group is present in the positions *ortho* or *para* to the halogen atom are, unlike orther aryl halides, decomposed with replacement of the halogen atom by a hydroxyl or alkoxyl group. Halogen compounds may thus be divided into two large classes, which comprise organic derivatives of all four halogens.

Organic compounds containing fluorine, chlorine and bromine may further be roughly subdivided according to their relative reactivity towards halogen-eliminating reagents.

On treatment with water, warming if the substance does not enter into solution, and decanting the aqueous portion, solutions of halogen hydracids will be present in the case of carboxylic acid halides and aliphatic halogen substituted ethers in which the halogen atom is attached to the same carbon atom as the alkoxyl group.

Carboxylic acid halides (pp. 69, 160) react with water, yielding the corresponding carboxylic acid and hydrochloric acid or one of its analogues.

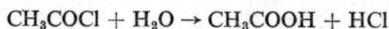

$$CH_3COCl + H_2O \rightarrow CH_3COOH + HCl$$

This reaction takes place rapidly in the cold with the halides of the lower fatty acids, but increasingly slowly with increasing molecular weight, possibly on account of decreasing solubility in water. On treatment with alcohols, in which they are soluble, they rapidly yield esters.

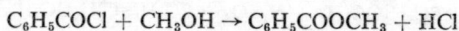

$$C_6H_5COCl + CH_3OH \rightarrow C_6H_5COOCH_3 + HCl$$

A similar reaction takes place, rather more slowly, on warming with phenols.

$$CH_3COCl + C_6H_5OH \rightarrow CH_3COOC_6H_5 + HCl$$

On slowly adding an acid halide to an excess of concentrated ammonia solution, with efficient agitation and cooling, the amide is formed

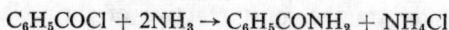

$$C_6H_5COCl + 2NH_3 \rightarrow C_6H_5CONH_2 + NH_4Cl$$

The amides of acids of high molecular weight, being insoluble in water, separate out during the reaction; those of low molecular weight, which dissolve more readily, frequently remain in solution, so that it is often convenient to carry out the reaction in the presence of a relatively large volume of ether; the ammonium chloride solution is separated and the ether distilled, when the amide remains as a residue. An analogous reaction takes place with amines such as aniline or diphenylamine.

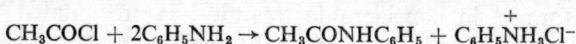

$$CH_3COCl + 2C_6H_5NH_2 \rightarrow CH_3CONHC_6H_5 + C_6H_5\overset{+}{N}H_3Cl^-$$

If the base be dissolved in ether, the amine hydrochloride is precipitated and may be filtered off, leaving the substituted amide in solution.

Similar in general behaviour are the *alkyl chloroformates* (or *chlorocarbonates*), (p. 160), which react slowly with cold water, yielding hydrochloric acid, carbon dioxide, and alcohols.

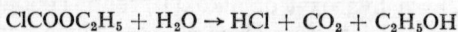

$$ClCOOC_2H_5 + H_2O \rightarrow HCl + CO_2 + C_2H_5OH$$

On warming with alcohols they yield alkyl carbonates and hydrogen chloride:

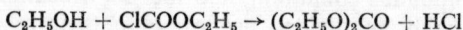

$$C_2H_5OH + ClCOOC_2H_5 \rightarrow (C_2H_5O)_2CO + HCl$$

and react similarly on shaking with cold alkaline solutions of phenols.

$$C_6H_5ONa + ClCOOC_2H_5 \rightarrow C_6H_5OCOOC_2H_5 + NaCl$$

When slowly added to cold concentrated ammonia, preferably covered with a large volume of ether, alkyl carbamates (urethanes) are formed and may be isolated by distilling the ethereal solution.

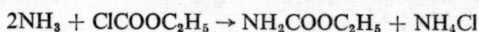

$$2NH_3 + ClCOOC_2H_5 \rightarrow NH_2COOC_2H_5 + NH_4Cl$$

An analogous reaction takes place with aniline, alkyl carbanilates being produced.

$$2C_6H_5NH_2 + ClCOOC_2H_5 \rightarrow C_6H_5NHCOOC_2H_5 + C_6H_5\overset{+}{N}H_3\overset{-}{Cl}$$

a-Halogen substituted aliphatic ethers (p. 153) on treatment with water, and in some cases with alcohol, yield the corresponding alcohols, aldehydes, and halogen hydrides.

$$CH_3OCH_2Cl + H_2O \rightarrow CH_2O + CH_3OH + HCl$$

The aldehyde produced may be identified by treatment of the reaction solution with a methanolic solution of 2,4-dinitrophenylhydrazine in the presence of concentrated hydrochloric acid, when its dinitrophenyl-hydrazone is precipitated.

On treatment with anhydrous potassium or sodium acetate, the halogen may be replaced by an acetoxyl group.

$$CH_3CH(OCH_3)Cl + CH_3COOK \rightarrow CH_3CH(OCH_3)OCOCH_3 + KCl$$

As a general rule, *halogen substituted ketones, acids and esters, in which the halogen atom is attached to the carbon atom next to the carbonyl group or similar group,* give rise to solutions containing considerable quantities of ionised halogen as shown by the formation of a distinct precipitate of silver halide, when treated as follows: about 0.5 g of substance dissolved in 20 cm³ of ethanol or aqueous ethanol, is refluxed with 0.5 cm³ of pyridine; chlorine compounds require about 30 minutes and bromine compounds about 5 minutes. The mixture is then cooled, diluted with distilled water, acidified with dilute nitric acid and treated with dilute silver nitrate solution.

Similarly, the halides of such radicals as *allyl* and *benzyl* part with their halogen fairly readily to pyridine.

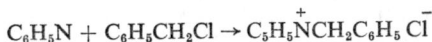

$$C_6H_5N + C_6H_5CH_2Cl \rightarrow C_5H_5\overset{+}{N}CH_2C_6H_5 \ \overset{-}{Cl}$$

Other classes of aliphatic chlorine and bromine compounds may yield a faint turbidity on the addition of silver nitrate, while aryl halides yield none whatever.

Simple *alkyl halides* (pp. 80, 145 et seq.) may be distinguished from other types of organic halogen compounds by their complete insolubility in concentrated sulphuric acid. Unsaturated and aromatic halogenated hydrocarbons may dissolve in hot sulphuric acid, with formation of acidic products, but this action takes place with increasing difficulty as the proportion of halogen in the molecule increases.

Chloroform, bromoform, iodoform, and chloral hydrate may be detected in very small amounts by the following test: to 4 cm³ of 20% sodium hydroxide solution are added enough pyridine to form a layer 2 mm deep and a drop of the liquid suspected to contain one of these compounds. The mixture is heated just to boiling and allowed to stand; should one of the above substances be present the pyridine layer assumes a red colour.

Aromatic halogen compounds (pp. 81, 85, 145–153, 156–159, 161–164) may be treated according to the methods suggested for the examination of aromatic hydrocarbons, namely nitration, sulphonation, oxidation, etc. (pp. 81–84), when characteristic derivatives can be prepared in the same ways. As stated above, halogen atoms directly united to an aromatic nucleus are incapable of undergoing the general metathetical reactions characteristic of alkyl halides; there is however one exception to this rule, and this is the formation of Grignard reagents with magnesium in dry ether. This reaction takes place as readily as it does with the aliphatic halogen compounds, and it also can best be adapted to identification by the formation of carboxylic acids through the action of carbon dioxide.

$$C_6H_5Br \rightarrow C_6H_5MgBr \rightarrow C_6H_5COOMgBr \rightarrow C_6H_5COOH$$

Iodine compounds differ from the corresponding chloro and bromo derivatives in their behaviour towards chlorine. Aliphatic iodides on treatment with chlorine are converted into the corresponding chlorides, with elimination of iodine. An analogous reaction takes place with bromine. Aromatic iodo compounds, on the other hand, lose no iodine but form yellow, crystalline dichlorides.

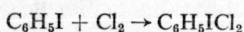

$$C_6H_5I + Cl_2 \rightarrow C_6H_5ICl_2$$

Types of Radical in which
CARBON, HYDROGEN, NITROGEN and
SULPHUR occur in conjunction

Under this head are included the thiocyanates, isothiocyanates, thioamides and thioureas, and sulphonamides. In other types of compounds in which carbon, hydrogen, nitrogen and sulphur have been detected, the groups involving the nitrogen and the sulphur atoms may be considered independently. Thus, for example, in the case of sulphanilic acid, the nitrogen may be demonstrated to be present in the form of an aromatic primary amine radical by the formation of a diazo solution and other characteristic reactions, while the sulphur may be shown to be in the form of a sulphonyl group by the fact that the substance forms neutral salts with alkalis, cannot be reduced to a sulphide or a mercaptan, and so on. The effect of the two groups upon each other is of course to be taken into consideration, as this frequently causes very considerable changes from the usual reactions of the individual types. Thus sulphanilic acid, on fusion with potash, instead of yielding potassium sulphite and an aminophenol, yields aniline and potassium sulphate.

In the types discussed in this section the sulphur and nitrogen atoms are so dependent upon each other that no selective examination of either is possible, the whole group being considered as one individual radical.

In the *aliphatic and aromatic thiocyanates* (p. 226) the fact that the alkyl or aryl radical is attached to sulphur is shown by reduction with zinc dust in hydrochloric acid.

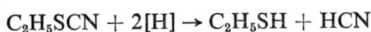

$$C_2H_5SCN + 2[H] \rightarrow C_2H_5SH + HCN$$

The resulting mercaptan should be isolated and examined, and hydrocyanic acid should be carefully tested for by passing the evolved gases into alkali and treating the solution for the 'Prussian blue' test.

On oxidation by boiling with nitric acid thiocyanates yield sulphonic acids.

$$C_2H_5SCN + 2[O] + H_2O \rightarrow C_2H_5SO_2OH + HCN$$

On boiling with alcoholic potash, potassium thiocyanate is formed:

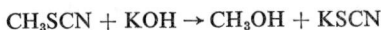

$$CH_3SCN + KOH \rightarrow CH_3OH + KSCN$$

which may be detected by diluting with water, acidifying with dilute nitric acid, and adding a drop of ferric chloride solution, when a red colouration of ferric thiocyanate will appear. Some thiocyanates whose boiling points lie above 160°C are converted partially or entirely into the corresponding isothiocyanates on distillation under atmospheric pressure.

Isothiocyanates (p. 226), or 'Mustard oils,' possess properties analogous to those of the isocyanates, though they are somewhat less reactive. Like the latter, they possess irritating odours. On long boiling with concentrated hydrochloric acid they break down into primary amines, with evolution of hydrogen sulphide, which should be detected by placing in the evolved vapours a piece of filter paper moistened with lead acetate solution.

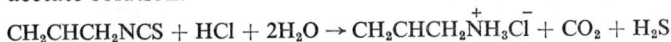

$$CH_2CHCH_2NCS + HCl + 2H_2O \rightarrow CH_2CHCH_2\overset{+}{N}H_3\overset{-}{Cl} + CO_2 + H_2S$$

On reduction with zinc dust in dilute mineral acid they yield primary amines with evolution of thioformaldehyde, which is recognisable by its leek-like odour.

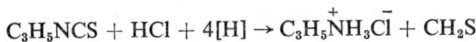

$$C_3H_5NCS + HCl + 4[H] \rightarrow C_3H_5\overset{+}{N}H_3\overset{-}{Cl} + CH_2S$$

On treatment with ammonia or with primary or secondary amines, derivatives of thiocarbamide are produced.

$$C_3H_5NCS + C_6H_5NH_2 \rightarrow C_3H_5NHCSNHC_6H_5$$

On warming in alcoholic solution with mercuric oxide or chloride, a derivative of urethane is produced, with precipitation of mercuric sulphide.

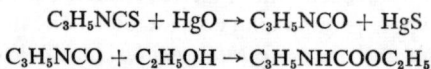

$$C_3H_5NCS + HgO \rightarrow C_3H_5NCO + HgS$$

$$C_3H_5NCO + C_2H_5OH \rightarrow C_3H_5NHCOOC_2H_5$$

Sulphonamides derived from ammonia (RSO_2NH_2) or from a primary amine (RSO_2NHR^1) are soluble in alkali; acidification of the solution regenerates the original substance (see Hinsberg's method, page 30).

When fused with potassium hydroxide as described under aromatic sulphonic acids (page 37) sulphonamides liberate ammonia or an amine: the evolution of sulphur dioxide on acidification of the aqueous extract is further evidence of the presence of a sulphonamide. Simple sulphonamides, on heating to about 155°C with 70–80% sulphuric acid for about 5 minutes yield the free sulphonic acid and ammonium sulphate,

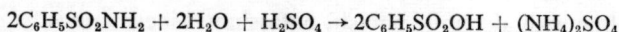

$$2C_6H_5SO_2NH_2 + 2H_2O + H_2SO_4 \rightarrow 2C_6H_5SO_2OH + (NH_4)_2SO_4$$

while primary and secondary substituted sulphonamides on hydrolysis yield the corresponding sulphonic acid and amine. If the mixture be heated to boiling, the parent hydrocarbon, sulphuric acid, and ammonia are formed.

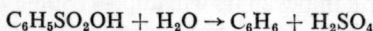

$$C_6H_5SO_2OH + H_2O \rightarrow C_6H_6 + H_2SO_4$$

Thioamides and thioureas are all solid compounds, more acidic than their oxygen analogues and so are soluble in alkali. When boiled with alkali, thiourea and simple thioamides ($RCSNH_2$) evolve ammonia, while substituted thioureas slowly liberate ammonia and/or amine. Acidification of the reaction mixture liberates hydrogen sulphide.

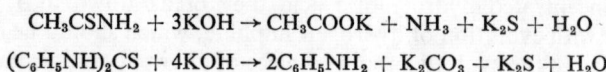

$$CH_3CSNH_2 + 3KOH \rightarrow CH_3COOK + NH_3 + K_2S + H_2O$$

$$(C_6H_5NH)_2CS + 4KOH \rightarrow 2C_6H_5NH_2 + K_2CO_3 + K_2S + H_2O$$

Treatment of the substance with a suspension of yellow mercuric oxide in ethanol and warming the mixture leads to the precipitation of black mercuric sulphide, together with formation of nitriles (from simple thioamides), cyanamides (from simple and monosubstituted thioureas) or substituted ureas (from symmetrically disubstituted thioureas).

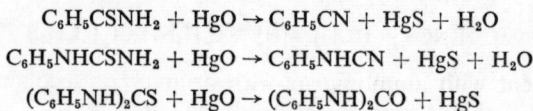

$$C_6H_5CSNH_2 + HgO \rightarrow C_6H_5CN + HgS + H_2O$$

$$C_6H_5NHCSNH_2 + HgO \rightarrow C_6H_5NHCN + HgS + H_2O$$

$$(C_6H_5NH)_2CS + HgO \rightarrow (C_6H_5NH)_2CO + HgS$$

The action of boiling dilute sulphuric acid decomposes thioamides and thioureas into carboxylic acids and guanidines, respectively, as the main organic products, hydrogen sulphide being liberated in all cases.

In the fusion test for sulphonic acids (q.v.) thioamides and thioureas evolve hydrogen sulphide.

Thiourea (thiocarbamide), on heating carefully until it melts, produces ammonium thiocyanate which may be readily detected by cooling the melt, dissolving it in water and adding aqueous ferric chloride: the blood-red colour of ferric thiocyanate appears.

Types of Radical Involving the Detection of CARBON and HYDROGEN only

In this section are set out the principles to be followed in the examination of the remaining portion of the molecule of substances in which elements other than carbon and hydrogen have been detected. Such examination can of course be carried out only after the allotment of the compound into its proper class with respect to the nitrogen, sulphur, or halogen contained therein. For the purposes in view, however, it is assumed throughout that in the types of compound described no elements other than carbon and hydrogen have been detected, for if the discussion is not restricted in this way, an unwieldy mass of exceptions and special cases would require to be described. All such exceptions and special cases will be found in the tables of compounds. The preliminary tests, if rigorously and completely carried out, will have greatly limited the range of possibilities, if they have not definitely established the class of the substance under examination; hence doubt should but rarely arise from this restriction of the field of discussion.

If the substance is readily soluble in water, the presence of oxygen in the molecule is certain. Hydrocarbons and ethers, and the majority of esters, acid anhydrides, and many substances of high molecular weight containing but few hydroxyl groups are insoluble or but sparingly soluble in water. Other types of compound, especially those of low molecular weight, such as alcohols, some phenols, aldehydes, ketones, carboxylic acids, sugars, and glycosides, frequently dissolve in cold water.

When the substance is insoluble in water, the action of metallic sodium should be tried upon a solution of the dry substance in absolutely dry ether, which should be drawn from a stock maintained permanently in contact with this metal. As a general rule, evolution of hydrogen ensues in all cases except those of hydrocarbons and ethers and some esters, such as ethyl benzoate, in which no replaceable hydrogen atom is present.

Hydrocarbons may be distinguished from ethers, and in fact from all oxygen containing compounds which are sufficiently soluble in benzene, by taking advantage of the observation that solutions of iodine in oxygen free liquids are violet in colour, while solutions of iodine in liquids containing oxygen (combined either in the solvent or in some other solute present in sufficient quantity) possess a brown tint. The test may be performed by mixing in a test-tube 5 cm^3 of a 0.005% solution of iodine in pure benzene with 5 cm^3 of a 5–10% solution in the same solvent of the substance under examination. The colour of the resulting mixture, viewed through the length of the tube, is compared with that of a mixture of the same volumes of the iodine solution and pure benzene.

Hydrocarbons (pp. 81, 145), unless unsaturated or aromatic, do not exhibit a wide scope of reactions. Saturated hydrocarbons may be distinguished from unsaturated and aromatic hydrocarbons by their failure to rise in temperature when treated with a cold mixture of concentrated nitric and sulphuric acids. They are also completely insoluble in concentrated sulphuric acid, while unsaturated hydrocarbons are as a rule either dissolved or polymerised to higher boiling products. Aromatic hydrocarbons vary among themselves in their reactivity towards strong sulphuric acid; benzene is only slowly sulphonated at its boiling point; toluene and ethylbenzene are dissolved rapidly on boiling, though slowly in the cold; *ortho* and *meta* xylenes are sulphonated rapidly in the cold, while *para* xylene requires long treatment at the boiling temperature for complete sulphonation; mesitylene and pseudocumene are very rapidly sulphonated in the cold; naphthalene and anthracene, on the other hand, are sulphonated only slowly in comparison with the highly methylated derivatives of benzene.

On account of the lack of reactivity of saturated hydrocarbons, their identification must generally rest upon determinations of boiling point and density.

Aromatic hydrocarbons may frequently be oxidised by different reagents such as nitric acid, chromic acid mixture,* a solution of chromic acid in glacial acetic acid or sulphuric acid of fairly high concentration, or potassium permanganate in acid, neutral, or alkaline solution. Different end-products are frequently obtained by varying the oxidising agent. As a general rule, side-chains are thus oxidised to aldehydic or carboxyl groups directly united to the aromatic nucleus. For example toluene or ethylbenzene yields benzoic acid, naphthalene yields phthalic acid, and *p*-xylene or cymene yields terephthalic acid.

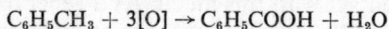

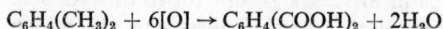

$$C_6H_5CH_3 + 3[O] \rightarrow C_6H_5COOH + H_2O$$
$$C_6H_4(CH_3)_2 + 6[O] \rightarrow C_6H_4(COOH)_2 + 2H_2O$$

The procedure for oxidation with alkaline permanganate, which is the method most generally applicable to all types of aromatic compounds, as well as hydrocarbons, is as follows: 1–2 g of the substance

---

* Prepared either by dissolving 10 g of chromium trioxide in 60 cm^3 of water, and adding 6 cm^3 of concentrated sulphuric acid; or by adding 11 cm^3 of sulphuric acid to a solution of 15 g of potassium dichromate in 60 cm^3 of water.

is boiled under reflux with very dilute alkali, some pieces of porous earthenware or capillary glass tubes sealed at one end being added to facilitate ebullition. A solution containing the exact amount of potassium permanganate, calculated from the equation

$$2KMnO_4 + H_2O \rightarrow 2MnO_2 + 2KOH + 3[O]$$

sufficient to produce the required effect, is gradually added to the boiling solution, and the heating continued until the original red colour of the permanganate has disappeared. The precipitated hydrated manganese dioxide is then filtered off, the filtrate concentrated to small bulk, filtered if necessary, acidified with hydrochloric acid, and the resulting acid filtered off or extracted with ether. It is inadvisable to employ more than the calculated quantity of permanganate, since the oxidation products themselves are frequently destroyed to some extent by the oxidising agent. Similarly, on oxidising with chromic acid mixture, the amount of oxidising agent is to be calculated from the equation

$$2CrO_3 + 3H_2SO_4 \rightarrow Cr_2(SO_4)_3 + 3H_2O + 3[O]$$

the action being continued until the original yellow colour has been entirely superseded by the green colour of the chromic salt. In addition to oxidation, sulphonation, bromination, and especially nitration, are of the greatest advantage for the purpose of preparing characteristic derivatives of aromatic hydrocarbons and other types of aromatic compounds.

Sulphonation may be effected either by warming with concentrated sulphuric acid or by the action of fuming sulphuric acid, according to the ease with which the compound is sulphonated. The resulting sulphonic acids frequently crystallise out on slightly diluting the sulphonation mixture at the end of the reaction, or the sodium salts may be thrown out by adding the mixture to a relatively large volume of saturated sodium chloride solution.

For the preparation of sulphonic derivatives the most satisfactory procedure, however, is to add the hydrocarbon gradually to about five times its weight of chlorosulphonic acid, with efficient cooling and agitation. On pouring the resulting mixture into crushed ice the corresponding sulphochloride (or mixture of isomeric sulphochlorides) separates as an oil or low melting solid, which can be extracted with chloroform and distilled under reduced pressure or filtered off and recrystallised from light petroleum.

$$(CH_3)_2C_6H_4 + 2ClSO_3H \rightarrow (CH_3)_2C_6H_3SO_2Cl + HCl + H_2SO_4$$

A small proportion of sulphone is frequently formed simultaneously;

such sulphones are less soluble in light petroleum than the sulpho-chlorides, or remain as residues on distillation. For further characteris-ation, the sulphochlorides may be converted into the corresponding sulphonamides by treatment with ammonia.

Bromination is best carried out by slowly adding the requisite amount of bromine to the hydrocarbon, to which a small quantity of iodine or iron filings has been added. After washing it with water, the product is fractionally distilled in order to separate unchanged hydrocarbon and dibromo derivatives from the principal monobromo product. If the bromine is added at 100–150°C in the absence of catalyst, bromination takes place in the side-chain; thus the xylenes yield xylyl bromides and xylylene bromides, and ethylbenzene yields a-bromo and $\alpha\beta$-dibromo derivatives. The preparation of these compounds is not recommended since they are often very strongly lachrymatory.

For the preparation of mononitro derivatives, a mixture of concen-trated nitric and sulphuric acids is added, with vigorous stirring, to the hydrocarbon alone or in solution in carbon tetrachloride, the temper-ature being held below 40°C. In order to obtain more highly nitrated derivatives, fuming (90–95%) nitric acid is substituted for the ordinary (65–70%) acid, and the temperature is raised towards the end of the reaction. Great care should be taken in nitrating unknown compounds, as the reaction is sometimes explosively violent and a splash of acid may cause serious injuries to the eyes or skin.

Many aromatic hydrocarbons, and aromatic compounds generally, form well-defined addition products with picric acid, produced by mixing saturated solutions of the substance under examination and of picric acid in 95% alcohol, warming, and allowing to cool slowly. The picrates, which crystallise out on cooling, may be recrystallised from the same solvent, and possess melting points characteristic of the hydrocarbon.

Unsaturated compounds, as stated in the chapter on examination for approximate constitution, yield addition products with bromine, and may be readily oxidised with formation of a variety of different end products, according to the degree of violence with which the oxidation is carried out. Unlike saturated and many aromatic hydrocarbons, they dissolve, at least partially, in concentrated sulphuric acid, from which, however, they cannot be recovered unchanged on dilution with water. The simpler aliphatic unsaturated hydrocarbons dissolve in cold 80–90% sulphuric acid, giving water-soluble alkyl sulphuric acids which on distillation with steam break up into secondary or tertiary alcohols.

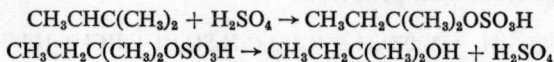

$$CH_3CHC(CH_3)_2 + H_2SO_4 \rightarrow CH_3CH_2C(CH_3)_2OSO_3H$$
$$CH_3CH_2C(CH_3)_2OSO_3H \rightarrow CH_3CH_2C(CH_3)_2OH + H_2SO_4$$

The higher the molecular weight of the hydrocarbon the more readily does this reaction take place and the lower the necessary concentration of the acid, but the higher members display an increasing tendency to polymerise to high boiling hydrocarbons which do not dissolve in the acid mixture.

Unsaturated, like aromatic, hydrocarbons react with fuming nitric acid, occasionally with considerable violence. This test serves to distinguish saturated paraffin hydrocarbons which alone are unattacked by this powerful reagent.

Ethers (pp. 79, 100) may be differentiated from saturated and aromatic hydrocarbons by the fact that they dissolve in cold concentrated sulphuric acid. On pouring the resulting solution very slowly into ice-cold water, the ether is thrown out of solution, and can be recovered unchanged. They may be oxidised by warming with a solution of chromic acid in concentrated sulphuric acid, with production, in the case of aliphatic or aliphatic-aromatic ethers, of carboxylic acids or of aldehydes. The aliphatic-aromatic ethers may be decomposed by boiling with constant boiling (48%) hydrobromic acid, which breaks them up into phenols and alkyl bromides.

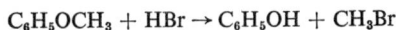

$$C_6H_5OCH_3 + HBr \rightarrow C_6H_5OH + CH_3Br$$

A similar effect may be brought about with concentrated hydriodic acid (cf. p. 252) or by carefully warming the ether with anhydrous aluminium chloride.

If the substance shows an acid reaction to litmus when examined in presence of water or aqueous alcohol, it may be a carboxylic acid or an acid anhydride. Some phenols and a few easily hydrolysable esters, such as methyl oxalate, may also show acid properties towards litmus. In the case of phenols, the acid reaction is generally extremely faint and often barely perceptible, while in the case of esters it is more the exception than the rule for an acid reaction to be obtained. In order to distinguish these classes, a small quantity, about 0.05 g of the substance, finely powdered if solid, is suspended or dissolved in pure water or aqueous alcohol, a drop of phenolphthalein solution added, and 0.1 Molar sodium or potassium hydroxide run in from a burette. Sharp end points will be given readily only by acids and very easily hydrolysable anhydrides and esters.

Carboxylic acids (pp. 77, 123, 128) are the most strongly acidic class of compound among substances containing only carbon, hydrogen, and oxygen, and their alkali salts are not decomposed by the action of carbon dioxide. The lower members of the series are miscible with water, and only a few acids exist which are insoluble in boiling water.

On warming on the water-bath with an excess of methyl or ethyl alcohol in presence of a small amount of concentrated sulphuric acid, or on passing a current of dry hydrogen chloride into a boiling solution of the acid in the alcohol, esters are produced.

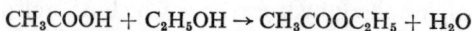

$$CH_3COOH + C_2H_5OH \rightarrow CH_3COOC_2H_5 + H_2O$$

When sulphuric acid has been employed for esterification, the mixture should be poured into cold water and ether added, the extract separated from the aqueous portion by means of a separating funnel, washed with dilute sodium carbonate solution and again with pure water, and finally distilled. When gaseous hydrogen chloride has been employed as a catalyst, the same process may be applied for isolation; but when the boiling point of the ester is high, the resultant mixture may be fractionally distilled without further treatment. Care must be taken, however, that the reaction has been given sufficient time to come to completion, otherwise the ester will be contaminated with unchanged acid. For this reason it is preferable, whenever possible, to wash the final product with sodium carbonate solution before distilling.

On warming on the water-bath at 50–60°C with phosphorus trichloride or with thionyl chloride, carboxylic acids yield the corresponding chlorides.

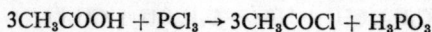

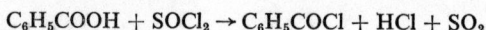

$$3CH_3COOH + PCl_3 \rightarrow 3CH_3COCl + H_3PO_3$$

$$C_6H_5COOH + SOCl_2 \rightarrow C_6H_5COCl + HCl + SO_2$$

These may be employed, without necessarily purifying before further conversion, for the preparation of esters, amides, or substituted amides.

On boiling an acid with an equivalent quantity of aniline or p-toluidine for several hours, the anilide or p-toluidide is produced.

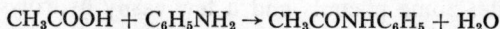

$$CH_3COOH + C_6H_5NH_2 \rightarrow CH_3CONHC_6H_5 + H_2O$$

The lower members of the fatty acid series may rapidly be distinguished by testing firstly with acid permanganate, which is decolourised by formic acid but by none of its homologues. The latter differ in the solubility of their ferric and cupric salts in organic liquids; 1–2% solution of the acid in water is accurately neutralised (to phenolphthalein) with 0.5 Molar alkali,* and 2 cm³ of the resulting solution is mixed with 1 cm³ of isoamyl alcohol (preferably diluted with half its volume of methyl alcohol), and one or two drops of 2% ferric chloride or copper sulphate are added. The mixture is well shaken and allowed to stand until it separates into layers. In the case of acetic acid (and formic acid) the aqueous layer is coloured and the isoamyl alcohol colourless; with propionic acid and higher homologues the colour is entirely in the alcohol. In the latter case the test is repeated, substituting ethyl ether for the isoamyl alcohol, when the colour is taken up by the ether only

---

* Sodium or potassium, not alkaline earth metal, salts of the acids must be employed.

with butyric and higher acids. These can be distinguished by employing benzene as the organic solvent, which becomes coloured only with valeric and higher acids. It is to be noted that the heavy metallic salts of butyric acid and its higher homologues are precipitated from aqueous solution, while with formic, acetic and propionic acids they are soluble in water.

The addition of ferric chloride solution serves roughly to distinguish simple acids from *aliphatic hydroxy acids* (p. 130). While simple acids form dull reddish colourations or precipitates when treated in aqueous solution with a few drops of this reagent, hydroxy acids, and in particular α-hydroxy acids, form intensely yellow coloured solutions on addition of a drop of ferric chloride. This last test may also be applied in the following manner: to a dilute aqueous solution of phenol a few drops of ferric chloride are added. On adding a small quantity of the resulting violet coloured solution to a solution of the α-hydroxy acid in water, the violet colouration is discharged, a deep yellow tint taking its place. The majority of aliphatic hydroxy acids are extremely soluble in water.

Many *aromatic hydroxy acids* (p. 131), in particular those in which the hydroxyl group is in the position *ortho* to the carboxyl group, give the violet or bluish colours produced by phenolic compounds on treatment with ferric chloride solution.

Acid anhydrides (pp. 69, 134) are generally sparingly soluble in water and are hydrolysed to the corresponding acids on boiling with water or dilute alkali. They react with hydroxylamine to produce hydroxamic acids (RCONHOH) which give highly coloured complex salts with ferric chloride solution: this reaction forms the basis of the *hydroxamic acid test* which is described under Esters, below.

A distinction may be made between esters and acid anhydrides by warming the substance, dissolved in benzene or chloroform, with a small volume of pure aniline for 1–2 minutes. The production of a precipitate of an anilide on cooling the solution indicates an anhydride.

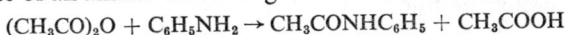

$$(CH_3CO)_2O + C_6H_5NH_2 \rightarrow CH_3CONHC_6H_5 + CH_3COOH$$

Anhydrides, when warmed with *n*-butyl alcohol give esters which may readily be detected by odour.

Esters (pp., 78, 135, 138) the majority of which are liquids insoluble in water, may be detected by the *hydroxamic acid test:* a preliminary test must be carried out by treating a solution of the substance in water or alcohol with aqueous ferric chloride—a red or violet colour indicates that the hydroxamic acid test is not applicable. Otherwise, 0.1 g of the compound are treated with 1 cm³ of a 5% solution of hydroxylamine hydrochloride in methanol and then with drops of methanolic potassium hydroxide until the mixture is alkaline to litmus. It is then boiled for 1 minute and the cooled mixture acidified with dilute hydrochloric acid. The addition of drops of ferric chloride solution produces a violet colouration due to a complex ferric salt.

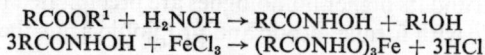

$$RCOOR^1 + H_2NOH \rightarrow RCONHOH + R^1OH$$
$$3RCONHOH + FeCl_3 \rightarrow (RCONHO)_3Fe + 3HCl$$

Esters may be hydrolysed by hot mineral acid, but it is usually advisable to carry out the hydrolysis by refluxing the ester (5 g) with an excess (40 cm^3) of a 30% solution of aqueous alkali, continuing the heating until all the ester has entered into solution (this may require from $\frac{1}{2}$ to 2 hr). The mixture is then distilled, and the alcohol which has passed over in the distillate is isolated by fractionation of the distillate and salting out the alcohol by the addition of solid potassium carbonate. The alcohol may, however, also be examined in solution. Alcohols are of course not present in the distillate obtained after hydrolysis of esters of most phenols and non-volatile alcohols such as glycols or glycerol; in such cases the reaction mixture should be saturated with carbon dioxide, when any phenol present may be isolated by distillation with steam or by extraction with ether, while polyhydroxylic alcohols can be separated only by evaporation to complete dryness on the water-bath and extraction of the residue with a mixture of alcohol and ether. The alkaline solution containing the salt of the acid should be acidified with dilute sulphuric acid and filtered, extracted with ether, or distilled, in order to isolate the free acid.

For aromatic esters a quicker method of hydrolysis is available: 0.3 g of pure potassium hydroxide is dissolved by warming gently with 5 cm^3 of diethylene glycol and 15 drops of water. The solution is cooled, 3 g of the ester is added and the whole refluxed gently, with stirring, until hydrolysis is complete (about 3–5 minutes usually). The alcohol and acid may be isolated as already described.

Many esters are converted into the amides on standing with concentrated ammonia solution.

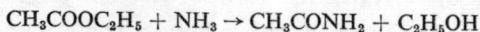

$$CH_3COOC_2H_5 + NH_3 \rightarrow CH_3CONH_2 + C_2H_5OH$$

Phenols (pp. 85, 108) exhibit, as stated above, slightly acid properties, being almost always soluble in solutions of alkali hydroxides, and occasionally in ammonia. The resultant alkali salts are, in contradistinction to the salts of carboxylic acids, decomposed by an excess of carbon dioxide.

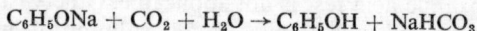

$$C_6H_5ONa + CO_2 + H_2O \rightarrow C_6H_5OH + NaHCO_3$$

Phenols may be recognised by the violet, blue, or green colours which are frequently imparted to their aqueous or alcoholic solutions by addition of one or two drops of ferric chloride. The permanence of such colourations varies greatly with individual phenols—some persisting indefinitely, some disappearing only after several minutes, and some changing in tint almost immediately. Some phenols, also, give colourations in alcoholic solution while giving none in water. It is to be observed that similar colourations are produced by ferric chloride with certain keto-enolic substances, such as acetoacetic esters. Such compounds also form solutions in dilute aqueous alkalis, and may be precipitated therefrom* by the action of carbon dioxide. They lack, however, that stability towards acid and alkaline hydrolytic reagents which is characteristic of phenols, and reflects their complete difference in class.

The majority of phenols are readily brominated on treatment with bromine water, with immediate formation of polybromo derivatives, which may be recrystallised from alcohol; they also may often be nitrated or sulphonated, with formation of characteristic derivatives. These reactions all take place with extreme readiness.

A sensitive test for phenols in very dilute solution is afforded by the development of a red colour on adding alkali and a small quantity of a cold solution of *p*-nitroaniline in dilute hydrochloric acid which has been decolourised (diazotised) by sodium nitrite. The following test is applicable to the detection of small quantities of phenols with the *para* position free: to 10 cm³ of the phenol solution is added one drop of a 10% solution of sodium nitrite, then 2–5 cm³ of concentrated sulphuric acid is run under the surface; a ring, green below and red above, forms at the zone of contact.

The remaining types included in this section exhibit no acid properties towards indicators.

Alcohols (pp. 70, 102) may be detected by one of the following tests:

(i) the compound is dissolved in water or pure dioxan and treated with a few drops of ceric nitrate solution in dilute nitric acid. A red colour indicates an alcohol, though it should be noted that the test is satisfactory only if the compound contains less than ten carbon atoms per molecule.

(ii) the substance (0.5 cm³) is treated with 0.3 cm³ acetyl chloride in a *dry* test-tube: alcohols (and other compounds having replaceable hydrogen) give a vigorous reaction. After about 2–3 minutes, the contents of the test-tube are poured into 3 cm³ of water, neutralised by careful addition of solid sodium bicarbonate and the residual product examined for properties of an ester.

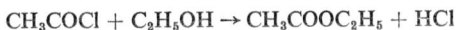

$$CH_3COCl + C_2H_5OH \rightarrow CH_3COOC_2H_5 + HCl$$

---

* Exceptions are the more highly acidic polynitro and polyhalogeno phenols.

Primary, secondary and tertiary alcohols may be differentiated by virtue of their behaviour towards chromic acid, which rapidly oxidizes primary and secondary alcohols to acids and ketones respectively, but does not oxidise tertiary alcohols under the test conditions. The oxidising agent is prepared by dissolving 1 g of chromium trioxide in 1 cm³ of concentrated sulphuric acid and pouring this solution carefully into 25 cm³ of distilled water. The cooled solution is used in the test procedure: to 1 cm³ of *pure* acetone is added one drop or 10 mg of the substance followed by one drop of the chromic acid reagent. The mixture is stirred immediately and the reaction of the suspected alcohol observed within *two seconds*. Primary and secondary alcohols react within this time to give a precipitate of a greenish-blue colour: tertiary alcohols give no sign of reaction and the solution retains its orange colour. It should be noted that positive results are also obtained from aldehydes and enols.

Simple primary secondary and tertiary alcohols may be differentiated by their reactivity towards strong hydrochloric acid. Primary alcohols as a rule are considerably more soluble in this reagent than in pure water but do not enter into reaction with it, even on long boiling; secondary alcohols yield the corresponding chlorides, but only on warming; while tertiary alcohols react so readily with concentrated hydrochloric acid that the chlorides separate in the cold. Exception to this rule must be made in the cases of alcohols of the benzyl type and of glycols, which are (although primary alcohols) converted into their chlorides on boiling with concentrated hydrochloric acid. The analogous reaction takes place rather more readily with hydrobromic acid and hydriodic acid. Addition of zinc chloride markedly promotes the esterification of alcohols by hydrochloric acid, so that by its use primary chlorides may be prepared in good yields.

Aldehydes and Ketones (pp. 75, 112, 116) are best detected by means of the 2,4-dinitrophenylhydrazine reagent, with which aldehydes and ketones, as well as *acetals*, give sparingly soluble dinitrophenyl-hydrazones which are usually yellow or red in colour.

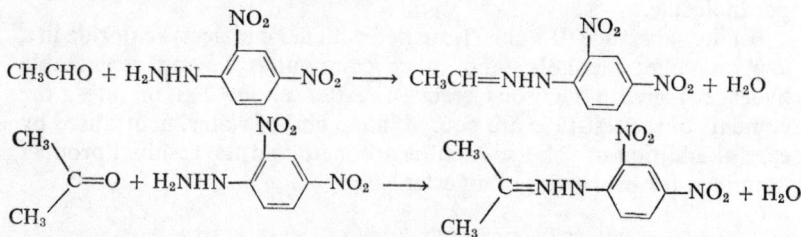

The reagent may be prepared as required thus: 0.5 g of 2,4-dinitro-phenylhydrazine is suspended in 25 cm³ of methanol and 1 cm³ of concentrated sulphuric acid is added cautiously. Heat is evolved and the solid usually dissolves: if necessary the mixture, when cool, may be filtered to produce a clear solution. To 3 cm³ of this reagent solution are added 2 drops (or 0.1 g) of the substance and the mixture is shaken. If a precipitate does not form at once, it may do so on allowing the mixture to stand for 15 minutes, or on gentle warming on a water-bath.

The majority of aldehydes and ketones react with a concentrated aqueous solution of sodium bisulphite, when a crystalline solid often separates out, its formation occasionally being accompanied by considerable evolution of heat. This precipitate consists of the sodium salt of an α-hydroxy sulphonic acid, usually termed an 'aldehyde or ketone bisulphite compound.'

$$C_6H_5CHO + NaHSO_3 \rightarrow C_6H_5CH(OH)SO_3Na$$

$$CH_3COCH_3 + NaHSO_3 \rightarrow (CH_3)_2C(OH)SO_3Na$$

It should be noted that the bisulphite compounds of many carbonyl compounds of these types are readily soluble in water, and may therefore produce no precipitate. An ethereal solution of an aldehyde or a ketone will nevertheless lose its solute on treatment with a saturated bisulphite solution, whether the resulting salt separates out or not. The tendency of aldehydes and of ketones in which the carbonyl group is attached to methyl or exists in a ring (as in cyclohexanone) to combine with bisulphites is so pronounced that they develop free alkali on shaking with neutral sodium sulphite solution.

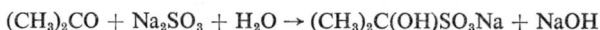

$$(CH_3)_2CO + Na_2SO_3 + H_2O \rightarrow (CH_3)_2C(OH)SO_3Na + NaOH$$

This can be observed by the formation of a red colour with phenolphthalein. On the other hand, ketones in which higher alkyl groups are attached to the carbonyl group react slowly with bisulphite or not at all.

Aldehydes and ketones containing a methyl group attached to the carbonyl group readily yield iodoform on treatment with iodine and dilute alkali.

$$CH_3CHO + 6I + 4NaOH \rightarrow CHI_3 + HCOONa + 3NaI + 3H_2O$$

This reaction is not specific for carbonyl compounds, as it is given by ethyl, isopropyl, and other alcohols which on oxidation yield the above type of aldehyde or ketone.

Solutions of many aldehydes and ketones develop characteristic red colours when treated with sodium nitroprusside and alkali. If, instead of alkali, piperidine be employed, blue colours result. On adding a dilute solution of m-phenylenediamine hydrochloride to aqueous or alcoholic solutions of aldehydes or ketones, a green fluorescence is developed. Alkaline solutions of some aldehydes and ketones give red colourations on addition of m-dinitrobenzene. These colour reactions are given only by ketones in which at least one methyl or ethyl radical is attached to the carbonyl group.

The simpler aldehydes and ketones, in extremely dilute aqueous solution, yield voluminous precipitates with Nessler's solution.

Aldehydes (pp. 75, 112) may be distinguished from ketones by testing with *Schiff's reagent*.* On treatment of the pure substance with this solution a distinct red colour is formed within a time limit of two minutes. In the case of water-soluble aldehydes the pink colour should appear almost instantly. The solution should never be warmed or treated with alkaline reagents, as both of these agencies restore the pink colour in the absence of any aldehyde. The presence of mineral acids, even too great an excess of the sulphurous acid, is to be avoided, as this tends to diminish the susceptibility of the reagent to aldehydes. Since many compounds may contain traces of substances which restore the colour of the fuchsine, the appearance of a pink tint after the expiration of the time limit is to be disregarded. Yellow to red colours are developed by aldehydes with benzidine in glacial acetic acid solution.

Aldehydes reduce *Fehling's solution* on warming. They also cause precipitation of metallic silver in the form of a mirror or in the amorphous condition when they are treated with *ammoniacal silver nitrate solution:*† it may be necessary to dissolve the substance in ethanol or to warm the reaction solution, but the latter must never be boiled.

On warming with concentrated potash solution, aliphatic aldehydes yield aldehyde resins with powerful odours, while aromatic aldehydes are converted into a mixture of the corresponding acid and alcohol (Cannizzaro Reaction):

$$2C_6H_5CHO + KOH \rightarrow C_6H_5COOK + C_6H_5CH_2OH$$

Aromatic aldehydes react with bromine in carbon tetrachloride or chloroform solution to form acid bromides:

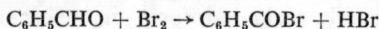

$$C_6H_5CHO + Br_2 \rightarrow C_6H_5COBr + HBr$$

while in aliphatic aldehydes the hydrogen atoms in the alkyl radical are substituted by bromine with formation of somewhat unstable bromo compounds.

Aromatic aldehydes, on warming with primary amines, yield condensation products (Schiff's bases) by elimination of water.

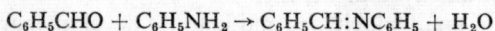

$$C_6H_5CHO + C_6H_5NH_2 \rightarrow C_6H_5CH:NC_6H_5 + H_2O$$

---

* Prepared by dissolving 0.2 g of rosaniline, or its acetate or hydrochloride (fuchsine), in 10 cm³ of a cold saturated solution of sulphur dioxide in water, and, after allowing to stand until the pink colour is entirely discharged, diluting to 100 cm³. This solution should be preserved away from light in a well-stoppered bottle.
† 5 % silver nitrate solution to which has been added ammonia solution until the first formed precipitate just redissolves.

Formaldehyde differs from other aliphatic aldehydes in some of its reactions. Thus on heating with concentrated alkali it yields methyl alcohol and a formate (but no aldehyde resin), while with dilute ammonia it forms neutral hexamethylenetetramine. Its oxime and phenylhydrazone are difficult to isolate, owing to their tendency to polymerise.

Paraldehydes and *acetals* (p. 112) behave like ethers towards alkalis, but are much more susceptible to the action of acids. Both types yield aldehydes on boiling with dilute mineral acids, the latter giving rise simultaneously to alcohols.

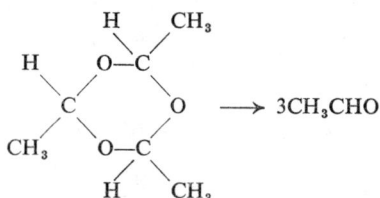

$$CH_3CH(OC_2H_5)_2 + H_2O \longrightarrow CH_3CHO + 2C_2H_5OH$$

They are not attacked by Fehling's solution nor by metallic sodium.

Ketones (pp. 75–116) are far less susceptible than aldehydes to the action of oxidising agents. They do not reduce Fehling's solution nor ammoniacal silver nitrate, and do not restore the colour to Schiff's reagent. Aliphatic and aliphatic-aromatic ketones may, however, be oxidised by boiling with chromic acid mixture, with formation of carboxylic acids.

$$C_6H_5COCH_3 + 4[O] \rightarrow C_6H_5COOH + CO_2 + H_2O$$

Aliphatic and aliphatic-aromatic ketones, on treatment with an equivalent quantity of bromine, are very readily brominated with substitution of bromine in the aliphatic group adjacent to the carbonyl.

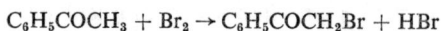

$$C_6H_5COCH_3 + Br_2 \rightarrow C_6H_5COCH_2Br + HBr$$

Glycosides and *carbohydrates* (pp. 75, 142), while properly belonging to the classes of alcohols, aldehydes, or ketones, may be discussed independently, inasmuch as they possess many properties peculiar to themselves. They may be detected by *Molisch's reaction*, which is stated to be so delicate that it is given by substances other than carbohydrates when contaminated by dust or by fibres from the filter paper: to a small portion (about 0.005 g) of the substance in 1 cm³ of water, 2 drops of a 10% solution of α-naphthol in chloroform are added. On carefully adding 1 cm³ of concentrated sulphuric acid from a finely pointed pipette so that it forms a separate layer below the water, a violet ring will be formed at the junction of the layers. On cautiously mixing the layers by shaking the test-tube in a stream of cold water, a deep purple solution results, which on dilution with cold water yields a violet precipitate. On shaking and adding a small quantity of the suspension to an excess of concentrated ammonia the colour is changed to dull brown.

On strongly heating alone, glycosides and carbohydrates, in common with some aliphatic hydroxy acids, yield the pungent odour attributed to 'burnt sugar.'

Glycosides may be distinguished from most carbohydrates by their failure to reduce Fehling's solution or react with phenylhydrazine until they have been boiled with dilute mineral acid, which splits them up into a sugar and an alcohol, phenol, etc.

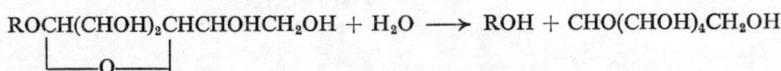

$$ROCH(CHOH)_2CHCHOHCH_2OH + H_2O \longrightarrow ROH + CHO(CHOH)_4CH_2OH$$

This inability to react with Fehling's solution before hydrolysis is also characteristic of the disaccharide sucrose (cane sugar), which is in fact a true glycoside.

Most of the complex natural glycosides yield intense colours on treatment with cold concentrated sulphuric acid.

Carbohydrates (pp. 74, 142) may be roughly subdivided into: (1) monosaccharides (pentoses and hexoses); (2) di- and tri-saccharides (such as saccharose, maltose, raffinose); (3) polysaccharides (such as starch, glycogen, inulin). Cellulose may also be included in the third division. All carbohydrates, on treatment with cold concentrated sulphuric acid, give only colourations varying between yellow, brown, and black, no members of this class giving red or purple colours.

Monosaccharides are mostly colourless crystalline solids, all readily soluble in water, yielding optically active solutions. The specific rotations of the sugars are important physical characteristics, and should in all cases be determined.

Monosaccharides are unchanged on boiling with dilute mineral acids, but decomposed with formation of indefinite products on warming with concentrated alkali. They are readily oxidised: by boiling with Fehling's solution, yielding a red precipitate of cuprous oxide, or by warming with a solution of silver nitrate in a mixture of concentrated ammonia and caustic soda solution, with precipitation of metallic silver in the amorphous condition or in the form of a mirror.

Ketoses may be distinguished by their resistance to the action of bromine water, their reducing power (towards Fehling's solution) being unaltered by treatment with this reagent. *Aldoses* on the other hand, readily decolourise bromine water, the resulting solution possessing a greatly diminished reducing power or none at all.

Pentoses give several specific reactions by means of which they may be distinguished from the hexoses. On distillation with 12% hydrochloric acid they yield furfural, which possesses a characteristic odour and develops a red colour with aniline acetate paper (rapid furfural test). On warming with 18% hydrochloric acid and a small quantity of phloroglucinol a purple colour is formed; while addition of a pentose to a boiling solution of orcinol in 18% hydrochloric acid containing a little ferric chloride causes the formation of a green colour.

Hexoses give a red colour on warming with resorcinol in hydrochloric acid; this test takes place more rapidly with ketohexoses than with aldohexoses. They may also be distinguished by Fenton's test: a small sample is moistened with water and warmed on the water-bath with one or two drops of phosphorus tribromide until the mixture begins to darken, whereupon it is cooled and mixed with a little alcohol and a few drops of ethyl malonate; on adding an excess of alcoholic potash and finally diluting with water a blue fluorescence appears.

A specific reaction for hexoses is the formation of levulinic acid. The sugar is heated on the water-bath for 15 to 20 hours with 4 to 5 parts of constant boiling (20%) hydrochloric acid under reflux; the mixture is cooled, filtered, and repeatedly extracted with ether. After distilling off the ether the residue is tested for levulinic acid by gently warming it with a solution of iodine in dilute sodium carbonate (iodoform test). Alternatively the levulinic acid may be identified as its phenylhydrazone.

Different hexoses may be distinguished by the formation of saccharic acid or of mucic acid on oxidation with nitric acid: a 2 g sample of the sugar is treated with 8–12 g of 25% nitric acid (sp. gr. 1.15) and the acid evaporated by heating on the water-bath until the syrupy residue begins to assume a brown colour. Water amounting to rather less in volume than the acid employed is then added. If a precipitate appears on thus diluting, it will in all probability consist of mucic acid, which, after cooling, should be filtered off and purified by dissolving in dilute alkali and reprecipitating by the addition of a slight excess of mineral acid. The filtrate from the mucic acid, or the clear solution if no precipitate has appeared, is then exactly neutralised in the warm by the cautious addition of potassium carbonate, and the resulting dark-coloured solution again acidified with acetic acid and evaporated to a syrup. After the further addition of a few drops of acetic acid and cooling, acid potassium saccharate separates out in those cases where saccharic acid is formed. This should be dried on a porous tile and recrystallised from a small quantity of water in order to remove any oxalic acid. The silver salt, which may be prepared by neutralising the acid salt with ammonia, boiling to remove any excess, adding silver nitrate solution, and filtering off and drying the precipitate at 110°C, should be ignited in order to determine the equivalent of the acid.

Di- and *Tri-saccharides* behave like glycosides in that when warmed with dilute mineral acids they are hydrolysed with formation of mono-saccharides. They vary, however, in their stability towards oxidising agents and phenylhydrazine. Thus, sucrose does not reduce Fehling's solution on boiling, and forms no compound when warmed with aqueous phenylhydrazine acetate; maltose and lactose, on the other hand, resemble the monosaccharides in their behaviour towards both of these reagents, with the difference that the osazones formed do not separate from solution until the reaction mixture is allowed to cool. Di- and tri-saccharides resemble the monosaccharides in their ready solubility in water.

Polysaccharides are far less soluble in cold water than the sugars, but solutions of some of them can be obtained on boiling with water. Such solutions are, however, never entirely clear. On hydrolysis by boiling with dilute mineral acids they yield monosaccharides. They do not react with Fehling's solution nor with phenylhydrazine. On treatment with a cold solution of iodine in aqueous potassium iodide characteristic colours are produced,* which are discharged on heating but reappear on cooling. Cellulose, on prolonged warming with concentrated hydrochloric acid, may be partially or entirely hydrolysed, with formation of dextrose. It is, however, insoluble in pure water, and develops no colour with iodine solution. Much information as to the nature and source of the cellulose is afforded by microscopical examin-ation of the fibres.

* e.g. starch: blue; glycogen: wine colour.

3 Separation of Mixtures of Organic Compounds

When it is required to identify the constituents of a mixture of organic substances, it will in all cases be necessary to separate each component from the mixture and to isolate it in a pure state before proceeding with the examination. To identify the constituents of a mixture without separating them is at best an extremely difficult feat.

Whilst specialised methods of separation such as the various types of chromatography and electrophoresis are now widely used in chemical laboratories they are not considered appropriate for discussion here.

Owing to the great number of possibilities, no definite rule for procedure can be laid down and advantage will have to be taken of any facts that emerge in the preliminary examination which may lead to a modification of the general scheme given below. The preliminary examination of the mixture is therefore of considerable importance and should include the physical state of the sample, its behaviour on distillation, determination of the elements present, behaviour on ignition, its solubility in water and whether an acid or an alkaline reaction is given.

Should the mixture appear to contain some volatile liquid, it may be heated on the water-bath in a distilling flask attached to a condenser until no more of the liquid passes over. The distillate should be re-distilled, employing a thermometer, in order to ascertain whether it is homogeneous. Should this not be the case, it must be further investigated with a view to separation by chemical methods. Fractional distillation in the ordinary way is generally not well adapted to the separation of mixtures available in only small amounts.

The residue in the flask or the original mixture if this was wholly solid, is treated with an excess of dry ether, any insoluble portion being filtered off and washed with the same solvent, the washings being added to the filtrate. By this means the majority of salts, carbohydrates and other polyhydroxylic compounds, sulphonic acids, and similar substances insoluble in ether, may be separated from the main portion. Such a residue is to be examined independently, extraction with cold methyl alcohol being carried out as a preliminary step towards further separation.

Should the mixture have been found to contain nitrogen the ethereal solution is shaken in a separating funnel with dilute sulphuric acid. In the absence of nitrogen this operation may be omitted. By this means basic substances are removed from the mixture on separating the aqueous and ethereal layers. The bases may be recovered by rendering the aqueous solution alkaline and again extracting with ether.

The ethereal solution after this treatment should be washed with a small quantity of water—the washings being discarded—and shaken with dilute caustic soda solution. This has the effect of removing all compounds of an acidic character. The treatment of the aqueous portion will be discussed below.

The ether now contains only neutral substances. From these any aldehydic and many ketonic compounds can be removed by shaking with a concentrated solution of sodium bisulphite. The aldehydes and ketones can be recovered from the resulting precipitate or aqueous solution by acidification with dilute sulphuric acid followed by distillation, extraction, or filtration.

The alkaline solution containing the acidic substances should be saturated with carbon dioxide and extracted with ether. By this procedure all phenolic compounds which contain no carboxyl or nitro groups, oximes, and similar weak acids are liberated and pass into the ether. On adding dilute sulphuric acid until evolution of carbon dioxide ceases, carboxylic acids and nitrophenols are liberated, and can be isolated by extraction, filtration, or distillation.

At this stage all ethereal solutions should be evaporated, the ether, if only for the sake of safety from conflagrations, being efficiently condensed and thus recovered. All residues should be tested afresh for constituent elements, and tests for homogeneity applied.

The results of the above operations are briefly summarised in the scheme shown at top of p. 67.

The various fractions, denominated I, II, etc., may contain the following types of compound.

I. Hydrocarbons, ethers, alcohols, ketones, esters, aliphatic halogen compounds; and conceivably aldehydes, acetals, nitriles, aliphatic amines, mercaptans, sulphides, and alkyl nitrates and nitrites whose boiling points lie below 100°C.

II. Metallic salts, salts of organic bases with mineral acids, carbohydrates and other polyhydroxylic compounds, amino acids, sulphonic acids of all types.

III. Aliphatic and aromatic primary, secondary, and tertiary amino compounds; possibly some amides.

IV. Aldehydic and ketonic compounds, containing no groups which would have placed them in another fraction.

Mixture.
|
Distilled on water-bath
|

Non-Volatile.
Treated with ether

Volatile.
(I)

Insoluble in ether.
(II)

Soluble in ether.
Treated with dilute H_2SO_4

Neutral or acidic.
Treated with dilute NaOH

Basic.
(III)

Neutral
Treated with $NaHSO_3$

Acidic.
Treated with CO_2

Aldehydes
and
Ketones.
(IV)

Neutral
(V)

Phenols.
(VI)

Acids
and
Nitrophenols.
(VII)

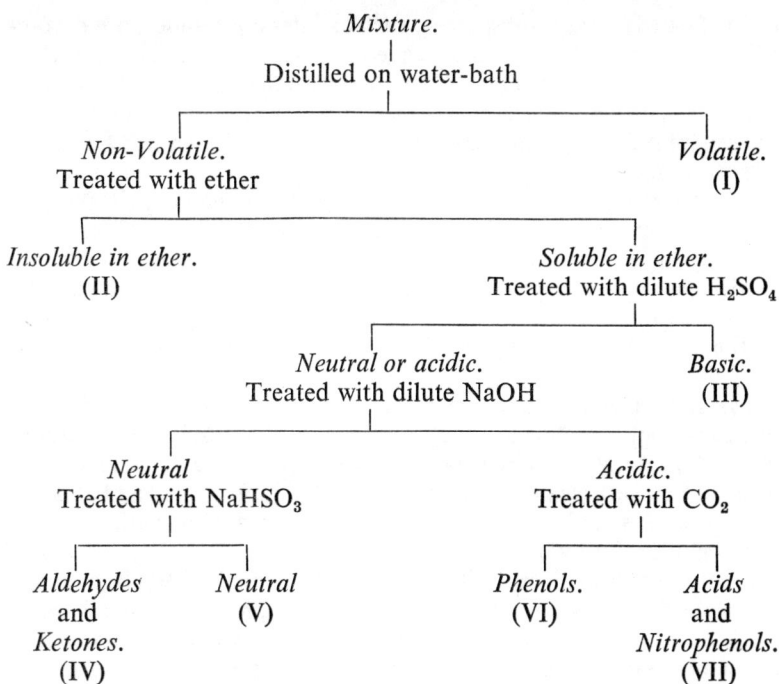

V. Hydrocarbons, ethers, alcohols, higher ketones, esters, and aliphatic or aromatic halogen compounds, not included in other fractions. Also nitriles; nitrohydrocarbons, nitro-ethers, nitro-alcohols; simple azoxy, azo, hydrazo compounds; sulphides, disulphides, sulphoxides, sulphones, sulphonamides and a few nonreactive anhydrides, ketones, and esters of mineral acids.

VI. Simple and substituted phenols, keto-enolic substances, mercaptans, some thioamides.

VII. Carboxylic acids; nitrophenols in which the nitro groups are present in either *ortho* or *para* position.

Any obvious decompositions produced by the action of water on the mixture, such as in the case of acid halides, will have been observed in the preliminary investigation of the mixture. Much information as to the nature of the constituents may be obtained by the mere performance of the above separations.

Steam distillation from acid and from alkaline solutions, while constituting an excellent means of separation in place of extraction with ether, has not been recommended in the above scheme on account of the fact that many substances, such as esters or amides, are decomposed by hot acid or alkaline aqueous solutions, but preserve their identity when treated in ethereal solution with these reagents in the cold.

For the further separation of the individual substances contained in the different fractions no definite scheme can be drawn up, but a few general suggestions may be offered.

Separation can in many cases be effected by treatment of the fraction with different liquids which exert selective solvent effects. Water, ethyl alcohol, ligroin, benzene, acetone, chloroform, or glacial or dilute acetic acid may be employed for this purpose. This method may also be applied by dissolving the mixture in one solvent and precipitating one of the components by the addition of another liquid in which it is insoluble.

Distillation with steam, when feasible, occasionally serves for the separation of substances—even of isomers—of differing volatility. Neutral, basic, and acidic substances which volatilise with steam may be separated very cleanly by alternate distillation from acid and alkaline solutions. This method should however be applied with caution, owing to possibilities of hydrolysis.

Finally, fractional crystallisation and fractional distillation under atmospheric or reduced pressure may be employed when all other methods have failed. Both of these operations are, however, often extremely tedious, and should be regarded only as a last resort. The criterion of purity in the case of fractional distillation is the isolation of a fraction of constant boiling point, while solid substances must be recrystallised until no rise in melting point is observable on further recrystallisation of any fraction.

4　Preparation of Derivatives

Acid anhydrides

(a) *Anilide* (from monobasic acids)

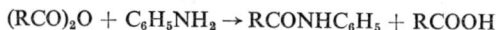

$$(RCO)_2O + C_6H_5NH_2 \rightarrow RCONHC_6H_5 + RCOOH$$

A mixture of 1 cm³ of aniline and 1 g of the anhydride are heated together almost to boiling and then cooled. About 3 cm³ of water are added and the whole stirred with a glass rod. The separated product is recrystallised from water or aqueous ethanol.

(b) *Anilic acid* (from dibasic acids)

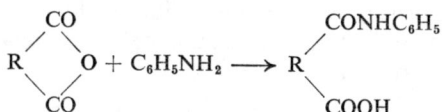

$$R\begin{array}{c} CO \\ \diagup \\ \diagdown \\ CO \end{array}O + C_6H_5NH_2 \longrightarrow R\begin{array}{c} CONHC_6H_5 \\ \diagup \\ \diagdown \\ COOH \end{array}$$

1 cm³ of aniline is heated gently with a solution of 1 g of anhydride in 30 cm³ of benzene. On cooling, the anilic acid separates and after washing it with dilute hydrochloric acid it is recrystallised from alcohol.

On heating at the melting point, some anilic acids decompose to give anils.

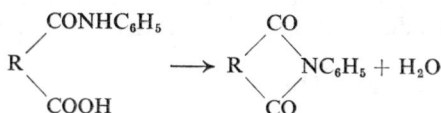

$$R\begin{array}{c} CONHC_6H_5 \\ \diagup \\ \diagdown \\ COOH \end{array} \longrightarrow R\begin{array}{c} CO \\ \diagup \\ \diagdown \\ CO \end{array}NC_6H_5 + H_2O$$

Acid chlorides

(a) *Amide*

$$RCOCl + NH_3 \rightarrow RCONH_2 + HCl$$

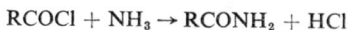

0.5 g of acid chloride is treated with 5 cm³ of concentrated ammonia solution (0.88 sp. gr.) and the mixture stirred. The amide is recrystallised from water or aqueous ethanol.

(b) *Anilide, p-toluidide*

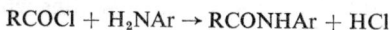

$$RCOCl + H_2NAr \rightarrow RCONHAr + HCl$$

1 g of the acid chloride is dissolved in 5 cm³ of pure ether or benzene and slowly treated with 1 g of aniline (or *p*-toluidine) dissolved in 10 cm³ of the same solvent until the odour of the acid chloride is no longer detectable. After vigorous stirring, the mixture is treated with

dilute hydrochloric acid to remove excess of amine and the ethereal (or benzene) solution of the product evaporated (CAUTION). The substituted amide is recrystallised from water or aqueous alcohol.

Alcohols

(a) *3,5-Dinitrobenzoate, p-nitrobenzoate*

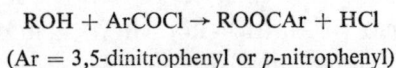

$$ROH + ArCOCl \rightarrow ROOCAr + HCl$$

(Ar = 3,5-dinitrophenyl or *p*-nitrophenyl)

The dry alcohol (0.5 g) is dissolved in 5 cm³ of dry pyridine and to the solution is added 1 g of 3,5-dinitrobenzoyl chloride or *p*-nitrobenzoyl chloride. The mixture is refluxed for about 30 minutes, cooled, and poured into 30 cm³ of dilute hydrochloric acid. The separated solid product is washed with dilute sodium carbonate solution and is recrystallised from ethanol, aqueous ethanol or benzene.

(b) *1-Naphthyl carbamate*

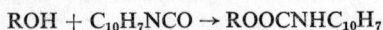

$$ROH + C_{10}H_7NCO \rightarrow ROOCNHC_{10}H_7$$

The reagent is best used for alcohols which are insoluble in water.

1 g of dry alcohol is shaken with 0.5 g of 1-naphthyl isocyanate in a dry test-tube which is loosely plugged with cotton wool. If the derivative does not separate, the mixture is heated to 100°C on an oil bath for 5 minutes and then cooled. The solid is extracted with boiling ligroin (b.p. 100–120°C) or with carbon tetrachloride and the solution cooled. The separated product is recrystallised from ligroin, ethanol or chloroform.

(c) *Hydrogen 3-nitrophthalate*

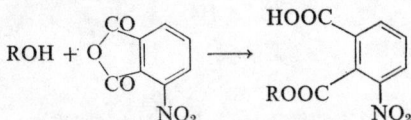

The anhydrous alcohol (1 g) is heated under reflux with 3-nitrophthalic anhydride (1 g) until the mixture liquifies and then for a further period of 5 minutes. If the alcohol has a boiling point in excess of 150°C, it is advisable to dissolve it in 5 cm³ of toluene and reflux this solution with the anhydride for 30 minutes after which the toluene is removed under reduced pressure. On cooling, the reaction product should solidify; otherwise it should be extracted with saturated sodium bicarbonate solution and the extract acidified with dilute hydrochloric acid. The product is recrystallised from aqueous ethanol. (The m.p. of 3-nitrophthalic acid is 218°C).

Amides and Imides

(a) *Xanthylamide*

$$RCONH_2 + $$

Xanthydrol   Xanthylamide

$$+ H_2O$$

To 7 cm³ of a 7% solution of xanthydrol in glacial acetic acid is added 0.5 g of the substance and the mixture is shaken and allowed to stand for 10 minutes. If the product does not separate in this time, the whole should be heated on a water bath for not more than 30 minutes and allowed to cool. The solid xanthylamide is filtered off and re-crystallised from aqueous dioxan or acetic acid.

(b) *Hydrolysis*

$$RCONH_2 + NaOH \rightarrow NH_3 + RCOONa \rightarrow RCOOH$$

The organic compound (0.5 g) is refluxed with an excess of 25% sodium hydroxide solution until all ammonia has been driven off. The resulting solution is acidified with concentrated hydrochloric acid and the liberated acid separated by filtration, distillation or ether extraction.

Amides, *N*-substituted

Hydrolysis

$$RCONHR' + H_2SO_4 + H_2O \rightarrow RCOOH + (R'NH_3)^+(HSO_4)^-$$

1 g of the substituted amide is treated with 10 cm³ of 70% sulphuric acid and refluxed for 30 minutes. The reaction mixture is cooled and diluted by careful addition of 5 cm³ of water and the organic acid removed by filtration, distillation or ether extraction. The residual liquor is rendered alkaline with sodium hydroxide solution and the amine removed by ether extraction. Both the amine and the carboxylic acid should be characterised.

Amines, Primary and Secondary

(a) *Acetyl derivative*

$$RNH_2 + (CH_3CO)_2O \rightarrow RNHCOCH_3 + CH_3COOH$$

$$RR'NH + (CH_3CO)_2O \rightarrow RR'NCOCH_3 + CH_3COOH$$

A mixture of 0.5 g of the amine with 0.5 cm³ of water, 0.5 cm³ of glacial acetic acid and 0.5 cm³ of acetic anhydride is shaken at room temperature and is heated gently for 15 minutes if reaction does not occur spontaneously. The cooled reaction mixture is filtered and the separated product recrystallised from water or aqueous alcohol.

For *ortho*-substituted derivatives of aromatic amines, a few drops of concentrated sulphuric acid should be added to the acetylating reagent which should contain 1.5 cm^3 of acetic anhydride in this case.

(b) *Benzoyl derivative*

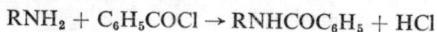

$$RNH_2 + C_6H_5COCl \rightarrow RNHCOC_6H_5 + HCl$$

$$RR'NH + C_6H_5COCl \rightarrow RR'NCOC_6H_5 + HCl$$

Schotten-Baumann method. About 0.5 g of the amine is suspended in 10 cm^3 of 5% sodium hydroxide solution in a small conical flask and treated dropwise with 1 cm^3 of benzoyl chloride, the vessel being securely stoppered and shaken between additions of the reagent. The mixture should be kept cool throughout the addition. Finally the reaction is completed by vigorous shaking for 5–10 minutes and the product filtered off, washed with water and recrystallised from alcohol.

(c) *p-Toluenesulphonyl derivative ('p-Tosyl')*

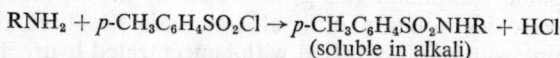

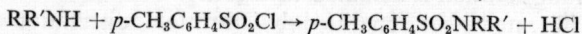

$$RNH_2 + p\text{-}CH_3C_6H_4SO_2Cl \rightarrow p\text{-}CH_3C_6H_4SO_2NHR + HCl$$
$$\text{(soluble in alkali)}$$

$$RR'NH + p\text{-}CH_3C_6H_4SO_2Cl \rightarrow p\text{-}CH_3C_6H_4SO_2NRR' + HCl$$

A solution of 1 g of *p*-toluenesulphonyl chloride and 0.5 g of the amine in a small volume of acetone is treated with 15 cm^3 of 5% sodium hydroxide solution and the mixture is shaken vigorously for 10–15 minutes, cooling as necessary. The derivative which precipitates is filtered off, washed with water and recrystallised from alcohol. It should be noted that the derivative formed from a primary amine is soluble in alkali and must be precipitated from the reaction mixture by acidification with hydrochloric acid.

(d) *2,4-Dinitrophenyl derivative*

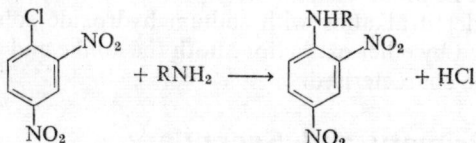

Note: the reagent is a skin irritant and must be handled with care.

1 g of the amine, 1 g of 2,4-dinitrochlorobenzene and 10 cm^3 of ethanol are treated with a slight excess of anhydrous potassium carbonate or anhydrous sodium acetate. The mixture is heated under reflux for 30 minutes, poured into cold water and the derivative filtered off, washed with water and recrystallised from alcohol.

(e) *Picrate* Picric acid combines with amines to give crystalline molecular complexes which usually have characteristic melting points and often have the composition 1 mole amine: 1 mole picric acid.

In the case of an amine which is soluble in water, its aqueous solution is treated with a slight excess of saturated aqueous solution of picric acid (solubility is approximately 1% in cold water). The solution is allowed to stand for a time, and is then shaken vigorously. The precipitated picrate is filtered off and recrystallised from water, ethanol or benzene.

Alternatively, a solution of the amine (0.5 g) in ethanol (2 cm³) is treated with a saturated ethanolic solution of picric acid (3 cm³) and warmed gently for one minute. After allowing the mixture to cool slowly, the product precipitates and is recrystallised from alcohol.

Amines, Tertiary

(a) *Methiodide*

$$RR'R''N + CH_3I \rightarrow RR'R''NCH_3^+I^-$$

A mixture of 0.5 g of the tertiary amine and 0.5 cm³ of colourless methyl iodide is allowed to stand for 5 minutes. If reaction has not occurred, the liquor is warmed under reflux on a water bath for 5 minutes and then cooled in ice water. If necessary to induce crystallisation, the sides of the vessel should be scratched with a glass rod. The solid product is recrystallised from absolute ethanol, ethyl acetate or ethanol-ether.

(b) *Picrate* Experimental details are given under Amines, primary and secondary (see above).

Amino Acids

(a) *Acetyl derivative*

$$HOOCRNH_2 + (CH_3CO)_2O \rightarrow HOOCRNHCOCH_3 + CH_3COOH$$

The derivative is prepared as described under Amines (p. 71).

(b) *Benzoyl, p-toluenesulphonyl derivatives*

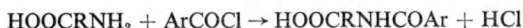

$$HOOCRNH_2 + ArCOCl \rightarrow HOOCRNHCOAr + HCl$$

These derivatives are prepared by the methods detailed under Amines, with the following modifications:

(i) for benzoyl derivatives, sodium carbonate solution should be used in place of sodium hydroxide solution, and when the reaction is over the derivative is precipitated by acidification with dilute hydrochloric acid.

(ii) for *p*-toluenesulphonyl derivatives, the reaction mixture should be shaken for 3 to 4 hours, and the derivative finally precipitated by acidification with dilute hydrochloric acid.

(c) *Phenylhydantoic acid, 1-naphthylhydantoic acid derivatives*

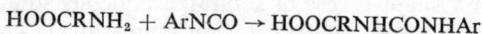

$$HOOCRNH_2 + ArNCO \rightarrow HOOCRNHCONHAr$$

1 g of phenyl isocyanate or 1-naphthyl isocyanate is added to a solution of 0.5 g of the amino acid in 25 cm³ of a 2% solution of sodium hydroxide and the mixture is shaken for 5 minutes and then allowed to stand for 25 minutes. After filtering to remove any insoluble diarylurea the solution is acidified to Congo Red with dilute hydrochloric acid. The precipitated hydantoic acid is filtered off, washed with cold water and recrystallised from water or alcohol.

Carbohydrates

(a) *Osazone*

$$
\begin{array}{ccc}
CHO & & CH{=}NNHC_6H_5 \\
| & & | \\
CHOH + 2C_6H_5NHNH_2 & \longrightarrow & C{=}NNHC_6H_5 \quad + 2H_2O \\
| & & | \\
R & & R \\
\end{array}
$$

$$
\left(
\begin{array}{c}
or\ CH_2OH \\
| \\
CO \\
| \\
R \\
\end{array}
\right)
$$

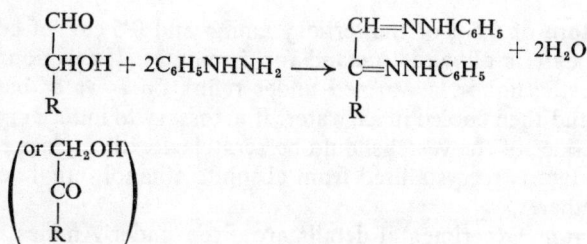

0.20 g of the carbohydrate, 0.40 g of pure (e.g. A.R. quality) phenylhydrazine hydrochloride, 0.60 g of crystalline sodium acetate and 4.0 cm³ of water are placed in a dry test-tube. The tube is then heated in a boiling water bath for 30 minutes, and allowed to cool in the water bath. The osazone may separate from the hot solution in some cases. The product is filtered off, washed with cold water and recrystallised from hot water or 60% ethanol.

(b) *Acetate* The β-acetate is normally obtained by the use of the following procedure, which completely acetylates all the hydroxy groups. 0.5 g of the carbohydrate, 0.5 g of powdered anhydrous sodium acetate and 3 cm³ of acetic anhydride are dissolved in 5 cm³ of glacial acetic acid by heating on a water bath under reflux and the solution is further heated for a period of two hours. The hot reaction mixture is poured cautiously into 25 cm³ of cold water and vigorously stirred to decompose the excess of acetic anhydride. The product is filtered off, washed with cold water and recrystallised from ethanol.

Carbonyl Compounds

(a) *2,4-Dinitrophenylhydrazone*

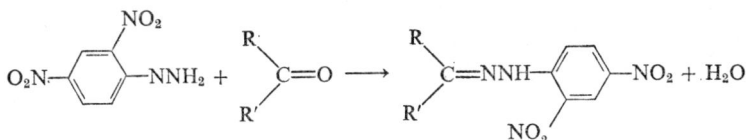

where R′ may be H.

(i) A solution is prepared by gently warming 0.25 g of 2,4-dinitrophenylhydrazine with 5 cm³ of methanol and 0.5 cm³ of concentrated sulphuric acid. The warm solution is clarified by filtration and treated with 0.1 g of the carbonyl compound dissolved in the minimum volume of methanol or ether. The mixture is boiled for 2 minutes, and then cooled. If the product does not separate, the solution should be cautiously treated with drops of dilute sulphuric acid. The filtered and washed dinitrophenylhydrazone is recrystallised from ethanol, acetic acid, nitrobenzene or xylene.

(ii) To 4 cm³ of a saturated solution of 2,4-dinitrophenylhydrazine in methyl cellosolve[1] (2-methoxyethanol) is added 0.1 g of the carbonyl compound (dissolved in a little methyl cellosolve if solid) followed by 3 drops of concentrated hydrochloric acid. If the derivative does not precipitate at once, the solution should be diluted carefully with drops of water. The product is filtered off, washed with ethanol and recrystallised as indicated in [a(i)] above.

Diethylene glycol dimethyl ether ('diglyme') may alternatively be used as a solvent in this reaction.[2]

(b) *Semicarbazone*

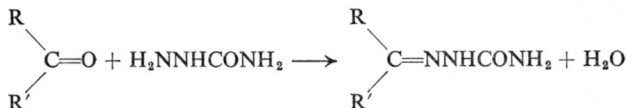

where R′ may be H.

To a solution of 0.8 g of hydrated sodium acetate and 0.5 g of semicarbazide hydrochloride in 5 cm³ of water is added 0.5 g of the compound. If after shaking the mixture it remains not completely clear, drops of pure methanol may be added to clarify. The solution is allowed to stand until the product separates (in some cases several hours may be required) or is heated gently on a water bath for up to 10 minutes followed by cooling in ice-water. The product is filtered

[1] Yeadon, A., S.S.R., 1969, 172, **50**, 594.
[2] Shine, H. J., J. Org. Chem., 1959, **24**, 252.

off, washed with a little cold water and recrystallised from water, aqueous ethanol, benzene or glacial acetic acid.

(c) *Oxime*

$$\begin{matrix} R \\ \diagdown \\ \diagup \\ R' \end{matrix} C{=}O + H_2NOH \longrightarrow \begin{matrix} R \\ \diagdown \\ \diagup \\ R' \end{matrix} C{=}NOH + H_2O$$

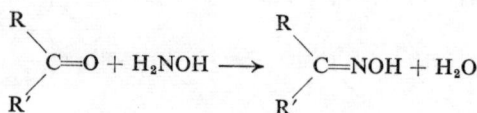

where R' may be H.

The method described for the preparation of semicarbazones ((b) above) may be used, substituting 1 g of hydroxylamine hydrochloride for the semicarbazide hydrochloride and raising the weight of sodium acetate employed to 2 g.

(d) *p-Nitrophenylhydrazone*

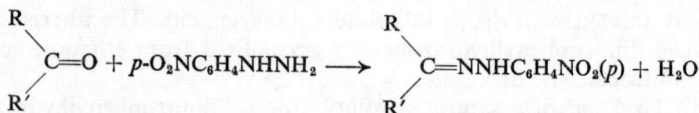

$$\begin{matrix} R \\ \diagdown \\ \diagup \\ R' \end{matrix} C{=}O + p\text{-}O_2NC_6H_4NHNH_2 \longrightarrow \begin{matrix} R \\ \diagdown \\ \diagup \\ R' \end{matrix} C{=}NNHC_6H_4NO_2(p) + H_2O$$

where R' may be H.

A mixture of 0.5 g of *p*-nitrophenylhydrazine, 15 cm³ of pure ethanol, 0.5 g of the carbonyl compound and 5 drops of glacial acetic acid are refluxed for 5 minutes, cooled and the precipitated product washed with ethanol and recrystallised from ethanol. (If there is difficulty in inducing precipitation of the product from the reaction liquor, reheat the latter, add water until turbid, then alcohol until clear and cool).

(e) *Dimethone*

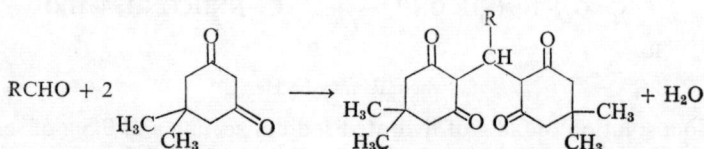

0.1 g of the aldehyde and one drop of piperidine are added to 4 cm³ of a 10% solution of dimedone in 50% aqueous ethanol. If the product does not form immediately, the reaction mixture may be warmed under reflux for 5 minutes; if the dimethone still has not separated, hot water should be added to the solution until a cloudiness appears and the mixture cooled to 5°C. The derivative is filtered off and recrystallised from aqueous methanol or ethanol.

Carboxylic Acids

(a) *Amide, anilide, p-toluidide*

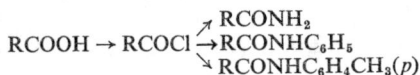

$$RCOOH \rightarrow RCOCl \begin{matrix} \nearrow RCONH_2 \\ \rightarrow RCONHC_6H_5 \\ \searrow RCONHC_6H_4CH_3(p) \end{matrix}$$

1 g of the acid (or its sodium salt) is refluxed with 3 cm³ of thionyl chloride in a fume chamber for 30 minutes or until no further reaction occurs. The excess of thionyl chloride is then removed by distillation.

(i) Amide

The cold acid chloride is treated cautiously with concentrated ammonia solution and warmed for a few moments. The solid product is recrystallised from aqueous alcohol.

(ii) Anilide, *p*-toluidide

The cold acid chloride is treated with a solution of 1 g of aniline (or *p*-toluidine) in benzene. The mixture is warmed carefully, cooled, and the product filtered off, washed with dilute hydrochloric acid to remove excess of amine and is finally recrystallised from water or ethanol. If the product does not crystallise from the benzene solution, the latter should be separated, washed with dilute hydrochloric acid and then with water and the solvent evaporated (CAUTION).

(b) *p-Bromophenacyl esters*

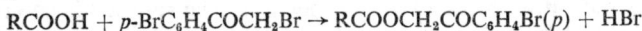

$$RCOOH + p\text{-}BrC_6H_4COCH_2Br \rightarrow RCOOCH_2COC_6H_4Br(p) + HBr$$

CAUTION. *p*-Bromophenacyl bromide is a skin and eye irritant.

0.5 g of the acid is dissolved in 5 cm³ of water in a small flask and is neutralised to phenolphthalein with 5% sodium hydroxide solution. The solution is rendered faintly acid by the addition of two drops of dilute hydrochloric acid and is then treated with a solution of 0.5 g of *p*-bromophenacyl bromide in 5 cm³ of ethanol. The reaction mixture is refluxed for one hour and is kept clear by the addition of ethanol if necessary. (Di- and tribasic acids require proportionately larger amounts of reagent and longer reaction times.) The product separates from the cooled mixture and is filtered and recrystallised from ethanol, aqueous ethanol or benzene.

(c) *p-Nitrobenzyl esters*

The procedure is as detailed for *p*-bromophenacyl esters, using *p*-nitrobenzyl bromide.

(d) *S-Benzylthiuronium salts*

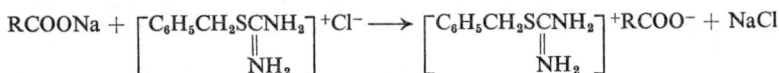

$$RCOONa + \begin{bmatrix} C_6H_5CH_2SCNH_2 \\ \| \\ NH_2 \end{bmatrix}^+ Cl^- \longrightarrow \begin{bmatrix} C_6H_5CH_2SCNH_2 \\ \| \\ NH_2 \end{bmatrix}^+ RCOO^- + NaCl$$

0.25 g of the acid is dissolved or suspended in 5 cm³ of warm water and neutralised to phenolphthalein with 4% sodium hydroxide solution.

The solution so formed (or a solution of 0.25 g of the sodium salt in water) is treated with 2 *drops* of dilute (0.1 M) hydrochloric acid and a solution of 1 g of *S*-benzylthiuronium chloride in 5 cm³ of water or 10 cm³ of ethanol. The mixture is cooled in ice-water until precipitation is complete. Recrystallisation (from aqueous alcohol) is sometimes difficult.

With some acids (e.g. succinic acid) better results are obtained by reversing the order of mixing, i.e. by adding the solution of the acid to the reagent solution.

Esters

Hydrolysis. The most general procedure for the characterisation of esters involves their hydrolysis to the parent carboxylic acid and alcohol (or phenol).

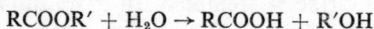

$$RCOOR' + H_2O \rightarrow RCOOH + R'OH$$

Various methods are available for such hydrolyses, the choice being dependent on the resistance of the ester to hydrolysis. Those given below should be suitable for the majority of esters likely to be encountered.

(i) Sodium methoxide

1 g of sodium is dissolved in 10 cm³ of pure methanol by warming in a dry flask fitted with a reflux condenser. The solution is cooled and treated with 1 g of the ester and 0.5 cm³ of water. The sodium salt of the parent carboxylic acid often separates at once* or else after boiling the mixture for a few minutes (occasionally 30–60 minutes is required). The product is filtered off, washed with a little cold methanol and may be identified by conversion to the *p*-bromophenacyl ester or other derivative.

The residual reaction liquor contains the alcoholic component* of the ester, and should be isolated by carefully distilling off the excess of methanol, and characterised as described under Alcohols.

(ii) Aqueous alkali

5 g of the ester is boiled under reflux with 40 cm³ of 30% potassium hydroxide (or sodium hydroxide) solution until hydrolysis is complete

---

* If the ester gives rise to both an acid and a phenol, the latter may be precipitated with the acid as the sodium salt or may be found in solution in the mother liquor. The acid and phenol may be separated as follows: the reaction mixture is made acid to litmus or Congo Red paper with dilute sulphuric acid and the carboxylic acid and phenol both extracted with ether. The acid is removed by washing the ethereal extract repeatedly with saturated sodium bicarbonate solution until effervescence ceases.

The carboxylic acid is isolated from the bicarbonate extract by acidification with dilute sulphuric acid followed by filtration or distillation, and characterised.

The phenol is isolated by careful evaporation of the ether and identified.

as indicated by a change in the odour or appearance of the mixture (often upwards of 60 minutes). A test portion (about 5 cm³) of the liquor is distilled off and saturated with potassium carbonate; if a distinct layer of an alcohol separates on standing, this layer should be removed, dried with anhydrous potassium carbonate and the alcohol identified. (Alternatively the alcohol may be removed from the test portion after saturating with potassium carbonate by ether extraction.)

A portion of the residue in the flask is first acidified with dilute sulphuric acid; if the carboxylic acid precipitates, the whole of the residual reaction liquor is acidified, the acid is filtered off, recrystallised and identified. If no precipitate is formed on acidification of the reaction liquor, the liquor is neutralised to phenolphthalein and the solution of the alkali salt of the acid is used to prepare a suitable derivative.

For separation of an acid and a phenol formed by hydrolysis of an ester, see the footnote to hydrolysis method (i).

Ethers

(a) *Sulphonamide*

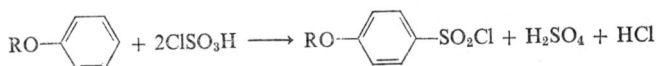

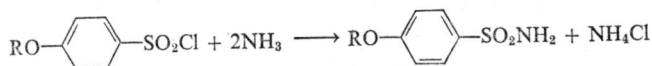

$$RO-\text{\Large\bigcirc} + 2ClSO_3H \longrightarrow RO-\text{\Large\bigcirc}-SO_2Cl + H_2SO_4 + HCl$$

$$RO-\text{\Large\bigcirc}-SO_2Cl + 2NH_3 \longrightarrow RO-\text{\Large\bigcirc}-SO_2NH_2 + NH_4Cl$$

1 g of the aryl ether is dissolved in 5 cm³ of dried chloroform in a dry test-tube, is cooled to 0°C and then treated with 3 cm³ of chlorosulphonic acid, added drop-wise. When the evolution of hydrogen chloride has subsided, the mixture is allowed to stand at room temperature for 20 minutes (or longer if reaction is very slow) and is then poured cautiously onto 25 g of crushed ice. The chloroform layer, which contains the arylsulphonyl chloride, is separated, washed with a little water, and added, with stirring, to 10 cm³ of concentrated ammonia solution. The mixture is allowed to stand for 10 minutes after which time the chloroform is evaporated on a water bath, the residue is cooled and treated with 5 cm³ of 10% aqueous sodium hydroxide in which the sulphonamide of the ether dissolves as the sodium salt $\left(RO-\text{\Large\bigcirc}-SO_2NHNa\right)$. The solution is filtered to remove unwanted insoluble products, such as sulphone, acidified with dilute hydrochloric acid and cooled in ice-water. The sulphonamide is filtered off and recrystallised from aqueous ethanol.

(b) *Picrate.* A solution of 0.5 g of the compound in the minimum volume of hot ethanol is added to a hot saturated solution of picric acid in ethanol (3–5 cm³) and the mixture is allowed to stand. The picrate complex is filtered off and may be recrystallised from ethanol, but it should be noted that picrates of aromatic ethers are relatively unstable and are liable to decompose during recrystallisation.

(c) *Bromo derivative.* Aryl ethers may be brominated by a solution of bromine in a variety of solvents; the number of bromine atoms introduced into the aromatic nucleus ranges commonly from one to three.

A solution of 1 g of the compound in 10–15 cm³ of glacial acetic acid is treated cautiously with a solution of 4 cm³ of liquid bromine in 10–15 cm³ of glacial acetic acid until the colour of bromine persists. The mixture is allowed to stand for 20 minutes, adding more bromine solution if the colour fades, and the liquor is then poured into 70 cm³ of water. The product is filtered off, washed with water and recrystallised from ethanol.

Alternative reaction solvents are carbon tetrachloride and chloroform, which after reaction are removed by distillation.

(d) *Nitro derivative.* Aryl ethers may be converted into solid derivatives by the process of nitration, using one of the methods detailed under Hydrocarbons.

(e) *Oxidation.* An aromatic ether possessing an aliphatic side chain may be converted into a suitable derivative by oxidation of the side chain, using one of the methods indicated under Hydrocarbons (d).

Halogenohydrocarbons

(1) Alkyl halides

(a) *S-Alkylthiuronium picrate*

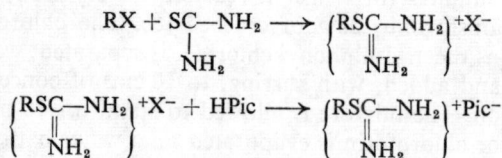

$$RX + \underset{\underset{NH_2}{|}}{SC}{-}NH_2 \longrightarrow \left\{\underset{\underset{NH_2}{\|}}{RSC}{-}NH_2\right\}^+X^-$$

$$\left\{\underset{\underset{NH_2}{\|}}{RSC}{-}NH_2\right\}^+X^- + HPic \longrightarrow \left\{\underset{\underset{NH_2}{\|}}{RSC}{-}NH_2\right\}^+Pic^-$$

where $HPic = 2,4,6\text{-}(NO_3)_3C_6H_2OH$

Alkyl chlorides often react slowly to give poor yields of the *S*-alkylthiuronium halides, but give better results if to the initial reaction mixture is added 0.5 g of potassium iodide followed by sufficient water or ethanol to give a clear solution at the boiling point. The derivative is not suitable for tertiary halides.

1 g of the alkyl halide, 1 g of powdered thiourea and 10 cm^3 of ethanol are refluxed for 10–20 minutes (if a primary bromide or iodide) or 2–3 hours (if a secondary bromide or iodide) or 30–60 minutes (if a polymethylene dihalide). Alkyl chlorides require 3–5 hours unless treated with potassium iodide (see note above) in which case heating for up to 60 minutes is often sufficient. To the reaction solution is added 0.5 g of picric acid, the mixture is boiled to obtain a clear solution, cooled and treated with drops of water if necessary to pre-cipitate the derivative, which is recrystallised from ethanol.

(b) *2-Naphthyl ether*

$$RX + 2\text{-}C_{10}H_7ONa \rightarrow 2\text{-}C_{10}H_7OR + NaX$$

A mixture of 2 g of 2-naphthol, 1 g of the compound, 1 g of potassium hydroxide and 10 cm^3 of ethanol is refluxed for 20 minutes. (For alkyl chlorides, the addition of 0.5 g of potassium iodide is advised.) To the reaction solution is added 2 g of potassium hydroxide and 20 cm^3 of water with stirring until the derivative separates. The product is filtered off, washed with water and recrystallised from ethanol.

(2) Halogenated aromatic hydrocarbons

(a) *Sulphonamide*

$$HArX + ClSO_2OH \rightarrow p\text{-}XArSO_2Cl \rightarrow p\text{-}XArSO_2NH_2$$

The procedure described under Ethers (a) should be modified by keeping the initial reaction mixture at 50°C for 15–30 minutes.

(b) *Nitro derivative*
The methods detailed under Hydrocarbons may be used.

(c) *Oxidation of side chain*

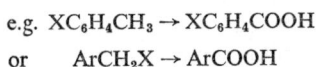

$$\text{e.g.} \quad XC_6H_4CH_3 \rightarrow XC_6H_4COOH$$

$$\text{or} \quad ArCH_2X \rightarrow ArCOOH$$

The oxidation of a side chain to the carboxylic acid group is con-veniently accomplished by means of potassium permanganate as described under Hydrocarbons (d).

Hydrocarbons

(a) *Nitro derivative*
Nitration of an aromatic hydrocarbon can produce a suitable derivative but the conditions necessary to achieve satisfactory results vary according to the ease with which the compound can be nitrated. The methods given below are illustrative of these variations. *All nitrations are potentially dangerous and should be undertaken with caution.*

(i) To a mixture of 4 cm³ each of concentrated sulphuric acid and concentrated nitric acid is added 0.5 g of the compound. The reaction liquor is heated on a water bath to 50°C for upwards of 10 minutes until reaction appears to be complete. The cooled mixture is then poured slowly into an excess of cold water and the product is filtered off, washed with water and recrystallised from dilute ethanol, ethanol or benzene.

(ii) The compound (0.5 g) is added slowly to a cold mixture of 5 cm³ of concentrated sulphuric acid and 3 cm³ of fuming nitric acid (CAUTION) with stirring. When the initial reaction has subsided, the reaction mixture is heated on a boiling water bath for 5–15 minutes, then cooled and poured into 50 cm³ of cold water. The product is filtered off (if an oil is obtained, vigorous stirring should cause crystallisation), washed well with cold water and recrystallised from aqueous ethanol, ethanol or benzene.

(iii) The organic compound (0.5 g) is dissolved in the minimum of glacial acetic acid and treated with a mixture of 2 cm³ of fuming nitric acid (CAUTION) and 2 cm³ of glacial acetic acid. The reaction mixture is heated under reflux for up to 10 minutes, allowed to stand until cold and then poured into 50 cm³ of cold water. The filtered product is washed with cold water and recrystallised from ethanol or benzene.

(b) *Sulphonamide*

$$ArH + ClSO_2OH \rightarrow ArSO_2Cl \rightarrow ArSO_2NH_2$$

The formation of the arylsulphonyl chloride should be carried out as described under Ethers (a).

The chloroform solution of the sulphonyl chloride may then be converted into the sulphonamide as follows: the chloroform layer is separated, washed with water, dried with anhydrous calcium chloride and the chloroform evaporated. The crude sulphonyl chloride is then heated in a fume chamber with 5 cm³ of concentrated ammonia solution for 10 minutes, the mixture is cooled, diluted with 10 cm³ of water, filtered and the solid product recrystallised from aqueous ethanol.

(c) *o-Aroylbenzoic acid*

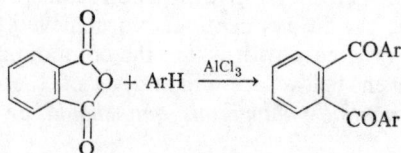

The process (carboxybenzoylation) involves the evolution of hydrogen chloride fumes and must be conducted in a fume cupboard or hood.

On a water bath, a mixture of 1 g of the hydrocarbon, 10 cm³ of dried methylene chloride or ethylene dichloride, 2.5 g of powdered anhydrous aluminum chloride and 1.2 g of pure phthalic anhydride is heated under a reflux condenser for 30 minutes or until no more hydrogen chloride fumes are evolved. The reaction mixture is cooled in ice-water and treated cautiously with 10 cm³ of concentrated hydrochloric acid with constant shaking, followed by 20 cm³ of water. All solid material should dissolve at this stage. The whole of the reaction mixture is extracted with ether (25 cm³) and the aqueous phase discarded. The ethereal layer is washed with 25 cm³ of dilute hydrochloric acid and is then (cautiously) extracted with 25 cm³ of dilute sodium carbonate solution, allowing the aqueous layer to run into 30 cm³ of dilute hydrochloric acid. The aroylbenzoic acid which is precipitated is filtered off, washed with cold water and recrystallised from dilute ethanol or acetic acid.

(d) *Oxidation of side chain*

e.g,
$$C_6H_5CH_3 \rightarrow C_6H_5COOH$$
$$o\text{-}C_6H_4(CH_3)_2 \rightarrow o\text{-}C_6H_4(COOH)_2$$
$$p\text{-}O_2NC_6H_4CH_3 \rightarrow p\text{-}O_2NC_6H_4COOH$$

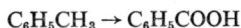

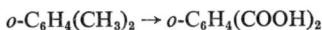

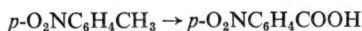

(i) Potassium permanganate
Under reflux, 1 g of the substance, 60 cm³ of water, 1.5 g of powdered sodium carbonate crystals and 4 g of finely powdered potassium permanganate are heated together until the purple colour of the permanganate has been discharged (up to 4 hours). The mixture is allowed to cool, is carefully acidified with dilute sulphuric acid and is then heated under reflux for 30 minutes before being cooled. Any residual manganese dioxide is removed by the addition of a little sodium bisulphite. The precipitated acid is filtered off and recrystallised from water, ethanol or a mixture thereof.

(ii) 'Chromic acid'
A suspension of 1 g of the compound, and 3 g of sodium dichromate in 10 cm³ of water is treated with 5 cm³ of concentrated sulphuric acid, added dropwise. The whole is then refluxed for up to 60 minutes, cooled and the carboxylic acid filtered off. The product is purified by dissolution in aqueous sodium carbonate from which it is precipitated by acidification and is then recrystallised from water or ethanol.

NOTE. For the oxidation of some polynuclear hydrocarbons e.g. anthracene, the above oxidising reagent mixture should be replaced by glacial acetic acid (20 cm³), chromium trioxide (2 g) and water

(4 cm³). The product is precipitated by addition of water to the reaction mixture and is recrystallised from acetic acid.

(e) *Picric acid derivative*

The method of preparation is described under Ethers (b); recrystallisation of the derivative may lead to its decomposition.

Mercaptans (Thiols)

(a) (*i*)*2,4-Dinitrophenyl thioether*

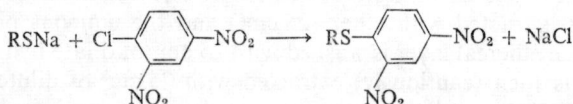

$$RSNa + Cl\!-\!\langle\text{aryl}\rangle\!-\!NO_2 \longrightarrow RS\!-\!\langle\text{aryl}\rangle\!-\!NO_2 + NaCl$$

To a solution of 0.5 g of the mercaptan in 15 cm³ of ethanol is added 2 cm³ of 10% aqueous sodium hydroxide followed by 1 g of 2,4-dinitrochlorobenzene dissolved in 5 cm³ of ethanol. The mixture is refluxed on a water bath for 10 minutes, rapidly filtered hot and the filtrate allowed to cool. The thioether crystallises on cooling and is recrystallised from ethanol.

(ii) *2,4-Dinitrophenyl sulphone*

The thioethers are readily oxidised to the corresponding sulphones which form valuable derivatives.

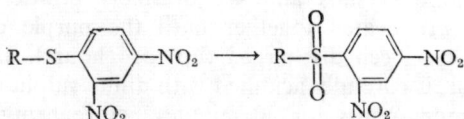

$$R\!-\!S\!-\!\langle\text{aryl}\rangle\!-\!NO_2 \longrightarrow R\!-\!\underset{O}{\overset{O}{S}}\!-\!\langle\text{aryl}\rangle\!-\!NO_2$$

To a solution of 1 g of the thioether in the minimum volume of warm glacial acetic acid is added 3% potassium permanganate solution, with shaking, as long as the colour is discharged. If the sulphide is precipitated during this addition, more acetic acid should be added. When reaction is complete, sulphur dixoide is passed into the reaction mixture until it is decolourised. Crushed ice is then added, the sulphone is filtered off and is recrystallised from ethanol.

(b) *Mercury salt* An excess of 10% aqueous mercuric cyanide is added to a solution of 6 drops (0.3 g) of the mercaptan in 2 cm³ of ethanol, and the mixture is shaken for a few minutes and then cooled in ice. The precipitated mercury mercaptide is filtered off and recrystallised from ethanol.

Nitriles

(a) *Hydrolysis to carboxylic acid*

$$2RCN + H_2SO_4 + 4H_2O \rightarrow 2RCOOH + (NH_4)_2SO_4$$

$$RCN + H_2O + NaOH \rightarrow RCOONa + NH_3$$

The nitrile (1 g) is refluxed with 5 cm³ of 30% aqueous sodium hydroxide (for alkyl cyanides) or 5 cm³ of 70% sulphuric acid (for aryl cyanides) for about 1–2 hours. The organic acid is liberated from the cooled mixture by the addition of an excess of hydrochloric acid (alkyl cyanides) or water (aryl cyanides) and identified.

(b) *Nitro derivative* Some aryl cyanides may be converted into a suitable derivative by nitration using method (a(i)) described under Hydrocarbons, keeping the temperature below 30°C throughout the reaction.

Nitrohydrocarbons, Halogenonitrohydrocarbons, Nitroethers

(a) *Reduction to amine*

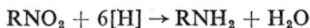

$$RNO_2 + 6[H] \rightarrow RNH_2 + H_2O$$

To a mixture of 1 g of the compound and 3 g of granulated tin are added 2 cm³ of ethanol and 10 cm³ of concentrated hydrochloric acid, with shaking until the initial reaction has subsided somewhat. The reaction mixture is then heated under reflux for 30 minutes or until the odour of the nitro compound is not detectable. The hot liquor is decanted, is cooled and made alkaline by cautious addition of 20% sodium hydroxide solution. The liberated amine is isolated by ether extraction (or steam distillation), the extract is dried with solid potassium hydroxide, the ether evaporated and the amine identified.

(b) *Oxidation of side chain* Aromatic nitro compounds containing a side chain may be oxidised to the corresponding carboxylic acids by the method (d(ii)) given under Hydrocarbons.

(c) *Nitration* Aromatic nitro compounds may be converted into polynitro derivatives by one of the methods described under Hydrocarbons (a).

Phenols

(a) *Benzoate*

$$ArOH + C_6H_5COCl \rightarrow ArOOCC_6H_5 + HCl$$

The Schotten-Baumann method described under Primary and Secondary Amines (b) should be followed.

(b) *p-Toluenesulphonate* (*'p-Tosyl'* *derivative*)

$$ArOH + p\text{-}CH_3C_6H_4SO_2Cl \rightarrow p\text{-}CH_3C_6H_4SO_2OAr$$

The derivative may be prepared either by the method given under Primary and Secondary Amines (c) or by that given under Alcohols (a), substituting *p*-toluenesulphonyl chloride for 3,5-dinitrobenzoyl chloride in the latter method.

(c) *3,5-Dinitrobenzoate, p-Nitrobenzoate*

$$ArOH + Ar'COCl \rightarrow ArOOCAr' + HCl$$

where Ar′ may be 3,5-dinitrophenyl or *p*-nitrophenyl. The procedure is detailed under Alcohols (a).

(d) *1-Naphthyl carbamate*

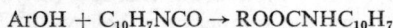

$$ArOH + C_{10}H_7NCO \rightarrow ROOCNHC_{10}H_7$$

The procedure given under Alcohols (b) may be followed, preferably modified by the addition of a few drops of an ethereal solution of triethylamine to the initial reaction mixture.

(e) *Aryloxyacetic acid*

$$ArOH + ClCH_2COOH \rightarrow ArOCH_2COOH + HCl$$

A mixture of 0.5 g of the phenol and 0.8 g of chloroacetic acid in 3 cm³ of 30% aqueous sodium hydroxide is treated with water until any solid formed is dissolved and the solution is then heated on a boiling water bath for 30–60 minutes. After cooling, the mixture is acidified to Congo Red with dilute hydrochloric acid and extracted with ether. The ethereal extract is washed with water and the aryloxy-acetic acid is then extracted by shaking with 15 cm³ of 5% sodium carbonate solution. The alkaline extract is acidified with dilute hydro-chloric acid, the precipitated derivative is filtered off, washed with cold water and recrystallised from hot water or dilute ethanol.

(f) *Bromo derivative* Phenols readily form bromo derivatives when treated by the method given under Ethers (c) using water or ethanol as solvent for the reaction.

Sulphides (Thioethers)

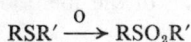

$$RSR' \xrightarrow{O} RSO_2R'$$

Sulphides may be characterised by oxidation to the corresponding sulphones as described under Mercaptans (a(ii)).

Sulphonic Acids

(a) *S-Benzylthiuronium salt* The experimental details are given under Carboxylic acids (d).

(b) *Amide*

$$RSO_2OH \xrightarrow{PCl_5} RSO_2Cl \xrightarrow{NH_3} RSO_2NH_2$$

In a fume chamber 1 g of the free acid or the anhydrous metallic salt is heated with 2 g of phosphorus pentachloride on an oil bath at 150°C for 30 minutes. The cooled mixture is treated with 15 cm³ of water, stirred and the washings decanted. The residual sulphonyl chloride is added to 5 cm³ of concentrated ammonia solution (0.880) with stirring and the mixture is then warmed on a water bath for 5 minutes. The sulphonamide usually crystallises from the cooled solution and is washed with water and recrystallised from water or aqueous ethanol.

(c) *Anilide*

$$RSO_2OH \xrightarrow{PCl_5} RSO_2Cl \xrightarrow{C_6H_5NH_2} RSO_2NHC_6H_5$$

The sulphonyl chloride is prepared as in (b) above and is then dissolved in acetone and treated with 0.5 cm³ of pure aniline dissolved in acetone together with 25 cm³ of dilute sodium hydroxide solution. The mixture is shaken for 10 minutes, extracted with ether to remove excess of aniline, acidified and the product filtered off and recrystallised from dilute ethanol.

Sulphonamides

(a) *Acetyl derivative*

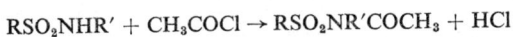

$$RSO_2NHR' + CH_3COCl \rightarrow RSO_2NR'COCH_3 + HCl$$

where R' may be H

The sulphonamide (1 g) is heated under reflux with 2.5 cm³ of acetyl chloride, adding up to 2 cm³ of glacial acetic acid if after 5 minutes solution is not complete. The excess of acetyl chloride is removed by distillation and the residue is poured into cold water. The product is collected, washed with water and recrystallised from aqueous ethanol.

(b) *Benzoyl derivative*

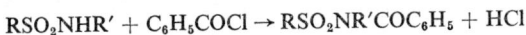

$$RSO_2NHR' + C_6H_5COCl \rightarrow RSO_2NR'COC_6H_5 + HCl$$

where R' may be H

The method given under Primary and Secondary Amines (b) may be suitably modified by acidifying the mixture when reaction is complete in order to precipitate the product.

(c) *Xanthyl derivative*

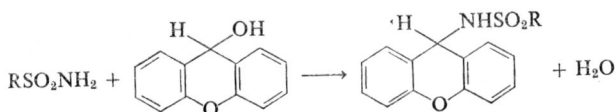

4

Primary sulphonamides may be converted into N-xanthylsulphon-amides by dissolving 0.25 g of the sulphonamide and 0.25 g of xanthy-drol in 10 cm³ of glacial acetic acid, warming gently if necessary. The solution is allowed to stand at room temperature until a solid separates (up to 90 minutes). The derivative is collected, recrystallised from dioxan-water (3:1) and dried at room temperature.

(d) *Hydrolysis*

$$ArSO_2NRR' + H_2O \xrightarrow{H_2SO_4} ArSO_2OH + RR'NH \cdot H_2SO_4$$

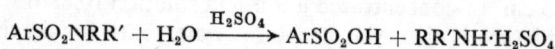

If $R = R' = H$, the main organic product of hydrolysis is a sulphonic acid. At the other extreme, if $R \neq R' \neq H$, the main products are a sulphonic acid and a secondary amine of the type $\begin{matrix} R \\ R' \end{matrix} \hspace{-4pt} >\hspace{-4pt} NH$.

In a large test-tube are placed 2 g of the sulphonamide and 3.5 cm³ of 80% (conc. acid:water = 3:1) sulphuric acid. The mixture is heated at 160°C and stirred continually with a thermometer until the solid goes into solution. The solution is allowed to cool and is poured into 25 cm³ of water. Dilute sodium hydroxide solution is added until the reaction solution is alkaline and any liberated amine is removed by ether extraction and identified. The alkaline aqueous residue contains the sodium salt of the sulphonic acid and after evaporation to small bulk may be used to prepare a derivative of the acid.

Sulphonyl Halides

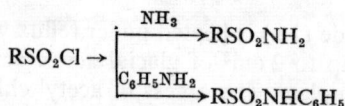

$$RSO_2Cl \rightarrow \begin{array}{l} \xrightarrow{NH_3} RSO_2NH_2 \\ \xrightarrow{C_6H_5NH_2} RSO_2NHC_6H_5 \end{array}$$

The amide or anilide should be prepared by following appropriate parts of the experimental procedures given under Sulphonic Acids.

5   The Use of Spectroscopic Methods

The value of spectroscopic methods in qualitative organic analysis has increased very greatly in recent years as a result of the development of instruments which make possible the convenient measurement of the infrared, ultraviolet, nuclear magnetic resonance and mass spectra of compounds using only very small amounts of substance.

Mass spectrometry can give valuable information concerning the molecular weight (see p. 258) of a substance as well as evidence for its molecular structure. Nuclear magnetic resonance spectroscopy has also emerged as a very powerful tool for structural investigations. For the present purpose however, discussion will be limited to a very brief account of the use of infrared and ultraviolet spectra in organic functional group determination.

Infrared spectroscopy

The bonds which hold together the atoms of an organic molecule may be likened to a system of springs which are continually undergoing stretching, bending and rotational motions. Each atom or group of atoms in the molecule oscillates about a point at which the attractive forces between electrons and nuclei balances the repulsive forces between the different nuclei and electron–electron repulsion. These natural oscillations have frequencies which depend upon the masses of the atoms and the strengths of the bonds. Provided that a change in the amplitude of motion results in a change in the molecular dipole moment any one of these molecular motions may be increased in amplitude by supplying energy in the form of infrared radiation of a frequency matching that of the natural molecular oscillation; the incident radiation will be absorbed by the molecule and its intensity is decreased on passing through the molecule. The intensity of absorption depends on the degree of change in dipole moment and the number of bonds participating in the absorption of radiation.

The organic chemist is mainly concerned with the absorption of infrared radiation having a wavelength between 2.5 μm and 15 μm (wave numbers 4000 cm^{-1} to 660 cm^{-1}) which gives information concerning the bending and stretching motions of a molecule.

It should be noted that infrared absorption band intensities are most often expressed as 'per cent transmittance' but also may be expressed as 'per cent absorbance'.

Some stretching and bending motions are not greatly affected by changes of structure in other parts of the molecule so that a knowledge of the characteristic group frequencies may enable the bands present in the spectrum of an unknown compound to be interpreted in terms of the corresponding functional groups. It must, however, be emphasized that shifts of characteristic group frequencies caused by electric and steric effects are quite common and although these shifts are often relatively small, much larger shifts are sometimes found and unless this is realised incorrect deductions may be made.

A detailed consideration of the typical infrared spectra of different classes of organic compounds is beyond the scope of this book but the following generalisations, taken in conjunction with the tables on pp. 90, 91 may be found useful.

Conjugation lowers the frequency of both bonds concerned, e.g. the stretching vibration of the $C\!=\!O$ group in saturated aliphatic ketones lies between 1725 and 1700 cm^{-1}, but conjugation with a $C\!=\!C$ group lowers the frequency to 1695–1660 cm^{-1} while the ethylenic group frequency is shifted downwards within its typical range 1690–1620 cm^{-1}.

Ring strain can cause an increase in the characteristic group frequency, e.g. the $C\!=\!O$ in succinic anhydride absorbs at 1865 cm^{-1} and 1782 cm^{-1}, and the $C\!=\!O$ absorption in cyclobutanone is 1780–1760 cm^{-1}.

Table 1. Some bond and group wave numbers (cm^{-1})

The ranges given are approximate.

OH and NH	Stretch	3200–3700
CH (aromatic)	Stretch	3000–3200
CH (aliphatic)	Stretch	2700–3300
S—H	Stretch	2550–2600
C≡C	Stretch⎫	2100–2300
C≡N	Stretch⎭	
C=O	Stretch	1650–1850
C=N	Stretch	1600–1700
C=C (aliphatic)	Stretch	1600–1700
C=C (aromatic)	Stretch	1585–1625
N—H	Bend	1550–1650
NO$_2$	Stretch	1500–1650
C—H	Bend	1300–1480
C—N (aromatic)	Stretch	1250–1350
O—H	Bend	1200–1450
C—N (aliphatic)	Stretch	1020–1220
C—O	Stretch	1050–1230
C=S	Stretch	1050–1235
C—F	Stretch	1000–1110
C—Cl	Stretch	700–750
S=O (sulphoxide)	Stretch	1030–1070
SO$_2$ (sulphone)	Stretch	⎧1120–1160 ⎩1300–1350
P=O	Stretch	1175–1350
P—O (aromatic)	Stretch	1180–1240
P—O (aliphatic)	Stretch	990–1050

Table 2. Some characteristic infrared spectra

Only those peaks are given where absorption about is 80% or greater. The range covered is 4000–400 cm⁻¹ (2.5–25.0 μm).
NOTE: Carbon–hydrogen stretching bands have been omitted.

Name	Peaks (cm⁻¹)	State	Cell path
n-Hexane	1465, 1460	Liquid	0.02 mm
	(merge as broad band)		
Nujol	none	Liquid	R.S.†
Cyclohexane	1445	Liquid	R.S.
Benzene	1480, 670*	Liquid	0.025 mm
Toluene	726, 692	Liquid	R.S.
m-Xylene	1616*, 1492*, 1460*, 767*, 690	Liquid	R.S.
Polystyrene	1602, 1495*, 1455*, 760*,		
	702* 547*	Film	
Indene	765	Liquid	R.S.
Methanol	3170*, 1030*	Liquid	R.S.
Ethanol	1070, 1060	Vapour	10 cm
Acetone	1713*, 1360, 1220	Liquid	0.015 mm
Methyl ethyl ketone	1715*	Liquid	R.S.
Cyclohexanone	1710*	Liquid	R.S.
Ethyl acetate	1740*, 1240	Liquid	0.015 mm
Diethyl phthalate	1725*, 1280	Liquid	R.S.
Ether	1120*	Liquid	0.025 mm
Dioxan	1250, 1120*, 885, 870	Liquid	R.S.
Chloroform	775*	Liquid	0.015 mm
Carbon tetrachloride	1545, 800*, 770*, 750*	Liquid	0.015 mm
Dichloroethane	710	Liquid	0.015 mm
Trichloroethylene	930*, 840*	Liquid	R.S.
Tetrachloroethylene	910*	Liquid	R.S.
Fluorolube	1195*, 1128*, 970	Liquid	R.S.
Carbon disulphide	1537, 1525	Vapour	10 cm
	(merge as broad band)		
Dimethyl sulphoxide	1050*	Liquid	R.S.

* Average values, given when the absorption yields a broad band.
† R.S. Between rock salt plates.

Hydrogen bonding decreases the stretching frequencies and increases the bending frequencies. As hydrogen bonding varies with the solvent used and the dilution, it is often instructive to measure the spectrum in different solvents, and at greater dilution which tends to reduce intermolecular hydrogen bonding.

α-Substitution of electron-withdrawing groups in compounds such as carboxylic acids and their derivatives, ketones, aldehydes, may lead to upward shifts of the C=O frequency e.g. in ethyl acetate, 1745 cm⁻¹, ethyl trichloroacetate, 1770 cm⁻¹.

Ultraviolet spectroscopy

The ultraviolet spectrum of a compound arises from the absorption of ultraviolet radiation by the molecule, resulting in transfer of electrons from lower orbitals to higher orbitals in the molecule. With few exceptions, only molecules containing multiple bonds have sufficiently stable excited states to give rise to absorption in the near ultraviolet region (that portion of the ultraviolet region which is most readily available, 200–400 nm) so that such spectra are diagnostic of unsaturation in the absorbing molecule. Thus many common types of compounds, e.g. alcohols, ethers, hydrocarbons are transparent in this region and although aldehydes, ketones and aliphatic nitro compounds have absorption maxima within the region these are of such low intensity that they are of very limited use for identification purposes. Monofunctional alkenes, alkynes, carboxylic acids, esters, and amides have absorption maxima just outside the near ultraviolet region.

If, however, two functions each containing a multiple bond are conjugated with each other the resulting interaction of π electrons (π–π conjugation) produces a strong absorption in the near ultraviolet region. Similarly, a change in the intensity of absorption (and in wavelength) occurs when a multiple bond is conjugated with a hetero atom carrying a non-bonding lone pair of electrons, such as —NR_2 or —OR (π–p conjugation). Consequently one of the most important uses of ultraviolet spectroscopy in organic analysis is the identification of conjugated groups.

On the other hand, if two absorbing groups are separated from each other by one or more insulating groups, such as —CH_2—, there is little interaction between the absorbing units and the ultraviolet spectrum equates approximately to the addition of the spectra of the absorbing groups (the additivity principle).

Unlike infrared spectra which commonly have upwards of ten maxima, ultraviolet spectra often contain only one or two, and these are usually broad bands. Consequently the identification of functional groups from a knowledge of the wavelengths of absorption maxima is much less simple than is the case with infrared spectra and generally prior knowledge of possible functional groups is necessary. However, a comparison of the spectrum of a compound whose structure is in doubt with that of a substance differing in structure only in a manner which should not affect the absorbing unit ('model' compound) is often a useful way of deciding between possible formulae. Thus, ethyl β-aminocrotonate may have the structure I or the tautomeric structure II.

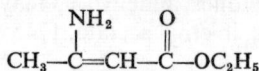

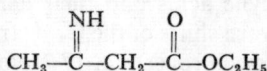

I   II

The essential difference, spectroscopically, is in the absorbing units

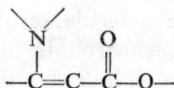

in I and

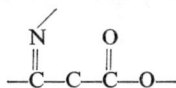

in II.

If ethyl β-dimethylaminocrotonate (III) is taken as a model compound having essentially the same system of conjugation as I it is

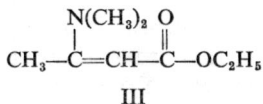

III

found that its ultraviolet spectrum is very similar to that of I (λ_{max} 275 nm and 267 nm respectively) and dissimilar to that of II which exhibits only slight absorption in the ultraviolet region, as would be expected since its molecule contains two absorbing units isolated by a —CH_2— group.

6 Classified Tables of the Commoner Organic Compounds

In the tables that follow, the compounds are listed according to the radicals present in them, the individuals of each class being arranged generally in progressive order of boiling or melting points. The order in which the various classes appear is dependent upon the elements present and in most cases the order of groups of compounds within these classes is that of increasing oxygenation e.g. hydrocarbons are followed by ethers, alcohols, phenols, aldehydes, ketones, carboxylic acids.

Abbreviations

The following abbreviations are used in the tables:

abs.	absolute
AcOH	acetic acid
addtn.	addition
alc.	alcoholic (ethanolic), alcohol
alk.	alkaline
anhyd.	anhydrous
approx. ca.	approximate(ly)
asym.	asymmetrical
aq.	aqueous
B.HCl	hydrochloride of base B
b.p.	boiling point*
c	concentration of a solution
calc.	calculated
cpd., cmpd.	compound
compd. with TNB	molecular complex formed with trinitrobenzene
conc.	concentrated
concn.	concentration
const.	constant
corr.	corrected
cryst.	crystalline, crystallised
d (before a figure)	density at specified temperature (superscript) relative to water at 4°C (subscript)
d (after a figure)	melts or boils with decomposition
decomp.	decomposition
deg.	degree

* Boiling points are given for N.T.P. or at reduced pressure expressed as mm of mercury (1 mm Hg = 13.5951 × 9.80665N m^{-2}).

deriv.	derivative
dil.	dilute
dist.	distilled
distln.	distillation
eqn.	equation
equiv.	equivalent
expt.	experiment(al)
htg.	heating
hyd.	hydrate(d)
i.r.	infrared
insol.	insoluble
liq.	liquid
max.	maximum
m.p.	melting point
min.	minimum
mod.	moderately
mol.	molecule, molecular
mol. wt.	molecular weight
n_D^t	refractive index for sodium D light at temperature $t°C$
obs.	observed
opt. act.	optically active
oxdtn.	oxidation
ppt.	precipitate
prep.	preparation
recryst.	recrystallised
S.A.T.P.	S-Alkylthiuronium picrate
S.B.T.	S-Benzylthiuronium salt
sol.	soluble
soln.	solution
sl.	slightly
spar.	sparingly
sp.	specific
subl.	sublimes
sym.	symmetrical
temp.	temperature
u.v.	ultraviolet
v.	very
vac.	vacuum
vol.	volume
wt.	weight
$[\alpha]_D^t$	specific rotation for sodium D light at temperature $t°C$

Temperatures are given to the nearest whole number and are expressed in degrees Celsius (Centigrade), $°C$.

HYDROCARBONS

LIQUID

B.P.°C

34° Isoprene C_5H_8, 2-Methyl buta-1,3-diene.
Unsaturated. d_4^{20} 0.6805. n_D^{20} 1.4216. Yields di- and tetrabromide with bromine. Thermal dgdtn. product of many terpenes.

36° n-Pentane C_5H_{12}
d_4^{20} 0.6263. n_D^{20} 1.3577

38° Trimethylethylene $(CH_3)_2$ C:CHCH_3, Amylene.
Unsaturated. d_4^{20} 0.6596. n_D^{20} 1.3876. Sol. in strong H_2SO_4 at 0° soln. yields t-amyl alc. on diltn. and distltn. Tech. 'amylene' usually a mixture, boiling range 25°–40°.

40° Cyclopentadiene C_5H_6
Unsaturated. d_4^{20} 0.798. n_D^{20} 1.4398. Stable only at low temp. Dimerises easily to dicyclopentadiene from which it can be regenerated on slow distln. Maleic anhydride adduct m.p. 164°.

49° Cyclopentane C_5H_{10}
d_4^{20} 0.7454. n_D^{20} 1.4065.

69° n-Hexane C_6H_{14}
d_4^{20} 0.6594. n_D^{20} 1.3750.

80° Benzene C_6H_6
m.p. 5.5°. d_4^{20} 0.8790. n_D^{20} 1.5011. Nitration → nitrobenzene b.p. 210°, m-dinitrobenzene m.p. 90°. $ClSO_3H$ → benzene sulphonyl chloride which with NH_3 gives sulphonamide m.p. 153°. Sulphonated slowly on boiling with conc. H_2SO_4. Picrate m.p. 84°. o-Aroyl benzoic acid deriv. m.p. 128°.

81° Cyclohexane C_6H_{12}
M.p. 6.5°. d_4^{20} 0.7783. n_D^{20} 1.4263. Vigorous oxdtn. → adipic acid m.p. 151°.

83° Cyclohexene C_6H_{10}
Unsaturated. d_4^{20} 0.811. n_D^{20} 1.4465. HNO_3 oxdtn. → adipic acid m.p. 151°. Nitrosochloride m.p. 152°. Reacts with N-bromo-succinimide.

98° n-Heptane C_7H_{16}
d_4^{20} 0.6837. n_D^{20} 1.3877.

101° Methylcyclohexane $C_6H_{11}CH_3$
d_4^{20} 0.7693. n_D^{20} 1.4231

110° Toluene $C_6H_5CH_3$
d_4^{20} 0.8670. n_D^{20} 1.4969. Oxdtn. with alk. $KMnO_4$ → benzoic acid m.p. 121°. 2,4-Dinitro deriv. m.p. 70° (2,4,6-trinitro deriv. m.p. 82°). Sulphonamide m.p. 137°. o-Aroyl benzoic acid m.p. 138°. p-Sulphonylchloride m.p. 69°.

115° Tropilidene C_7H_8, Cycloheptatriene
Unsaturated. d_4^{20} 0.887. n_D^{20} 1.5243. Maleic anhydride adduct m.p. 103°. Forms dibromide.

B.P.°C

136° Ethyl benzene $C_6H_5CH_2CH_3$
d_4^{20} 0.8671. n_D^{20} 1.4959. Oxdtn. with alk. $KMnO_4 \rightarrow$ benzoic acid
m.p. 121°. Sulphonamide m.p. 109°. o-Aroyl benzoic acid m.p.
128°.

138° p-Xylene $C_6H_4(CH_3)_2$
d_4^{20} 0.8610. n_D^{20} 1.4958. Oxdtn. with alk. $KMnO_4 \rightarrow$ terephthalic
acid (q.v.). Sulphonamide m.p. 147°. o-Aroyl benzoic acid m.p.
132°.

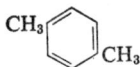

139° m-Xylene $C_6H_4(CH_3)_2$
d_4^{20} 0.8641. n_D^{20} 1.4972. Oxdtn. with alk. $KMnO_4 \rightarrow$ isophthalic
acid (q.v.). Sulphonamide m.p. 137°. o-Aroyl benzoic acid m.p.
142°.

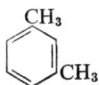

142° Phenylacetylene $C_6H_5C \vdots CH$
Unsaturated. Gives acetophenone with dil. H_2SO_4. Gives white
ppt. with Tollen's reagent.

144° o-Xylene $C_6H_4(CH_3)_2$
d_4^{20} 0.8801. n_D^{20} 1.5052. Oxdtn. with alk. $KMnO_4 \rightarrow$ phthalic acid
(q.v.). Sulphonamide m.p. 144°.

146° Styrene $C_6H_5CH:CH_2$, Phenylethylene
Unsaturated. d_0^{23} 0.9038. Polymerises with a drop of conc. H_2SO_4
to glassy mass. Oxdtn. with acid $KMnO_4 \rightarrow$ benzaldehyde, and
by alk. $KMnO_4$ to benzoic acid m.p. 121°. Dibromide m.p. 73°.
Nitrosochloride m.p. 97°.

152° Cumene $C_6H_5CH(CH_3)_2$, Isopropylbenzene
d_4^{20} 0.8618. n_D^{20} 1.4912. Oxdtn. with alk. $KMnO_4 \rightarrow$ benzoic acid
m.p. 121°. Sulphonamide m.p. 106°. o-Aroyl benzoic acid m.p.
133°.

156° α-Pinene $C_{10}H_{16}$
Unsaturated. Main constituent of turpentine oil. d_4^{15} 0.862. n_D^{12}
1.4650. $[\alpha]_D^{20}$ + 49°. Oxdtn. yields pinonic acid m.p. 103–104°
(racemic form).

165° Mesitylene $C_6H_3(CH_3)_3$
d_4^{20} 0.8651. n_D^{20} 1.4994. Oxdtn. with alk. $KMnO_4 \rightarrow$ trimesic acid.
Readily sulphonated and nitrated. Cold conc. HNO_3/conc.
$H_2SO_4 \rightarrow$ trinitro deriv. m.p. 230°. Sulphonamide m.p. 141°.
o-Aroyl benzoic acid m.p. 212°.

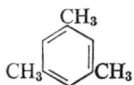

169° t-Butylbenzene $C_6H_5C(CH_3)_3$
d_4^{20} 0.8665. n_D^{20} 1.4926.

169° Pseudocumene $C_6H_3(CH_3)_3$
d_4^{20} 0.8762. n_D^{20} 1.5049. Fuming $HNO_3 \xrightarrow{0°C}$ 5-nitro deriv. m.p. 71°.
Warm conc. HNO_3/conc. $H_2SO_4 \rightarrow$ 3,5,6-trinitro deriv. m.p.
185°. Sulphonamide m.p. 181°.

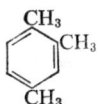

B.P.°C

170°(d)   Dicyclopentadiene $C_{10}H_{12}$
Unsaturated. See cyclopentadiene.

177°   *p*-Cymene $CH_3C_6H_4CH(CH_3)_2$
d_4^{20} 0.8573. n_D^{20} 1.4909. Oxdtn. with alk. KMnO$_4$ → terephthalic acid (*q.v.*). Sulphonamide m.p. 115°. *o*-Aroyl benzoic acid m.p. 124°. Frequent dgdtn. product of monoterpenes, e.g. camphor + P$_2$O$_5$ → *p*-cymene, on heating.

177°   Limonene $C_{10}H_{16}$
Unsaturated. d_4^{20} 0.8411. $[\alpha]_D^{20}$ + 127°. Tetrabromide m.p. 104–105°.

178°   Dipentene $C_{10}H_{16}$
Unsaturated. d_4^{20} 0.840. n_D^{20} 1.4727. Tetrabromide m.p. 124°. Racemic form of limonene.

182°   Indene C_9H_8
d_4^{20} 0.9968. n_D^{20} 1.577. Oxdtn. with boiling 30% HNO$_3$ → phthalic acid (*q.v.*). Yields Na deriv. on heating with sodium. Boiling gl. AcOH → dimer m.p. 57–58°. Gradually polymerises in air/light. Picrate m.p. 98° (explosive).

187°/195°   Decalin $C_{10}H_{18}$, Decahydronaphthalene
Cis and trans forms. Commercial decalin is a mixture. Trans-b.p. 187°, d_4^{20} 0.870, n_D^{20} 1.4695. Cis-b.p. 195°, d_4^{20} 0.896, n_D^{20} 1.4810.

207°   Tetralin $C_{10}H_{12}$, Tetrahydronaphthalene
d_4^{20} 0.971. n_D^{20} 1.5392. Undergoes autoxdtn. in air, turning yellow; deposits colourless crystals of hydroperoxide m.p. 56°. Oxdtn. by aq. KMnO$_4$ → Phthalonic acid (*q.v.*).

245°   1-Methyl naphthalene $C_{10}H_7CH_3$
d_4^{20} 1.025. n_D^{20} 1.6174. Oxdtn. → 1-naphthoic acid m.p. 162°. Picrate m.p. 141–142°. *o*-Aroyl benzoic acid m.p. 169°.

288°   n-Hexadecane $C_{16}H_{34}$, n-Cetane
d_4^{20} 0.722. n_D^{20} 1.434. Used in gas–liquid chromatography as a non-polar stationary phase.

375°   Squalane $C_{30}H_{56}$ (2,6,10,15,19,23-Hexamethyltetracosane)
d_4^{20} 0.807. n_D^{20} 1.4515. Used in gas–liquid chromatography as a non-polar stationary phase.

SOLID

M.P.°C

13°   *p*-Xylene. See liquid hydrocarbons.

M.P.°C

26° Diphenylmethane $C_6H_5CH_2C_6H_5$
B.p. 265°. CrO_3 oxdtn. → benzophenone m.p. 48°. Conc. HNO_3/ conc. H_2SO_4 → tetranitro deriv. m.p. 172°.

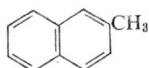

32° 2-Methylnaphthalene $C_{10}H_7CH_3$
B.p. 241°. Picrate m.p. 115°. o-Aroyl benzoic acid, m.p. 190°. Cold conc. HNO_3 → 1-nitro deriv. m.p. 81°.

51° Camphene $C_{10}H_{16}$
B.p. 159°. Unsaturated. $[\alpha]_D^{20} \pm 104°$. Dibromo m.p. 89°. Hydrochloride m.p. 149°.

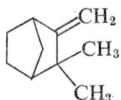

52° Dibenzyl $C_6H_5CH_2CH_2C_6H_5$
B.p. 284°. Easily oxidised to benzoic acid m.p. 121°. Cold fuming HNO_3 → dinitro deriv. m.p. 180°.

40–60° Benzyldiphenyl $C_6H_5CH_2C_6H_4C_6H_5$.
(range) Commerical product is a mixture of o- and p-isomers (approx. ratio 60:40). Used in gas–liquid chromatography as a slightly polar stationary phase.

62° Diphenylacetylene $C_6H_5C \vdots CC_6H_5$, Tolan
Unsaturated. Adds two or four halogen atoms. Hydrated by mineral acids to deoxybenzoin m.p. 60°. Absptn. max. at 279 and 298 nm. Picrate m.p. 111°.

70° Diphenyl $C_6H_5C_6H_5$
B.p. 256°. Cold conc. HNO_3 → 4-nitro deriv. m.p. 114°. Warm fuming HNO_3 → 4,4′-dinitro m.p. 235°. CrO_3/gl. AcOH oxdtn. → benzoic acid m.p. 121°. o-Aroylbenzoic acid m.p. 225°. Sulphonamide m.p. 228°.

80° Naphthalene $C_{10}H_8$
B.p. 218°. Picrate m.p. 150°. Sublimes readily. Steam volatile. Oxdtn. by HNO_3 or $KMnO_4$ yields benzoic acid m.p. 121°. Oxdtn. by CrO_3/AcOH yields 1,4-naphthaquinone (q.v.). Gives green colour with dry $AlCl_3$ in $CHCl_3$ soln.

93° Triphenylmethane $(C_6H_5)_3CH$
Cold fuming HNO_3 → trinitro deriv. m.p. 206°. Redtn. by Zn/ AcOH → triamino deriv. m.p. 208°. Oxdtn. with PbO_2/HCl gives intense red colour of rosaniline. Oxdtn. by CrO_3 or hot conc. HNO_3 yields triphenyl carbinol m.p. 162°.

95° Acenaphthene $C_{12}H_{10}$
B.p. 278°. Picrate m.p. 161°. CrO_3 oxdtn. → acenaphthene-quinone m.p. 261°. o-Aroyl benzoic acid m.p. 199°.

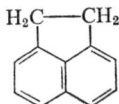

99° Azulene $C_{10}H_8$
Deep blue colour. Strong absptn. maxima 534 → 697 nm region. Dissolves in conc. acids, not recovered on diltn. Isomerises to naphthalene on heating above 270°. 1,3,5-trinitrobenzene adduct m.p. 167°.

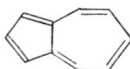

M.P.°C

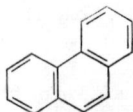

100°   Phenanthrene $C_{14}H_{10}$
Picrate m.p. 143°. $CrO_3/AcOH$ oxdtn. → phenanthraquinone
m.p. 205°. 1,3,5-Trinitrobenzene adduct m.p. 145°.

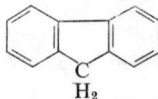

114°   Fluorene $C_{13}H_{10}$
$CrO_3/AcOH$ oxdtn. → fluorenone m.p. 83° (separate product
from unreacted fluorene on alumina column). Dissolves in warm
conc. H_2SO_4 → blue colour. o-Aroyl benzoic acid m.p. 228°.
Picrate unstable.

124°   Stilbene $C_6H_5CH:CHC_6H_5$, Trans 1,2-diphenylethylene.
Unsaturated, but decolourises bromine slowly. Dibromide m.p.
237°. Acid $KMnO_4$ oxdtn. → benzaldehyde.

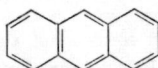

217°   Anthracene $C_{14}H_{10}$
B.p. 340°. Sublimes. Picrate m.p. 138°. $CrO_3/AcOH$ oxdtn. →
anthraquinone m.p. 280°. Br_2/CCl_4 → dibromo m.p. 221°.
Maleic anhydride adduct m.p. 263°.

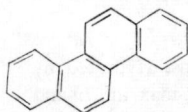

255°   Chrysene $C_{18}H_{12}$
CrO_3 oxdtn. → chrysaquinone m.p. 240°. Does not form maleic
anhydride adduct. Picrate unstable.

ETHERS

LIQUID

B.P.°C

11°   Ethylene oxide C_2H_4O, Oxirane
d_4^0 0.897. n_4^{20} 1.3614. Miscible with H_2O, alcohol, ether. Reduces
Tollen's reagent. Gives iodoform reaction.

35°   Diethyl ether $(C_2H_5)_2O$
d_4^{20} 0.7135. n_D^{20} 1.3526. Sl. sol. in H_2O. Dissolves in cold conc.
H_2SO_4; recovered unchanged on diltn.

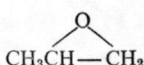

35°   Propylene oxide C_3H_6O
d_4^0 0.859. Misicble with H_2O, alcohol, ether. Hot H_2O → propylene glycol.

45°   Methylal. See Aldehydes and Acetals.

65°   Tetrahydrofuran C_4H_8O
d_4^{20} 0.889. n_D^{20} 1.407. Excellent solvent for Grignard reactions.
Very sol. in H_2O.

85°   Ethylene glycol dimethyl ether $CH_3OCH_2CH_2OCH_3$
d_4^{20} 0.866. n_D^{20} 1.3796. H_2O soluble.

89°   Ethylal.   See Aldehydes and Acetals.

B.P.°C

102°   Dioxan $C_4H_8O_2$

d_4^{20} 1.0336. n_D^{20} 1.4232. Miscible with H_2O, alcohol.

103°   Acetal. See Aldehydes and Acetals.

121°   Ethylene glycol diethyl ether $C_2H_5OCH_2CH_2OC_2H_5$
d_4^{20} 0.848.

124°   Paraldehyde. See Aldehydes and Acetals.

135°   Ethylene glycol monoethyl ether $C_2H_5OCH_2CH_2OH$, 2-Ethoxyethanol.
d_4^{20} 0.9297. n_D^{20} 1.408. Miscible with water, alcohol, ether. 3,5-Dinitrobenzoate m.p. 75°.

142°   Di n-butyl ether $(C_4H_9)_2O$
d_4^{20} 0.7683. n_D^{20} 1.3989. Sl. sol. in H_2O.

154°   Anisole $C_6H_5OCH_3$, Methyl phenyl ether.
d_4^{20} 0.994. n_D^{20} 1.522. Picrate m.p. 81°. Sulphonamide m.p. 111°.

162°   Diethylene glycol dimethyl ether $CH_3OCH_2CH_2OCH_2CH_2OCH_3$, Diglyme.
d_{20}^{20} 0.944. n_D^{20} 1.4099. Miscible with water.

170°   Benzyl methyl ether $C_6H_5CH_2OCH_3$
d_4^{20} 0.965. n_D^{20} 1.5008. Insol. H_2O. Warm conc. H_2SO_4/gl.AcOH → benzyl acetate (q.v.). Picrate 116°.

172°   Phenetole $C_6H_5OC_2H_5$ Ethyl phenyl ether
d_4^{20} 0.966. n_D^{20} 1.5080. Picrate m.p. 92°. Sulphonamide m.p. 150°.

188°   Diethylene glycol diethyl ether $(C_2H_5OCH_2CH_2)_2O$
d_4^{20} 0.906. n_D^{20} 1.411.

206°   Veratrole $C_6H_4(OCH_3)_2$, Catechol dimethyl ether.
M.p. 22°. d_{25}^{25} 1.084. n_D^{22} 1.5287. Dibromo m.p. 92°. Fuming HNO_3 → nitro deriv. m.p. 95°. Sulphonamide m.p. 136°. Picrate m.p. 57°.

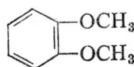

217°   Resorcinol dimethyl ether $C_6H_4(OCH_3)_2$
n_D^{20} 1.4233. Dibromo m.p. 140°. Sulphonamide m.p. 166°.

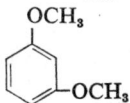

232°   Safrole.   4-Allyl-1,2-methylenedioxybenzene.
d_4^{20} 1.10. n_D^{20} 1.5383. Excess Br_2 → pentabromo m.p. 169°. Alk $KMnO_4$ oxdtn. → piperonylic acid m.p. 228°.

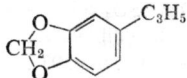

235°   Anethole $C_3H_5C_6H_4OCH_3$, p-Propenyl anisole.
M.p. 22°. n_D^{20} 1.558. Picrate m.p. 70°(d). CrO_3 oxdtn. → anisic acid m.p. 184°.

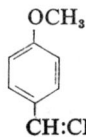

B.P.°C
246°   Isosafrole. 4-Propenyl 1,2-methylenedioxybenzene
       d_4^{15} 1.125. n_D^{20} 1.578. $Br_2/CS_2 \rightarrow$ tribromo m.p. 109°. Alk. $KMnO_4$
       oxdtn. $\rightarrow$ piperonylic acid m.p. 228°.
254°   Eugenol. See Phenols.
271°   1-Naphthyl methyl ether $C_{10}H_7OCH_3$
       d_4^{20} 1.092. n_D^{20} 1.6256. Picrate m.p. 130°.
298°   Dibenzyl ether $(C_6H_5CH_2)_2O$
       d_4^{20} 1.043. Picrate m.p. 78°.

SOLID

M.P.°C

27°   Diphenyl ether $(C_6H_5)_2O$
       B.p. 259°. Picrate m.p. 110°. Disulphonamide m.p. 159°.
31°   Guaiacol. See Phenols.
56°   Hydroquinone dimethyl ether $C_6H_4(OCH_3)_2$

CH_3O⟨⟩OCH_3

       B.p. 213°. Yellow soln. in conc. H_2SO_4, Br_2/gl. AcOH (boiling) $\rightarrow$
       dibromo m.p. 142°. Sulphonamide m.p. 148.°
72°   2-Naphthyl methyl ether $C_{10}H_7OCH_3$, Nerolin.

OCH_3

       B.p. 273°. Picrate m.p. 117°. Sulphonamide m.p. 151°.

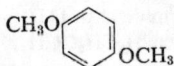

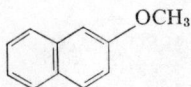

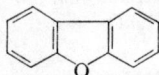

86°   Dibenzofuran $C_{12}H_8O$, Diphenylene oxide.
       B.p. 288°. Picrate m.p. 94°.

ALCOHOLS

LIQUID

B.P.°C
65°   Methyl alcohol CH_3OH
       d_4^{20} 0.791. Miscible with H_2O. Oxidised to formaldehyde by
       dipping hot Cu wire into it. Cold dil. $K_2Cr_2O_7/H_2SO_4$ soln. $\rightarrow$
       CH_2O. 3,5-Dinitrobenzoate m.p. 109°. 1-Naphthyl carbamate
       m.p. 124°. 3-Nitro hydrogen phthalate m.p. 152°.
78°   Ethyl alcohol C_2H_5OH
       d_4^{20} 0.789. Miscible with H_2O. Hot dil. $K_2Cr_2O_7/H_2SO_4$ soln. $\rightarrow$
       CH_3CHO. Alk. $KMnO_4$ oxdtn. $\rightarrow$ acetic acid. Iodine/dil. NaOH
       soln. $\rightarrow$ iodoform CHI_3 m.p. 119°. 3,5-Dinitrobenzoate m.p. 93°.
       1-Naphthyl carbamate m.p. 79°. 3-Nitro hydrogen phthalate
       m.p. 157°.
82°   Isopropyl alcohol $(CH_3)_2CHOH$
       d_4^{20} 0.785. Miscible with H_2O. Hot dil. $K_2Cr_2O_7/H_2SO_4$ soln. $\rightarrow$
       acetone. With I_2 in cold Na_2CO_3 soln. $\rightarrow$ iodoform m.p. 119°.
       3,5-Dinitrobenzoate m.p. 122°. 1-Naphthyl carbamate m.p. 106°.
       3-Nitro hydrogen phthalate m.p. 153°.

B.P.°C
83°   t-Butyl alcohol $(CH_3)_3COH$
     M.p. 25°. Miscible with H_2O. Yields t-butyl chloride b.p. 52°
     with excess cold conc. HCl rapidly. 3,5-Dinitrobenzoate m.p.
     142°. 1-Naphthyl carbamate m.p. 101°. p-Nitrobenzoate m.p.
     116°.
97°   n-Propyl alcohol $C_2H_5CH_2OH$
     d_4^{20} 0.804. Miscible with H_2O. Hot dil. $K_2Cr_2O_7/H_2SO_4$ soln→
     propionaldehyde. Alk. $KMnO_4$ oxdtn. → propionic acid. 3.5-
     Dinitrobenzoate m.p. 74°. 1-Naphthyl carbamate m.p. 80°.
     3-Nitro hydrogen phthalate m.p. 142°.
97°   Allyl alcohol $CH_2:CHCH_2OH$
     d_4^{20} 0.854. Miscible with H_2O. Irritating odour. Unsaturated. Dil.
     $K_2Cr_2O_7/H_2SO_4$ soln. → acrolein (very pungent). 3,5-Dinitro-
     benzoate m.p. 50°. 1-Naphthyl carbamate m.p. 109°. 3-Nitro
     hydrogen phthalate m.p. 124°.
99°   s-Butyl alcohol $CH_3CHOHC_2H_5$
     d_4^{20} 0.807. Sol. in 6 vols. cold H_2O. Hot dil. $K_2Cr_2O_7/H_2SO_4$
     soln. → methyl ethyl ketone. I_2/Na_2CO_3 soln. in cold → iodo-
     form. 3,5-Dinitrobenzoate m.p. 76°. 1-Naphthyl carbamate m.p.
     97°. 3-Nitro hydrogen phthalate m.p. 131°.
102° t-Amyl alcohol $(CH_3)_2C(OH)C_2H_5$
     d_4^{20} 0.809. Sol. in 6 vols. cold H_2O. Yields t-amyl chloride b.p. 86°
     with excess cold conc. HCl. Dehydrated on warming with oxalic
     acid → amylene (q.v.). 3,5-Dinitrobenzoate m.p. 117°. 1-
     Naphthyl carbamate m.p. 72° (poor yield). p-Nitrobenzoate m.p
     85°.
108° Isobutyl alcohol $(CH_3)_2CHCH_2OH$
     d_4^{20} 0.802. Sol. in 8 vols. cold H_2O. Boiling dil. $K_2Cr_2O_7/H_2SO_4$
     soln. → isobutyraldehyde. 3,5-Dinitrobenzoate m.p. 88°. 1-
     Naphthyl carbamate m.p. 104°. 3-Nitro hydrogen phthalate m.p.
     183°.
117° n-Butyl alcohol $CH_3(CH_2)_2CH_2OH$
     d_4^{20} 0.810. Almost insol. cold H_2O. Boiling dil. $K_2Cr_2O_7/H_2SO_4$ →
     n-butyraldehyde. 3,5-Dinitrobenzoate m.p. 63°. 1-Naphthyl
     carbamate m.p. 72°. 3-Nitro hydrogen phthalate m.p. 147°.
119° s-Amyl alcohol $C_2H_5CH_2CHOHCH_3$
     Opt. act. Racemate props. d_4^{20} 0.809. 3,5-Dinitrobenzoate m.p.
     62°. 1-Naphthyl carbamate m.p. 74°. 3-Nitro hydrogen phthalate
     m.p. 103°.
124° 2-Methoxyethanol $CH_3OCH_2CH_2OH$, Ethylene glycol mono-
     methyl ether. 'Methyl Cellosolve'.
     d_4^{20} 0.965. Miscible with H_2O. 1-Naphthyl carbamate m.p. 113°.
     3-Nitro hydrogen phthalate m.p. 129°. p-Nitrobenzoate m.p. 51°.
129° Active amyl alcohol $C_2H_5CH(CH_3)CH_2OH$
     Opt. act. Racemate props. Sl. sol. H_2O. d_4^{20} 0.819. 3,5-Dinitro-
     benzoate m.p. 70°. 1-Naphthyl carbamate m.p. 82°. 3-Nitro
     hydrogen phthalate m.p. 158°.

B.P.°C

132° Isoamyl alcohol $(CH_3)_2CHCH_2CH_2OH$
      d_4^{20} 0.809. Almost insol. H_2O. Boiling dil. $K_2Cr_2O_7/H_2SO_4$
      soln. → isovaleraldehyde. 3,5-Dinitrobenzoate m.p. 62°. 1-
      Naphthyl carbamate m.p. 68°. 3-Nitro hydrogen phthalate m.p.
      166°.

135° 2-Ethoxyethanol $C_2H_5OCH_2CH_2OH$, Ethylene glycol mono-
      ethyl ether. 'Ethyl Cellosolve'.
      d_4^{20} 0.930. Miscible with H_2O. 3,5-Dinitrobenzoate m.p. 76°.
      1-Naphthyl carbamate m.p. 67°. 3-Nitro hydrogen phthalate m.p.
      121°.

138° n-Amyl alcohol $CH_3(CH_2)_3CH_2OH$
      d_4^{20} 0.815. Sl. sol. in H_2O. 1-Naphthyl carbamate m.p. 68°. 3-
      Nitro hydrogen phthalate m.p. 136°. Hydrogen phthalate m.p.
      75°.

140° Cyclopentanol C_5H_9OH
      d_4^{20} 0.947. Sl. sol. in H_2O. 3,5-Dinitrobenzoate m.p. 115°. 1-
      Naphthyl carbamate m.p. 118°. Phenyl carbamate m.p. 132°.

157° n-Hexyl alcohol $CH_3(CH_2)_4CH_2OH$
      d_4^{20} 0.819. 3,5-Dinitrobenzoate m.p. 61°. 1-Naphthyl carbamate
      m.p. 59°.

160° Cyclohexanol $C_6H_{11}OH$
      M.p. 25°. Almost insol. in cold H_2O. Boiling dil. $K_2Cr_2O_7/H_2SO_4$
      soln. → cyclohexanone. Hot conc. HNO_3 oxdtn. → adipic acid
      m.p. 151°. Distln. with a trace of H_3PO_4 (or conc. H_2SO_4) yields
      cyclohexene. 3,5-Dinitrobenzoate m.p. 113°. 1-Naphthyl carba-
      mate m.p. 128°. 3-Nitro hydrogen phthalate m.p. 160°.

166° Diacetone alcohol. See Ketones (Liquid).

170° Furfuryl alcohol $(C_4H_3O)CH_2OH$
      d_4^{20} 1.130. Sl. Sol. in cold H_2O. Resinifies on warming with acids.
      Green colour with conc. HCl. 3,5-Dinitrobenzoate m.p. 80°.
      1-Naphthyl carbamate m.p. 129°. Hydrogen phthalate m.p. 85°.

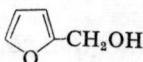

176° n-Heptyl alcohol $CH_3(CH_2)_5CH_2OH$
      d_4^{20} 0.822. Insol. in H_2O. 1-Naphthyl carbamate m.p. 60°. 3-
      Nitro hydrogen phthalate m.p. 127°.

177° Tetrahydrofurfuryl alcohol $(C_4H_7O)CH_2OH$
      d_4^{20} 1.054. Miscible with H_2O. 3,5-Dinitrobenzoate m.p. 84°.
      1-Naphthyl carbamate m.p. 90°.

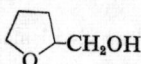

179° s-Octyl alcohol $CH_3(CH_2)_5CHOHCH_3$
      Opt. act. Racemate props. d_4^{20} 0.820. Almost insol. in H_2O.
      Boiling dil. $K_2Cr_2O_7/H_2SO_4$ soln. → methyl n-hexyl ketone.
      1-Naphthyl carbamate m.p. 63°. Hydrogen phthalate m.p. 55°.

188° Propylene glycol $CH_3CHOHCH_2OH$, Propan-1,2-diol.
      Opt. act. Racemate props. d_4^{20} 1.040. Miscible with H_2O. Heating

B.P.°C

with ZnCl$_2$ → propionaldehyde. 3,5-Dinitrobenzoate (di) m.p.
147°. *p*-Nitrobenzoate (di) m.p. 127°. Phenylcarbamate (di) m.p.
153°.

197° Ethylene glycol   CH$_2$OHCH$_2$OH
d_4^{20} 1.114. Miscible with H$_2$O. Conc. HNO$_3$ oxdtn. → oxalic acid.
Heating with KHSO$_4$ → acetaldehyde. 3,5-Dinitrobenzoate (di)
m.p. 169°. 1-Naphthyl carbamate (di) m.p. 176°. *p*-Nitrobenzoate
(di) m.p. 140°. Monoethyl ether b.p. 135°, see 2-Ethoxyethanol.

199° Linalool   C$_{10}$H$_{18}$O

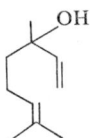

Opt. act. Unsaturated. Properties of (−)-form: [α]$_D^{20}$ −3° to
−17°. d_4^{20} 0.862. Dil. K$_2$Cr$_2$O$_7$/H$_2$SO$_4$ soln. → citral. *p*-Nitro-
benzoate m.p. 70°. Phenylcarbamate m.p. 65°.

206° Benzyl alcohol   C$_6$H$_5$CH$_2$OH
d_4^{20} 1.045. Sol. in 25 vols. cold H$_2$O. Oxdtn. by CrO$_3$ or acid
KMnO$_4$ → benzaldehyde. Alk. KMnO$_4$ oxdtn. → benzoic acid.
3,5-Dinitrobenzoate m.p. 112°. 1-Naphthyl carbamate m.p. 134°.
3-Nitro hydrogen phthalate m.p. 183°.

216° Trimethylene glycol   CH$_2$OHCH$_2$CH$_2$OH, Propan 1,3-diol.
d_4^{20} 1.054. Miscible with H$_2$O. 3,5-Dinitrobenzoate (di) m.p. 178°.
1-Naphthyl carbamate (di) m.p. 164°. *p*-Nitrobenzoate (di) m.p.
119°.

220° 2-Phenylethyl alcohol   C$_6$H$_5$CH$_2$CH$_2$OH
d_4^{20} 1.023. Almost insol. in H$_2$O. Rose oil odour. Dil. K$_2$Cr$_2$O$_7$/
H$_2$SO$_4$ soln. → phenylacetic acid m.p. 76°. 3,5-Dinitrobenzoate
m.p. 108°. 1-Naphthyl carbamate m.p. 119°. 3-Nitro hydrogen
phthalate m.p. 123°.

222° (+)-Citronellol   C$_{10}$H$_{20}$O
[α]$_4^{20}$ +4°. d_4^{17} 0.857. Pleasant odour. Oxdtn. → β-methyl adipic
acid m.p. 89°.

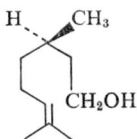

229° Geraniol   C$_{10}$H$_{18}$O
d_4^{20} 0.889. Unsaturated. Pleasant odour. 3,5-Dinitrobenzoate
m.p. 63°. 3-Nitro hydrogen phthalate m.p. 117°.

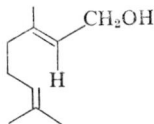

235° Hydrocinnamyl alcohol   C$_6$H$_5$CH$_2$CH$_2$CH$_2$OH
d_4^{20} 1.008. Sl. sol. in H$_2$O. Boiling dil. K$_2$Cr$_2$O$_7$/H$_2$SO$_4$ soln. →
hydrocinnamic acid m.p. 48°. Alk. KMnO$_4$ oxdtn. → benzoic
acid. 3,5-Dinitrobenzoate m.p. 92°. 3-Nitro hydrogen phthalate
m.p. 117°.

B.P.°C

237°   2-Phenoxyethyl alcohol $C_6H_5OCH_2CH_2OH$
d_4^{22} 1.102. Sl. sol. in H_2O. 3-Nitro hydrogen phthalate m.p. 112°.
p-Tosyl m.p. 80°.

245°   Diethylene glycol $(HOCH_2CH_2)_2O$
d_4^{20} 1.118. Miscible with H_2O. 3,5-Dinitrobenzoate (di) m.p. 149°.
p-Nitrobenzoate (di) m.p. 149°. Monoethyl ether b.p. 202°
(Carbitol).

285°   Triethylene glycol $(HOCH_2CH_2OCH_2)_2$
d_4^{20} 1.125. Miscible with H_2O. Dimethyl ether b.p. 216° (Triglyme)

290°d   Glycerol $CH_2OHCHOHCH_2OH$, Propan-1,2,3-triol.
M.p. 18°. d_4^{20} 1.260. Miscible with H_2O. Heating with $KHSO_4$ (or
conc. H_2SO_4) → acrolein (very pungent). 1-Naphthyl carbamate
(tri) m.p. 192°. p-Nitrobenzoate (tri) m.p. 188°. Benzoate (tri)
m.p. 71°.

SOLID

M.P.°C

20°   1-Phenylethanol $CH_3CH(C_6H_5)OH$, Methyl phenyl carbinol.
B.p. 203°. Oxdtn. → acetophenone. Dehydration (acid cat.) →
styrene. 3,5-Dinitrobenzoate m.p. 95°. 1-Naphthyl carbamate
m.p. 106°.

24°   Lauryl alcohol $CH_3(CH_2)_{10}CH_2OH$, n-Dodecanol.
B.p. 259°. Insol. in H_2O. 3,5-Dinitrobenzoate m.p. 60°. 1-
Naphthyl carbamate m.p. 80°. 3-Nitro hydrogen phthalate m.p.
124°.

25°   Cyclohexanol. See Alcohols (Liquid).

25°   t-Butyl alcohol. See Alcohols (Liquid).

25°   Anisyl alcohol $CH_3OC_6H_4CH_2OH$

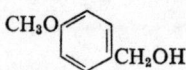

B.p. 259°. Insol. in H_2O. $KMnO_4$ oxdtn. → anisic acid m.p. 184°.
1-Naphthyl carbamate m.p. 126°.

33°   Cinnamyl alcohol $C_6H_5CH:CHCH_2OH$
B.p. 257°. Unsaturated. Sl. sol. in H_2O. CrO_3 oxdtn. → cinnamic
acid m.p. 133°. 3,5-Dinitrobenzoate m.p. 121°. 1-Naphthyl
carbamate m.p. 114°. p-Nitrobenzoate m.p. 78°. Dibromo m.p.
74°.

35°   Terpineol $C_{10}H_{17}OH$
Opt. act. (+) or (−). Unsaturated. B.p. 220°. Insol. in H_2O.
3,5-Dinitrobenzoate m.p. 79°. 1-Naphthyl carbamate m.p. 147°.
p-Nitrobenzoate m.p. 139°.

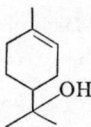

40°   Pinacol $(CH_3)_2C(OH)C(OH)(CH_3)_2$, Tetramethyl glycol.
M.p. of anhydrous. Hydrate m.p. 46°. B.p. 172°. Sl. sol. in cold
H_2O, easily in hot. On cooling, hot aq. soln. deposits hydrated
crystals. Gives iodoform test. CrO_3 oxdtn. → acetone. Boiling
with strong H_2SO_4 → pinacolone b.p. 106°. Phenyl carbamate
(di) m.p. 215°. Acetate (di) m.p. 65°.

M.P.°C

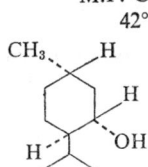

42° (−)-Menthol $C_{10}H_{19}OH$
B.p. 216°. $[\alpha]_D^{20}$ −59.6° (alcohol). Peppermint odour. Properties of natural (−) form. Boiling dil. $K_2Cr_2O_7/H_2SO_4$soln. → (−) menthone b.p. 207°. 3,5-Dinitrobenzoate m.p. 153°. 1-Naphthyl carbamate m.p. 120°. p-Nitrobenzoate m.p. 62°.

50° Cetyl alcohol $CH_3(CH_2)_{14}CH_2OH$
$CrO_3/AcOH$ oxdtn. (or KOH fusion) → palmitic acid m.p. 62°. 3,5-Dinitrobenzoate m.p. 66°. 1-Naphthyl carbamate m.p. 82°. 3-Nitro hydrogen phthalate m.p. 122°.

55° 2-Butyn 1,4-diol $CH_2OHC:CCH_2OH$
CrO_3/H_2SO_4 → acetylene dicarboxylic acid.

69° Benzhydrol $(C_6H_5)_2CHOH$, Diphenyl carbinol.
Readily oxidised to benzophenone. 3,5-Dinitrobenzoate m.p. 141°. 1-Naphthyl carbamate m.p. 136°. p-Nitrobenzoate m.p. 131°.

87° Saligenin. See Phenols (Solid).

111° D-Sorbitol $CH_2OH(CHOH)_4CH_2OH$
M.p. of anhydrous form. $[\alpha]_D^{20}$ −1.9°. Reduction product from D-glucose. Hexa-acetate m.p. 100°. Tribenzylidene m.p. 103°.

137° Benzoin. See Ketones (Solid).

148° Cholesterol $C_{27}H_{46}O$
Opt. act. Acetate m.p. 115°. 1-Naphthyl carbamate m.p. 160°. Benzoate m.p. 145°. p-Nitrobenzoate m.p. 190°. Hydrogen phthalate m.p. 161°.

162° Triphenyl carbinol $(C_6H_5)_3COH$
Intense yellow colour in conc. H_2SO_4. With conc. HCl (or CH_3COCl) → chloride m.p. 108°. Acetate m.p. 83°. Benzoate m.p. 167°.

165° Ergosterol $C_{28}H_{44}O$
$[\alpha]_D^{20}$ −130°. Acetate m.p. 180°. Benzoate m.p. 168°. 3,5-Dinitrobenzoate m.p. 202°. 1-Naphthyl carbamate m.p. 202°. λ_{max}. 280 nm.

166° D-Mannitol $CH_2OH(CHOH)_4CH_2OH$
Opt. act. but aq. soln. shows only sl. rotation. Reduction product from D-mannose. Sol. in 7 parts cold H_2O. Almost insol. in alcohol. Gives ppt. with $CuSO_4/NH_4OH$ soln. Hexa-acetate m.p. 125°. Hexabenzoate m.p. 149°.

208° (+)-Borneol $C_{10}H_{17}OH$
$[\alpha]_D^{20}$ +37° (alcohol). B.p. 212°. Odour like camphor. Boiling HNO_3 → camphor. 3,5-Dinitrobenzoate m.p. 154°. 1-Naphthyl carbamate m.p. 131°. p-Nitrobenzoate m.p. 137°.

225° *meso*-Inositol $(CHOH)_6$, Hexahydroxycyclohexane.
Opt. inact. Fairly sol. in H_2O. Sweet taste. $KMnO_4$ oxdtn. → CO_2 only. CrO_3 oxdtn. (cold) → CO_2,HCOOH. Boiling HNO_3 → oxalic acid. $(CH_3CO)_2O$/Na acetate → hexa-acetate m.p. 216°. Hexabenzoate m.p. 258°,

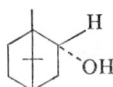

M.P.°C

258°   Pentaerythritol   C(CH$_2$OH)$_4$
Sl. sol. in cold H$_2$O. Tetra-acetate m.p. 83°. Tetrabenzoate m.p. 99°.

PHENOLS

LIQUID

B.P.°C

196°   Salicylaldehyde. See Aldehydes and Acetals.
202°   m-Cresol   CH$_3$C$_6$H$_4$OH

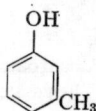

d_4^{20} 1.034. Blue-violet colour with neutral FeCl$_3$ soln. Blue colour in phthalein test (see Phenol). Br$_2$ water → tribromo m.p. 84°. 3,5-Dinitrobenzoate m.p. 165°. Aryloxyacetic acid m.p. 103°. 1-Naphthyl carbamate m.p. 128°.

211°   1,2,4-Xylenol. See Phenols (Solid).
223°   Methyl salicylate. See Esters of Hydroxy Acids.
234°   Ethyl salicylate. See Esters of Hydroxy Acids.
237°   Carvacrol   (CH$_3$)$_2$CHC$_6$H$_3$(CH$_3$)OH

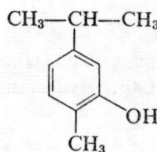

d_4^{20} 0.976. Transient green colour with neutral alc. FeCl$_3$ soln. 3,5-Dinitrobenzoate m.p. 78°. Aryloxyacetic acid m.p. 151° 1-Naphthyl carbamate m.p. 116°.

237°   Isopropyl salicylate. See Esters of Hydroxy Acids.
240°   n-Propyl salicylate. See Esters of Hydroxy Acids.
243°   m-Methoxy phenol   CH$_3$OC$_6$H$_4$OH, Resorcinol monomethyl ether.
d_4^4 1.07. Faint violet colour with neutral alc. FeCl$_3$ soln. Br$_2$/AcOH (boiling) → tribromo m.p. 104°. Aryloxyacetic acid m.p. 117°. 1-Naphthyl carbamate m.p. 129°.

254°   Eugenol   C$_3$H$_5$C$_6$H$_3$(OCH$_3$)OH, 4-Allyl-2-methoxy phenol.

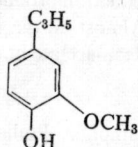

d_4^{20} 1.066. Unsaturated. Odour of cloves. Blue colour with neutral alc. FeCl$_3$ soln. Fusion with KOH → protocatechuic acid m.p. 200°. Excess Br$_2$ (ether) → tetrabromo m.p. 118°. Aryloxyacetic acid m.p. 100°. 3,5-Dinitrobenzoate m.p. 131°. 1-Naphthyl carbamate m.p. 122°. p-Tosyl m.p. 85°.

267°   Isoeugenol   C$_3$H$_5$C$_6$H$_3$(OCH$_3$)OH, 4-Propenyl-2-methoxy phenol.
d_4^{20} 1.085. Unsaturated. Cis/trans forms. Green colour with neutral alc. FeCl$_3$ soln. trans-Acetate m.p. 79°. Benzoate m.p. 103°. p-Nitrobenzoate m.p. 109°. Aryloxyacetic acid m.p. 116°. 3,5-Dinitrobenzoate m.p. 158°.

268°   n-Butyl salicylate. See Esters of Hydroxy Acids.

B.P.°C

283° *m*-Acetoxy phenol $CH_3COOC_6H_4OH$, Resorcinol monoacetate.
Extremely viscous. Saponification → resorcinol. Acetic anhydride → diacetate b.p. 278°.

320° Benzyl salicylate. See Esters of Hydroxy acids.

SOLID

M.P.°C

26° 1,2,4-Xylenol $(CH_3)_2C_6H_3OH$, 2,4-Dimethyl phenol.

OH
CH₃ (structure)
CH₃

B.p. 211°. Transient blue-violet colour with neutral aq. $FeCl_3$ soln. (Green-blue in alcohol). 3,5-Dinitrobenzoate m.p. 165°. Aryloxyacetic acid m.p. 141°. 1-Naphthyl carbamate m.p. 135°.

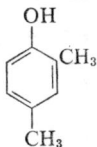

28° *o*-Hydroxy acetophenone. See Ketones (Solid).

28° Guaiacol $CH_3OC_6H_4OH$

OH
OCH₃ (structure)

B.p. 205°. Green-blue colour with neutral alc. $FeCl_3$ soln. (red in aq. soln.). Heated with Zn dust → anisole. 3,5-Dinitrobenzoate m.p. 141°. Aryloxyacetic acid m.p. 116°. 1-Naphthyl carbamate m.p. 118°. *p*-Tosyl m.p. 85°. Tribromo m.p. 116°.

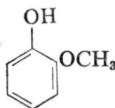

31° *o*-Cresol $CH_3C_6H_4OH$
B.p. 191°. Violet colour with neutral aq. $FeCl_3$ soln. Red colour in phthalein test (see Phenol). Br_2/H_2O → dibromo m.p. 56°. 3,5-Dinitrobenzoate m.p. 138°. Arloxyacetic acid m.p. 152°. 1-Naphthyl carbamate m.p. 142°.

36° *p*-Cresol $CH_3C_6H_4OH$
B.p. 202°. Blue colour with neutral aq. $FeCl_2$ soln. 3,5-Dinitrobenzoate m.p. 188°. Aryloxyacetic acid m.p. 136°. 1-Naphthyl carbamate m.p. 146°. *p*-Tosyl m.p. 70°. Benzoate m.p. 71°.

42° Phenol C_6H_5OH
B.p. 182°. Violet colour with neutral aq. $FeCl_3$ soln. Br_2/H_2O → tribromo m.p. 95°. Warm with phthalic anhydride and drop of conc. H_2SO_4 → phenolphthalein, turns red in alkali. 3,5-Dinitrobenzoate m.p. 146°. Aryloxyacetic acid m.p. 99°. 1-Naphthyl carbamate m.p. 133°. *p*-Tosyl m.p. 96°. Benzoate m.p. 68°.

42° Phenyl salicylate. See Esters of Hydroxy Acids.

49° 1,2,6-Xylenol $(CH_3)_2C_6H_3OH$, 2,6-Dimethyl phenol.
Aryloxyacetic acid m.p. 139°. 3,5-Dinitrobenzoate m.p. 159°. 1-Naphthyl carbamate m.p. 176°.

50° Thymol $(CH_3)_2CHC_6H_3(CH_3)OH$

CH₃–CH–CH₃
OH (structure)
CH₃

B.p. 233°. Transient green colour with neutral alc. $FeCl_3$ soln. Gives red colour on gentle warming with one drop KOH soln. and I_2/KI soln. 3,5-Dinitrobenzoate m.p. 103°. Aryloxyacetic acid m.p. 148°. 1-Naphthyl carbamate m.p. 160°. *p*-Tosyl m.p. 71°.

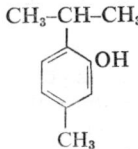

53° *p*-Methoxy phenol $CH_3OC_6H_4OH$, Hydroquinone monomethyl ether.

M.P.°C

B.p. 243°. Not steam volatile. Reduces Tollen's reagent on warming. Aryloxyacetic acid m.p. 111°. Benzoate m.p. 87°.

58°   Orcinol (hydrated) $CH_3C_6H_3(OH)_2$
Loses H_2O at 100° to give anhydrous form m.p. 107°. Blue-violet colour with neutral aq. $FeCl_3$ soln. Reduces Tollen's reagent. 3,5-Dinitrobenzoate m.p. 190°. Aryloxyacetic acid m.p. 217°. 1-Naphthyl carbamate m.p. 160°. Benzoate m.p. 88°. Tribromo m.p. 104°.

58°   o-Phenyl phenol $C_6H_5C_6H_4OH$, 2-Hydroxy biphenyl.
(67° corr.)   B.p. 275°. Benzoate m.p. 76°. p-Tosyl m.p. 65°. Acetate m.p. 62°.

62°   1,3,4-Xylenol $(CH_3)_2C_6H_3OH$, 3,4-Dimethyl phenol.
B.p. 225°. Aryloxyacetic acid m.p. 163°. 3,5-Dinitrobenzoate m.p. 182°. 1-Naphthyl carbamate m.p. 142°.

68°   1,3,5-Xylenol $(CH_3)_2C_6H_3OH$, 3,5-Dimethyl phenol.
3,5-Dinitrobenzoate m.p. 195°. Aryloxyacetic acid m.p. 111°. p-Tosyl m.p. 83°. Tribromo m.p. 166°.

70°   Methyl m-hydroxy benzoate. See Esters of Hydroxy Acids.
71°   Pseudocumenol $(CH_3)_3C_6H_2OH$, 2,4,5-Trimethyl phenol.
B.p. 232°. No colour with $FeCl_3$. Benzoate m.p. 63°. Aryloxyacetic acid m.p. 132°.

73°   Ethyl m-hydroxy benzoate. See Esters of Hydroxy Acids.
75°   1,2,5-Xylenol $(CH_3)_2C_6H_3OH$, 2,5-Dimethyl phenol.
B.p. 212°. No colour with $FeCl_3$. Dibromo m.p. 79°. Tribromo m.p. 179°. 3,5-Dinitrobenzoate m.p. 137°. 1-Naphthyl carbamate m.p. 172°. p-Nitrobenzoate m.p. 87°. Aryloxyacetic acid m.p. 118°.

75°   1,2,3-Xylenol $(CH_3)_2C_6H_3OH$, 2,3-Dimethyl phenol.
Aryloxyacetic acid m.p. 187°. p-Nitrobenzoate m.p. 126°.

81°   Vanillin. See Aldehydes and Acetals.
87°   Saligenin $HOC_6H_4CH_2OH$, o-Hydroxy benzyl alcohol.
Aryloxyacetic acid m.p. 120°. Benzoyl (di) m.p. 85°. 1-Naphthyl carbamate m.p. 283°.

94°   1-Naphthol $C_{10}H_7OH$
B.p. 278°. White ppt. with neutral aq. $FeCl_3$ soln. (no colour). Green colour in phthalein test (see Phenol). Warm with dil. NaOH, Cu powder and $CCl_4 \rightarrow$ blue colour (distinguish from 2-naphthol). Violet colour with I_2/KI soln. and excess NaOH (2-naphthol–colourless). Violet ppt. with NaOBr soln. 3,5-Dinitrobenzoate m.p. 217°. Aryloxyacetic acid m.p. 192°.

M.P.°C

1-Naphthyl carbamate m.p. 152°. *p*-Tosyl m.p. 88°. Br₂/AcOH →
dibromo m.p. 105°.

95°   2-Naphthyl salicylate. See Esters of Hydroxy Acids.

104°   Catechol $C_6H_4(OH)_2$

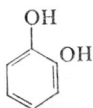

B.p. 245°. Sol. in H_2O. Green colour with neutral aq. FeCl₃
soln.; addtn. of Na₂CO₃ → red, characteristic of phenols with 2
ortho OH gps. Easily oxidised. Reduces Fehling's soln. White
ppt. with Pb acetate soln. Br₂/CCl₄ → tetrabromo m.p. 192°.
Dibenzoate m.p. 84°. 3,5-Dinitrobenzoate m.p. 152°. *p*-Nitro-
benzoate m.p. 169°.

109°   *p*-Hydroxy acetophenone. See Ketones (Solid).

110°   Resorcinol $C_6H_4(OH)_2$

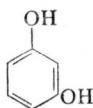

B.p. 276°. Sol. in H_2O. Blue-violet colour with neutral aq. FeCl₃
soln. In phthalein test (see Phenol) → fluorescein (orange soln.
in NaOH with strong yellow-green fluorescence). No ppt. with
Pb acetate soln. 3,5-Dinitrobenzoate m.p. 201°. Dibenzoate m.p.
117°. *p*-Tosyl m.p. 80°. Dibromo m.p. 112°. Aryloxyacetic acid
m.p. 195°.

116°   Ethyl *p*-hydroxy benzoate. See Esters of Hydroxy Acids.

117°   *p*-Hydroxy benzaldehyde. See Aldehydes and Acetals.

122°   2-Naphthol $C_{10}H_7OH$

B.p. 286°. No colour with neutral aq. FeCl₃ soln. (white opales-
cence). In phthalein test (see Phenol) → faint green colour (some
fluorescence). Warm conc. KOH soln. → blue colour with
CHCl₃ (CCl₄ no colour). Yellow colour with NaOBr soln.
3,5-Dinitrobenzoate m.p. 210°. 1-Naphthyl carbamate. m.p.
157°. Aryloxyacetic acid m.p. 154°. *p*-Tosyl m.p. 125°. Benzoate
m.p. 107°. Bromo m.p. 84°. Acetate m.p. 70°.

124°   Toluhydroquinone $CH_3C_6H_3(OH)_2$

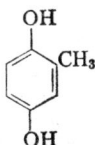

Properties like hydroquinone. Mild oxdtn. → toluquinone m.p.
68°. Dibenzoate m.p. 119°. *p*-Tosyl m.p. 125°.

131°   Methyl *p*-hydroxy benzoate. See Esters of Hydroxy Acids.

133°   Pyrogallol $C_6H_3(OH)_3$

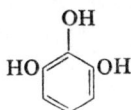

Sol. in H_2O. Reddish colour with neutral aq. FeCl₃ soln. Very
easily oxidised. Alk. soln. darkens rapidly in air. Blue ppt. with
FeSO₄ soln. 3,5-Dinitrobenzoate (tri) m.p. 205°. Aryloxyacetic
acid (tri) m.p. 198°. Tribenzoate m.p. 90°. *p*-Nitrobenzoate (tri)
m.p. 230°. Dibromo m.p. 158°. Triacetate m.p. 165°.

140°   Hydroxyhydroquinone $C_6H_3(OH)_3$

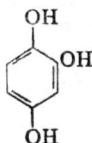

Tribenzoate m.p. 120°. Triacetate m.p. 96°.

M.P.°C

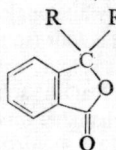

147° Resacetophenone. See Ketones (Solid).
158° Salicyclic acid. See Aromatic Hydroxy Acids (Solid).
165° p-Phenyl phenol $C_6H_5C_6H_4OH$, p-Hydroxy biphenyl.
       B.p. 306°. Benzoate m.p. 149°. p-Tosyl m.p. 179°. Acetate m.p.
       87°.
169° Hydroquinone $C_6H_4(OH)_2$, Quinol.
       Sol. in H_2O. Insol. in C_6H_6. Transient blue colour with neutral
       aq. $FeCl_3$ soln. Mild oxdtn. → benzoquinone (yellow) and/or
       quinhydrone (dark green). Gives ppt. with Pb acetate soln. (in
       presence of NH_3). Aryloxyacetic acid m.p. 250°. p-Tosyl m.p.
       159°. Acetate m.p. 123°. Benzoate m.p. 205°. Bromo (di) m.p.
       186°.
200° m-Hydroxy benzoic acid. See Aromatic Hydroxy Acids (Solid).
213° p-Hydroxy benzoic acid. See Aromatic Hydroxy Acids (Solid).
218° Phloroglucinol $C_6H_3(OH)_3$
       Dihydrate. Loses H_2O at 100°. Sol. in H_2O. Transient violet
       colour with neutral aq. $FeCl_3$ soln. Br_2 → tribromo m.p. 151°.
       Keto-enol props. e.g. yields tri-oxime with NH_2OH. 3,5-Dinitro-
       benzoate m.p. 162°. Benzoate m.p. 185°. Acetate m.p. 104°.
254° Phenolphthalein $C_{20}H_{14}O_4$
(261°) R = —$C_6H_4OH(p)$. Insol. in H_2O. Sol. in alc. Acid/base indi-
       cator; red soln. in dil. alkali; colour discharged by large excess
       conc. alkali. Dibenzoate m.p. 169°. Diacetate m.p. 143°.

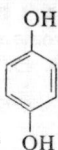

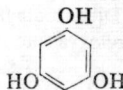

289° Alizarin. See Quinones.

ALDEHYDES AND ACETALS

LIQUID

B.P.°C

Gas Formaldehyde CH_2O
     B.p. −21°. Formalin is a 40% (approx.) aq. soln. b.p. 97–98°.
     Pungent odour. Strong reducing agent. V. dil. aq. soln. →
     red-brown ppt. with Nessler's soln. With NH_4OH gives hexa-
     methylene tetramine $(CH_2)_6N_4$ (q.v.). No colour with Na nitro-
     prusside. 2,4-Dinitrophenylhydrazone* m.p. 168°. p-Nitro-
     phenylhydrazone m.p. 182°. Dimethone m.p. 189°.
21° Acetaldehyde CH_3CHO
     Miscible with H_2O. Strong apple-like odour. V. dil. aq. soln
     → yellow ppt. with Nessler's soln. Warm dil. NaOH → resinifies.
     Red colour with Na nitroprusside (blue in presence of piperidine

* The preparation of aryl hydrazones is described in great detail
by E. Enders, 'Methoden der Organischen Chemie', Houben-Weyl,
4th Edn. (1967).

B.P.°C

or diethylamine). 2,4-Dinitrophenylhydrazone m.p. 168°. *p*-Nitrophenylhydrazone m.p. 128°. Dimethone m.p. 140°. Semicarbazone m.p. 163°.

45° Methylal $CH_2(OCH_3)_2$
Miscible with H_2O. Boiling dil. HCl → CH_2O and CH_3OH.

49° Propionaldehyde C_2H_5CHO
Sol. in 5 vols. cold H_2O. 2,4-Dinitrophenylhydrazone m.p. 155°. *p*-Nitrophenylhydrazone m.p. 124°. Dimethone m.p. 155°.

50° Glyoxal $(CHO)_2$
Usually as aq. soln. Phenylhydrazone m.p. 180°. Oxime m.p. 178°. Semicarbazone m.p. 270°.

52° Acrolein $CH_2:CHCHO$
Sol. in 2–3 vols. cold H_2O. Very pungent odour. Unsaturated. 2,4-Dinitrophenylhydrazone m.p. 165°. *p*-Nitrophenylhydrazone m.p. 151°. Dimethone m.p. 192°. Semicarbazone m.p. 171°.

63° Isobutyraldehyde $(CH_3)_2CHCHO$
Sol. in 9 vols. cold H_2O. Polymerises with a trace of H_2SO_4 in cold to para form m.p. 59°. 2,4-Dinitrophenylhydrazone m.p. 187° (182°). *p*-Nitrophenylhydrazone m.p. 130°. Dimethone m.p. 154°. Semicarbazone m.p. 125°.

64° Dimethyl acetal $CH_3CH(OCH_3)_2$
Sl. sol. in H_2O. Boiling dil. HCl → CH_3CHO and CH_3OH.

74° n-Butyraldehyde $C_2H_5CH_2CHO$
Sl. sol. in cold H_2O. 2,4-Dinitrophenylhydrazone m.p. 123°. *p*-Nitrophenylhydrazone m.p. 91°. Dimethone m.p. 135°. Semicarbazone m.p. 104° (77°).

75° Pivaldehyde $(CH_3)_3CCHO$, Trimethyl acetaldehyde.
2,4-Dinitrophenylhydrazone m.p. 210°. *p*-Nitrophenylhydrazone m.p. 119°. Semicarbazone m.p. 190°.

89° Ethylal $CH_2(OC_2H_5)_2$
Sl. sol. in cold H_2O. Boiling dil. HCl → CH_2O and C_2H_5OH.

92° Isovaleraldehyde $(CH_3)_2CHCH_2CHO$
2,4-Dinitrophenylhydrazone m.p. 123°. *p*-Nitrophenylhydrazone m.p. 110°. Dimethone m.p. 155°. Semicarbazone m.p. 107° (132°)

103° Acetal $CH_3CH(OC_2H_5)_2$
Sl. sol. in H_2O. Boiling dil. HCl → CH_3CHO and C_2H_5OH.

103° n-Valeraldehyde $CH_3(CH_2)_3CHO$
2,4-Dinitrophenylhydrazone m.p. 107°. *p*-Nitrophenylhydrazone m.p. 74°. Dimethone m.p. 108°.

103° Crotonaldehyde $CH_3CH:CHCHO$
Fairly sol. in H_2O. Very pungent odour. Unsaturated. Forms soluble adduct with $NaHSO_3$, not regenerated by Na_2CO_3. 2,4-Dinitrophenylhydrazone m.p. 190°. *p*-Nitrophenylhydrazone m.p. 185°. Dimethone m.p. 193° (186°). Semicarbazone m.p. 201°. Oxime m.p. 119°.

124° Paraldehyde $(CH_3CHO)_3$
Sl. sol. in H_2O. Cyclic trimer of acetaldehyde which is regenerated by gentle warming with a trace of H_2SO_4.

B.P.°C

131° Caproaldehyde $CH_3(CH_2)_4CHO$, n-Hexaldehyde.
2,4-Dinitrophenylhydrazone m.p. 104°. Dimethone m.p. 108°.
Semicarbazone m.p. 106°.

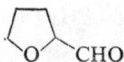

143° Tetrahydrofurfural $(C_4H_7O)CHO$
Warming with conc. HCl → bright red colour. 2,4-Dinitrophenyl-
hydrazone m.p. 134°. Semicarbazone m.p. 166°. Dimethone
m.p. 123°.

155° n-Heptaldehyde $CH_3(CH_2)_5CHO$, Oenanthal.
2,4-Dinitrophenylhydrazone m.p. 108°. p-Nitrophenylhydrazone
m.p. 73°. Dimethone m.p. 103°. Semicarbazone m.p. 109°.

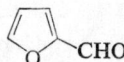

161° Furfural $(C_4H_3O)CHO$
Colourless liq. turns brown in air. Bran-like odour. Paper
moistened with aniline acetate turns red in its vapour. $KMnO_4$
oxdtn. → furoic acid m.p. 133°. Alc. KOH gives Cannizzaro
reaction. 2,4-Dinitrophenylhydrazone m.p. ca. 200°. p-Nitrophenylhydrazone m.p. 152° (127°). Dimethone m.p. 162°d.
Semicarbazone m.p. 203°. Phenylhydrazone m.p. 97°.

169° Succindialdehyde $(CH_2CHO)_2$
2,4-Dinitrophenylhydrazone m.p. 280°. Di-oxime m.p. 172°.

179° Benzaldehyde C_6H_5CHO
Spar. sol. in H_2O. Odour of almonds. Often contaminated with
benzoic acid from aerial oxdtn. Alk. $KMnO_4$ oxdtn. → benzoic
acid. Warming with strong KOH gives Cannizzaro reaction
$(C_6H_5COOH + C_6H_5CH_2OH)$. Alc. KCN → benzoin condensation. 2,4-Dinitrophenylhydrazone m.p. 237°. p-Nitrophenylhydrazone m.p. 192° (262°). Dimethone m.p. 195°. Semicarbazone
m.p. ca. 227°. Dibenzalacetone m.p. 112° (see Acetone). 0.88
NH_3 → hydrobenzamide m.p. 101°.

193° Phenylacetaldehyde. See Aldehydes and Acetals (Solid).

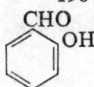

196° Salicylaldehyde HOC_6H_4CHO
Spar. sol. in H_2O. Violet colour with neutral aq. $FeCl_3$ soln.
Alk. $KMnO_4$ oxdtn. → salicyclic acid m.p. 158°. Shaken with
acetic anhydride and a drop of conc. H_2SO_4 → triacetate m.p.
100°. 2,4-Dinitrophenylhydrazone m.p. 252°d. p-Nitrophenylhydrazone m.p. 227°. Dimethone m.p. 211°. Semicarbazone m.p.
232°d. Phenylhydrazone m.p. 142°. p-Nitrobenzoate m.p. 128°.
Aryloxyacetic acid m.p. 133°.

199° m-Tolualdehyde $CH_3C_6H_4CHO$
Alk. $KMnO_4$ oxdtn. → m-toluic acid m.p. 110°. 2,4-Dinitro-
phenylhydrazone m.p. 212° (194°). p-Nitrophenylhydrazone m.p.
157°. Dimethone m.p. 172°. Semicarbazone m.p. 224° (216°).
Phenylhydrazone m.p. 91°.

200° o-Tolualdehyde $CH_3C_6H_4CHO$
Alk. $KMnO_4$ oxdtn. → o-toluic acid m.p. 104°. 2,4-Dinitro-
phenylhydrazone m.p. 194°. p-Nitrophenylhydrazone m.p. 222°.
Dimethone m.p. 167°. Semicarbazone m.p. ca. 212°. Phenylhydrazone m.p. 105°.

205° p-Tolualdehyde $CH_3C_6H_4CHO$

B.P.°C

Alk. KMnO$_4$ oxdtn. → p-toluic acid m.p. 178°. 2,4-Dinitro-phenylhydrazone m.p. 234°. p-Nitrophenylhydrazone m.p. 200°. Semicarbazone m.p. 234°. Phenylhydrazone m.p. 112°. Oxime m.p. 80°.

207°   Citronellal C$_{10}$H$_{18}$O, Rhodinal.

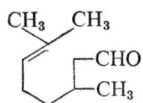 [α]$_D^{20}$ ±11°. Unsaturated. Takes up 1 mole bromine, product yields cymene on heating. 2,4-Dinitrophenylhydrazone m.p. 78°. Dimethone m.p. 78°. Semicarbazone m.p. 84°.

222°   Benzaldehyde diethyl acetal C$_6$H$_5$CH(OC$_2$H$_5$)$_2$
Hydrolysis → C$_6$H$_5$CHO + C$_2$H$_5$OH

224°   Hydrocinnamaldehyde C$_6$H$_5$CH$_2$CH$_2$CHO
Mild oxdtn. → hydrocinnamic acid m.p. 48°. 2,4-Dinitrophenyl-hydrazone m.p. 149°. p-Nitrophenylhydrazone m.p. 123°. Semicarbazone m.p. 127°. Oxime m.p. 94°.

228°d   Citral a C$_{10}$H$_{16}$O, Geranial.

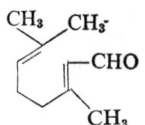 Opt. inact. Odour of lemon. Br$_2$ → tetrabromo. λ_{max} 235 nm, $\varepsilon = 40,000$. 2,4-Dinitrophenylhydrazone m.p. 110°. Semi-carbazone m.p. 164°.

247°   Anisaldehyde CH$_3$OC$_6$H$_4$CHO

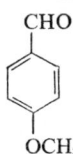 Insol. in H$_2$O. Mild oxdtn. → anisic acid m.p. 184°, 2,4-Dinitro-phenylhydrazone m.p. 253°d. p-Nitrophenylhydrazone m.p. 161°. Dimethone m.p. 145°. Semicarbazone m.p. 210°. Phenyl-hydrazone m.p. 120°. Oxime m.p. 92°.

252°d   Cinnamaldehyde C$_6$H$_5$CH:CHCHO
Insol. in H$_2$O. Unsaturated. Odour of cinnamon. Alk. KMnO$_4$ oxdtn. → benzoic acid. Warming with aniline → cinnamalaniline m.p. 109°. Alc. NH$_3$ → hydrocinnamide m.p. 106°. 2,4-Dinitro-phenylhydrazone m.p. 255°d. p-Nitrophenylhydrazone m.p. 195°. Dimethone m.p. 219°(213°). Phenylhydrazone m.p. 168°. Semicarbazone m.p. 216°(208°).

SOLID

M.P.°C

34°   1-Naphthaldehyde C$_{10}$H$_7$CHO

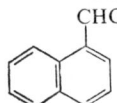 Mild oxdtn. → 1-naphthoic acid m.p. 162°. p-Nitrophenyl-hydrazone m.p. 224°(234°). Semicarbazone m.p. 221°. Oxime m.p. 98° (90°).

34°   Phenylacetaldehyde C$_6$H$_5$CH$_2$CHO
B.p. 193°. Mild oxdtn. → phenylacetic acid m.p. 76°. 2,4-Dinitrophenylhydrazone m.p. 121°. Dimethone m.p. 165°. Semi-carbazone m.p. 156°. Oxime m.p. 100°.

37°   Piperonal (CH$_2$O$_2$)C$_6$H$_3$CHO,3,4-Methylenedioxybenzaldehyde.

M.P.°C

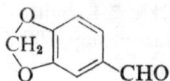

B.p. 263°. Spar. sol. in H_2O. Odour of heliotrope. Oxdtn. → piperonylic acid. m.p. 229°. 2,4-Dinitrophenylhydrazone m.p. 265°d. p-Nitrophenylhydrazone m.p. 200°. Dimethone m.p. 178° (193°). Semicarbazone m.p. 230°. Phenylhydrazone m.p. 102°. Oxime m.p. 110°.

60°   2-Naphthaldehyde $C_{10}H_7CHO$

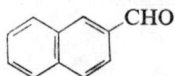

Mild oxdtn. → 2-naphthoic acid m.p. 185°. 2,4-Dinitrophenylhydrazone m.p. 270°. p-Nitrophenylhydrazone m.p. 230°. Semicarbazone m.p. 245°. Oxime m.p. 156°.

81°   Vanillin $CH_3O(OH)C_6H_3CHO$

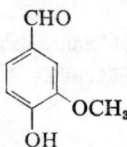

Sl. sol. in H_2O. Odour of vanilla. Red to violet colour with neutral aq. $FeCl_3$ soln. 2,4-Dinitrophenylhydrazone m.p. 271°. p-Nitrophenylhydrazone m.p. 229°. Dimethone m.p. 197°. Oxime m.p. 117°. Phenylhydrazone m.p. 105°. Acetate m.p. 160°. Aryloxyacetic acid m.p. 189°.

96°   Glycolaldehyde $HOCH_2CHO$, Hydroxyacetaldehyde.
Readily sol. in H_2O. Br_2/H_2O → glycollic acid. Dil. alk. → mixture of hexoses. Phenylhydrazone m.p. 162°.

104°   m-Hydroxy benzaldehyde HOC_6H_4CHO
Reddish colour with neutral aq. $FeCl_3$ soln. Mild oxdtn. → m-hydroxy benzoic acid m.p. 200°. 2,4-Dinitrophenylhydrazone m.p. 260°d. p-Nitrophenylhydrazone m.p. 222°. Semicarbazone m.p. 199°. Oxime m.p. 88°.

115°   Metaldehyde $(CH_3CHO)_n$
Sublimes. Insol. in H_2O. Polymer of acetaldehyde which is regenerated by action of dil. acids.

117°   p-Hydroxy benzaldehyde HOC_6H_4CHO
Spar. sol. in cold H_2O. Water soluble $NaHSO_3$ adduct. Faint violet colour with neutral aq. $FeCl_3$ soln. Yields p-hydroxy benzoic acid m.p. 213° only with difficulty on oxdtn. but readily on KOH fusion. Warming with aniline → p-hydroxybenzal aniline m.p. 190°. 2,4-Dinitrophenylhydrazone m.p. 280°d*. p-Nitrophenylhydrazone m.p. 266°. Dimethone m.p. 188°. Semicarbazone m.p. 223°. Phenylhydrazone m.p. 178°.

170°–180°   Polyoxymethylene $HO(CH_2O)_nH$, Paraformaldehyde.
Linear high polymer of formaldehyde which is regenerated by heating or action of dil. alkali.

KETONES

LIQUID

B.P.°C

56°   Acetone CH_3COCH_3
Miscible with H_2O. With iodine + dil. alkali → iodoform m.p.

* From acetic acid, purple colour; from H_2O monohydrate m.p. 260°, red colour.

B.P.°C

119°. With benzaldehyde and alkali in alc. soln. → dibenzalacetone m.p. 112°. Dil. aq. soln. + Na nitroprusside → red
colour. 2,4-Dinitrophenylhydrazone m.p. 126°. *p*-Nitrophenyl-
hydrazone m.p. 148°. Semicarbazone m.p. 188°. Oxime m.p. 59°.

80° Methyl ethyl ketone $CH_3COC_2H_5$, Butan-2-one.
Miscible with H_2O. Gives iodoform reaction. Red colour with
Na nitroprusside. 2,4-Dinitrophenylhydrazone m.p. 115° (111°).
p-Nitrophenylhydrazone m.p. 128°. Semicarbazone m.p. 135°
(148°).

88° Diacetyl $CH_3COCOCH_3$, Butan-2,3-dione.
With *o*-phenylenediamine → quinoxaline deriv. m.p. 106°. 2,4-
Dinitrophenylhydrazone (di) m.p. 315°. *p*-Nitrophenylhydrazone
(mono) m.p. 230°. Semicarbazone (di) m.p. 278°.

94° Methyl isopropyl ketone $CH_3COCH(CH_3)_2$
Gives iodoform reaction. Red colour with Na nitroprusside.
2,4-Dinitrophenylhydrazone m.p. 117°. *p*-Nitrophenylhydrazone
m.p. 109°. Semicarbazone m.p. 112°. Oxime m.p. 109°.

102° Diethyl ketone $C_2H_5COC_2H_5$, Pentan-3-one.
Sl. sol. in H_2O. 2,4-Dinitrophenylhydrazone m.p. 156°. *p*-
Nitrophenylhydrazone m.p. 144°. Semicarbazone m.p. 139°.

102° Methyl n-propyl ketone $CH_3COCH_2C_2H_5$, Pentan 2-one.
Gives iodoform reaction. Red colour with Na nitroprusside.
2,4-Dinitrophenylhydrazone m.p. 142°. *p*-Nitrophenylhydrazone
m.p. 117°. Semicarbazone m.p. 110°.

106° Pinacolone $CH_3COC(CH_3)_3$, Methyl t-butyl ketone.
Sl. sol. in H_2O. Camphor-like odour. CrO_3 oxdtn. → pivalic acid
(*q.v.*). Gives iodoform reaction. Red colour with Na nitroprusside. 2,4-Dinitrophenylhydrazone m.p. 125°. Semicarbazone
m.p. 157°. Oxime m.p. 75°.

117° Methyl isobutyl ketone $CH_3COCH_2CH(CH_3)_2$
Gives iodoform reaction. Red colour with Na nitroprusside.
2,4-Dinitrophenylhydrazone m.p. 95°. *p*-Nitrophenylhydrazone
m.p. 79°. Semicarbazone m.p. 132°.

127° Methyl n-butyl ketone $CH_3CO(CH_2)_3CH_3$, Hexan-2-one.
Gives iodoform reaction. Red colour with Na nitroprusside.
2,4-Dinitrophenylhydrazone m.p. 106°. *p*-Nitrophenylhydrazone
m.p. 88°. Semicarbazone m.p. 122°.

130° Mesityl oxide $CH_3COCH:C(CH_3)_2$, Isopropylidene acetone.
Sl. sol. in H_2O. Unsaturated. Odour like peppermint. Boiling
dil. H_2SO_4 → acetone. 2,4-Dinitrophenylhydrazone m.p. 203°.
p-Nitrophenylhydrazone m.p. 133°. Phenylhydrazone m.p. 142°.
Semicarbazone m.p. 164°(133°).

130° Cyclopentanone C_5H_8O
2,4-Dinitrophenylhydrazone m.p. 146°. *p*-Nitrophenylhydrazone
m.p. 154°. Semicarbazone m.p. 206°.

B.P.°C

139°   Acetyl acetone $CH_3COCH_2COCH_3$, Pentan-2,4-dione.
Sol. in 8 vols. cold H_2O. Keto-enolic properties. Orange-red
colour with neutral aq. $FeCl_3$ soln. Sol. in alkali, on boiling →
acetone and acetic acid. Light blue ppt. of Cu salt with Cu
acetate soln. (Similarly many transition metals). Gradual addtn.
to excess aq. NH_2OH soln. → dioxime m.p. 149°. With 2,4-
dinitrophenylhydrazine → deriv. m.p. 209° or pyrazole deriv.
m.p. 122°. p-Nitrophenylhydrazine → pyrazole deriv. m.p. 100°.
Semicarbazide → pyrazole deriv. m.p. 107°.

148°   Acetoin $CH_3CHOHCOCH_3$, 3-Hydroxy butan-2-one.
M.p. 15°. Opt. act. 2,4-Dinitrophenylhydrazone m.p. 318°.
Semicarbazone m.p. 185° (202°). Acetate b.p. 173°.

151°   Methyl n-amyl ketone $CH_3CO(CH_2)_4CH_3$, Heptan-2-one.
2,4-Dinitrophenylhydrazone m.p. 89°. Phenylhydrazone m.p.
207°. Semicarbazone m.p. 123°.

156°   Cyclohexanone $C_6H_{10}O$
Sl. sol. in cold H_2O. Conc. HNO_3 oxdtn. → adipic acid m.p. 151°.
Phenylhydrazone m.p. 82° gives tetrahydrocarbazole m.p. 118°
when heated with 3 parts acetic acid for 5 minutes. 2,4-Dinitro-
phenylhydrazone m.p. 160°. p-Nitrophenylhydrazone m.p. 146°.
Semicarbazone m.p. 166°. Oxime m.p. 91° gives ε-caprolactam
when heated with strong H_2SO_4.

164°   Diacetone alcohol $(CH_3)_2C(OH)CH_2COCH_3$
Miscible with H_2O, alc. and ether. Distil with dil. NaOH →
acetone. Distil with trace of iodine → mesityl oxide. 2,4-Dinitro-
phenylhydrazone m.p. 202°. Oxime m.p. 56° (104°). 3,5-Dinitro-
benzoate m.p. 55°.

165°   2-Methylcyclohexanone $C_7H_{12}O$
2,4-Dinitrophenylhydrazone m.p. 136°. p-Nitrophenylhydrazone
m.p. 132°. Semicarbazone m.p. 197°.

172°   Methyl n-hexyl ketone $CH_3CO(CH_2)_5CH_3$, Octan-2-one
2,4-Dinitrophenylhydrazone m.p. 58°. p-Nitrophenylhydrazone
m.p. 93°. Semicarbazone m.p. 123°.

181°   Cycloheptanone $C_7H_{12}O$, Suberone.
2,4-Dinitrophenylhydrazone m.p. 148°. p-Nitrophenylhydrazone
m.p. 137°. Semicarbazone m.p. 163°.

194°   Acetonyl acetone $CH_3COCH_2CH_2COCH_3$, Hexan-2,5-dione.
2,4-Dinitrophenylhydrazone (di) m.p. 257°. p-Nitrophenylhydrazone (di) m.p. 210°. Semicarbazone (di) m.p. 223°. Oxime
(di) m.p. 138°. Phenylhydrazone (di) m.p. 120°.

202°   Acetophenone $C_6H_5COCH_3$
M.p. 20°. Spar. sol. in H_2O. Alk. $KMnO_4$ oxdtn. → benzoic acid.
Gives iodoform reaction. 2,4-Dinitrophenylhydrazone m.p.

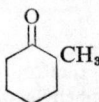

B.P.°C

240°(249°). *p*-Nitrophenylhydrazone m.p. 184°. Semicarbazone m.p. 198°.

207° (−)-Menthone $C_{10}H_{18}O$

$[\alpha]_D^{20}$ − 28°. Insol. in H_2O. Peppermint odour. Na/Hg amalgam redtn. → menthol. 2,4-Dinitrophenylhydrazone m.p. 146°. Semicarbazone m.p. 184°.

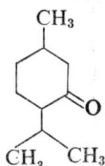

215° Isophorone $C_9H_{14}O$

Characteristic odour. Unsaturated. Semicarbazone m.p. 200° (190°). Oxime m.p. 76°.

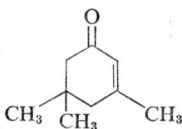

218° Propiophenone $C_6H_5COC_2H_5$

M.p. 19°. 2,4-Dinitrophenylhydrazone m.p. 191°. *p*-Nitrophenylhydrazone m.p. 147°. Semicarbazone m.p. 178°.

222° n-Butyrophenone $C_6H_5CO(CH_2)_3CH_3$

(228°) 2,4-Dinitrophenylhydrazone m.p. 190°. Semicarbazone m.p. 186°

230° (+)-Carvone $C_{10}H_{14}O$

$[\alpha]_D^{20}$ + 63°. Spearmint odour. Unsaturated. Forms tetrabromo. λ_{max} 235 nm, ε = 19,000. Acids → carvacrol b.p. 237°. 2,4-Dinitrophenylhydrazone m.p. 191°. *p*-Nitrophenylhydrazone m.p. 175°. Semicarbazone m.p. 162° (142°).

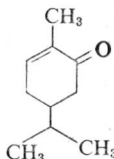

140°/18 mmHg β-Ionone $C_{13}H_{20}O$

Cedarwood odour. *p*-Nitrophenylhydrazone m.p. 173°. Semicarbazone m.p. 148°. λ_{max} 293.5 nm, ε = 8700.

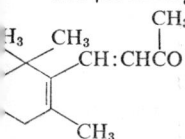

SOLID

M.P.°C

19° Propiophenone. See Ketones (Liquid).

20° Acetophenone. See Ketones (Liquid).

27° Benzyl methyl ketone $C_6H_5CH_2COCH_3$, Phenylacetone. B.p. 216°. 2,4-Dinitrophenylhydrazone m.p. 156°. *p*-Nitrophenylhydrazone m.p. 145°. Semicarbazone m.p. 198° (188°).

M.P.°C

28°   *p*-Methyl acetophenone $CH_3C_6H_4COCH_3$
B.p. 226°. 2,4-Dinitrophenylhydrazone m.p. 258°. *p*-Nitro-
phenylhydrazone m.p. 198°. Semicarbazone m.p. 205°. Oxime
m.p. 88°.

28°   *o*-Hydroxy acetophenone $HOC_6H_4COCH_3$
B.p. 215°. Red to violet colour with neutral aq. $FeCl_3$ soln.
Phenylhydrazone m.p. 110°. Semicarbazone m.p. 210°. Oxime
m.p. 117°. Acetate m.p. 89°.

28°   Phorone $(CH_3)_2CH:CHCOCH:CH(CH_3)_2$,   Diisopropylidene
acetone.
B.p. 198°. Unsaturated. $Br_2/CCl_4 \rightarrow$ tetrabromo m.p. 88°.
Boiling very dil. $H_2SO_4 \rightarrow$ acetone. 2,4-Dinitrophenylhydrazone
m.p. 112°. Semicarbazone m.p. 221°.

34°   Dibenzyl ketone $(C_6H_5CH_2)_2CO$
Gives deep violet colour with benzil in alc. KOH soln. on
warming. 2,4-Dinitrophenylhydrazone m.p. 100°. Semicarbazone
m.p. 146°. Oxime m.p. 124°.

34°   Methyl 1-naphthyl ketone $C_{10}H_7COCH_3$
Phenylhydrazone m.p. 146°. Oxime m.p. 139°. Picrate m.p. 116°.

38°   *p*-Methoxy acetophenone $CH_3OC_6H_4COCH_3$
2,4-Dinitrophenylhydrazone m.p. 220° (227°). *p*-Nitrophenyl-
hydrazone m.p. 195°. Semicarbazone m.p. 196° (181°). Oxime
m.p. 87°.

42°   Benzalacetone $C_6H_5CH:CHCOCH_3$, Benzylidene acetone.
Yellow. Insol. in H_2O. Unsaturated. Acid $KMnO_4$ oxdtn. $\rightarrow$
benzaldehyde. $Br_2/CCl_4 \rightarrow$ dibromo m.p. 125°. 2,4-Dinitro-
phenylhydrazone m.p. 227°. *p*-Nitrophenylhydrazone m.p. 166°.
Semicarbazone m.p. 186°. Oxime m.p. 115°.

42°   1-Indanone C_9H_8O
2,4-Dinitrophenylhydrazone m.p. 258°. *p*-Nitrophenylhydrazone
m.p. 235°. Semicarbazone m.p. 233°. Oxime m.p. 145°.

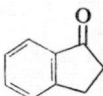

48°   Benzophenone $C_6H_5COC_6H_5$
Insol. in H_2O. Yellow soln. in conc. H_2SO_4. Zn/alc. NaOH
redtn. $\rightarrow$ benzhydrol m.p. 69°. 2,4-Dinitrophenylhydrazone m.p.
238°. *p*-Nitrophenylhydrazone m.p. 154°. Semicarbazone m.p.
164°. Oxime m.p. 142°.

53°   Methyl 2-naphthyl ketone $C_{10}H_7COCH_3$
Phenylhydrazone m.p. 171°. 2,4-Dinitrophenylhydrazone m.p.
262°d. Oxime m.p. 145°. Picrate m.p. 82°. Semicarbazone m.p.
235°.

58°   Chalcone $C_6H_5CH:CHCOC_6H_5$, Benzalacetophenone.
$Br_2/CH_3COOH \rightarrow$ dibromo m.p. 158°. Phenylhydrazone m.p.
118°. 2,4-Dinitrophenylhydrazone m.p. 244°d. Semicarbazone
m.p. 168° (180°).

60°   Deoxybenzoin $C_6H_5CH_2COC_6H_5$, Benzyl phenyl ketone.
2,4-Dinitrophenylhydrazone m.p. 204°. *p*-Nitrophenylhydrazone
m.p. 163°. Semicarbazone m.p. 148°. Oxime m.p. 98°.

M.P.°C

61° Benzoylacetone $C_6H_5COCH_2COCH_3$
B.p. 261°. Keto-enol properties. 2,4-Dinitrophenylhydrazone m.p. 151°. *p*-Nitrophenylhydrazone m.p. 100°.

83° Fluorenone $C_{13}H_8O$
Yellow. 2,4-Dinitrophenylhydrazone m.p. 283°. *p*-Nitrophenylhydrazone m.p. 269°. Phenylhydrazone m.p. 151°. Oxime m.p. 196°.

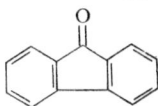

95° Benzil $C_6H_5COCOC_6H_5$
Yellow. Insol. in H_2O. Boiling alc. KOH → benzilic acid m.p. 150°. Alk. $KMnO_4$ oxdtn. → benzoic acid. With *o*-phenylenediamine (AcOH) → quinoxaline deriv. m.p. 126°. 2,4-Dinitrophenylhydrazone (di) m.p. 189°. Oxime (di) m.p. 237°. Semicarbazone (di) m.p. 243°.

109° *p*-Hydroxy acetophenone $HOC_6H_4COCH_3$
Red to violet colour with neutral aq. $FeCl_3$ soln. 2,4-Dinitrophenylhydrazone m.p. 261°. Semicarbazone m.p. 199°. Oxime m.p. 145°. Benzoate m.p. 134°.

112° Dibenzalacetone $C_6H_5CH:CHCOCH:CHC_6H_5$
Yellow. 2,4-Dinitrophenylhydrazone m.p. 180°. *p*-Nitrophenylhydrazone m.p. 173°. Semicarbazone m.p. 190°. Oxime m.p. 143°.

113° Anisoin $CH_3OC_6H_4CHOHCOC_6H_4OCH_3$, *pp'*-dimethoxybenzoin.
Semicarbazone m.p. 185°. Acetate m.p. 95°.

137° Benzoin ($\pm$) $C_6H_5CHOHCOC_6H_5$
Insol. in H_2O. Oxdtn. by Fehling's soln. (alcohol) → benzil. 2,4-Dinitrophenylhydrazone m.p. 234°. Oxime m.p. 151°. Benzoate m.p. 125°. Acetate m.p. 83°. Semicarbazone m.p. 205°.

139° Furoin $(C_4H_3O)CHOHCO(C_4H_3O)$
F·CHOHCO·F F = ⟨furan⟩ . 2,4-Dinitrophenylhydrazone m.p. 216°. Oxime m.p. 161°. Benzoate m.p. 93°. Mild oxdtn. → furil m.p. 162°.

147° Resacetophenone $(HO)_2C_6H_3COCH_3$
Red to violet colour with neutral aq. $FeCl_3$ soln. Phenylhydrazone m.p. 159°. Benzoate (di) m.p. 81°.

162° Furil $(C_4H_3O)COCO(C_4H_3O)$
F·COCO·F F = ⟨furan⟩ . Yellow. With *o*-phenylenediamine → quinoxaline deriv. m.p. 134°. 2,4-Dinitrophenylhydrazone m.p. 215°. *p*-Nitrophenylhydrazone m.p. 199°.

179° (+)-Camphor $C_{10}H_{16}O$
Sublimes. B.p. 209°. Opt. act. $[\alpha]_D^{20}$ + 44° (alcohol). Characteristic odour. Soln. unchanged in conc. H_2SO_4. $NaBH_4/CH_3OH$ redtn. → isoborneol m.p. 212° (sealed tube). SeO_2 oxdtn. →

M.P.°C

camphorquinone m.p. 199°. 2,4-Dinitrophenylhydrazone m.p. 177°. *p*-Nitrophenylhydrazone m.p. 217°. Oxime m.p. 118°. Semicarbazone m.p. 247° (237°).

218° Phloroglucinol. See Phenols (Solid).

QUINONES

SOLID

M.P.°C

45° Thymoquinone $C_{10}H_{12}O_2$

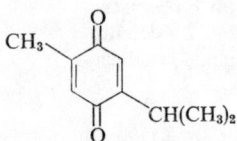

B.p. 232°. Yellow. Steam volatile. Spar. sol. in H_2O. SO_2/H_2O redtn. → hydrothymoquinone m.p. 140°. 2,4-Dinitrophenylhydrazone (mono) m.p. 179°. Oxime (mono) m.p. 161°. Semicarbazone (mono) m.p. 204°.

60–70° *o*-Benzoquinone $C_6H_4O_2$

Red. Unstable. Liberates I_2 from KI soln. SO_2 redtn. → catechol. Dioxime m.p. 142°.

69° Toluquinone $C_7H_6O_2$

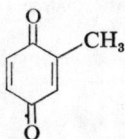

Yellow. Spar. sol. in cold H_2O. Sharp odour. SO_2/H_2O redtn. → toluhydroquinone m.p. 124°. Acetic anhydride + conc. H_2SO_4 (few drops) → 2,4,5-triacetoxytoluene m.p. 114°. 2,4-Dinitrophenylhydrazone (di) m.p. 269°.

106° 2-Methyl 1,4-naphthaquinone. Menadione. Vitamin K_3

Yellow. Redtn. → quinol deriv. m.p. 170°.

115° *p*-Benzoquinone $C_6H_4O_2$

Sublimes. Yellow. Spar. sol. in cold H_2O. Sharp odour. Steam volatile. Reduces Tollen's reagent. Liberates I_2 from KI soln. $FeSO_4$/dil. H_2SO_4 redtn. → quinhydrone m.p. 171°. Reductive acetylation (acetic anhydride, Na acetate, Zn dust) → hydroquinone diacetate m.p. 123°. 2,4-Dinitrophenylhydrazone m.p. 186°.

116°d 1,2-Naphthaquinone $C_{10}H_6O_2$ 2-Naphthaquinone.

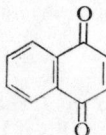

Red. Insol. in H_2O. Odourless. Not steam volatile. Green soln. in conc. H_2SO_4. $KMnO_4$ oxdtn. → phthalic acid. Warming with alc. $NH_2OH\cdot HCl$ soln. → 2-nitroso 1-naphthol m.p. 163°d. Reductive acetylation → diacetate deriv. m.p. 105°. With acetic anhydride and conc. H_2SO_4 (few drops) → 1,2,4-triacetoxy naphthalene m.p. 134°.

125° 1,4-Naphthaquinone $C_{10}H_6O_2$

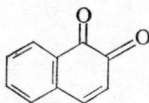

Yellow. Spar. sol. in H_2O. Sharp odour. Steam volatile. Soln. unchanged in conc. H_2SO_4. Not reduced by cold SO_2/H_2O. Reductive acetylation → diacetate deriv. m.p. 128°. 2,4-Dinitrophenylhydrazone (mono) m.p. 278°. Semicarbazone m.p. 247°d.

M.P.°C

171° Quinhydrone $C_6H_4O_2 \cdot C_6H_4(OH)_2$
Molecular compd. of benzoquinone and hydroquinone. Dark green. Spar. sol. in H_2O. Boiling H_2O → benzoquinone and hydroquinone. Reduces Tollen's reagent. Zn/HCl redtn. → hydroquinone. $K_2Cr_2O_7$/dil. H_2SO_4 oxdtn. → benzoquinone.

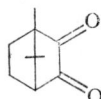

199° Camphorquinone $C_{10}H_{14}O_2$
Yellow. Steam volatile. Spar. sol. in cold H_2O, benzene. Sol. in alcohol. 2,4-Dinitrophenylhydrazone (di) m.p. 190°. Semicarbazone m.p. 236°.

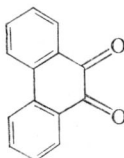

205°ca Phenanthraquinone $C_{14}H_8O_2$
Yellow-orange. Spar. sol. in H_2O. Dull green colour in conc. H_2SO_4. CrO_3 or $KMnO_4$ oxdtn. → diphenic acid m.p. 228°. Reduced by warm SO_2/alcohol soln. → hydrophenanthraquinone m.p. 148°. Reductive acetylation → diacetate deriv. m.p. 202°. With o-phenylenediamine → quinoxaline deriv. m.p. 222°. p-Nitrophenylhydrazone m.p. 245°.

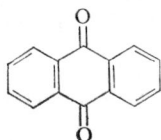

284°ca Anthraquinone $C_{14}H_8O_2$
Pale yellow. Spar. sol. in benzene, ether, alcohol. Sol. unchanged in conc. H_2SO_4. Red colour on heating with Zn dust/NaOH soln. Reductive acetylation → diacetate deriv. m.p. 260°. KOH fusion → benzoic acid. Sn/AcOH redtn. → anthrone m.p. 155°.

289° Alizarin $C_{14}H_8O_4$, 1,2-Dihydroxy anthraquinone.
Orange. Insol. in H_2O. Purple soln. in alkali or conc. H_2SO_4. Red colour with Al and Sn salts. Acetic anhydride/Na acetate → monoacetate m.p. 201°; prolonged treatment → diacetate m.p. 186°.

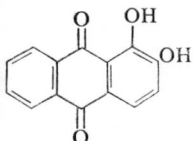

SATURATED CARBOXYLIC ACIDS

LIQUID

B.P.°C

101° Formic HCOOH
Miscible with H_2O. Sharp odour. Heat with conc. H_2SO_4 → CO. Decolourises acid $KMnO_4$ soln. Yellow colour with conc. $NaHSO_3$ soln. Reduces $HgCl_2$ to Hg_2Cl_2. Equiv. wt. 46. SBT* m.p. 146°. p-Bromophenacyl ester m.p. 140°.

118° Acetic CH_3COOH
M.p. 16°. Miscible with H_2O. Equiv. wt. 60. Amide m.p. 82°. Anilide m.p. 114°. p-Toluidide m.p. 148°. p-Nitrobenzyl ester m.p. 78°. SBT m.p. 134°. p-Bromophenacyl ester m.p. 85°.

* S-Benzyl thiuronium salt.

B.P.°C

141° Propionic C_2H_5COOH
Miscible with H_2O. Equiv. wt. 74. Amide m.p. 79°. Anilide m.p. 104°. *p*-Toluidide m.p. 124°. SBT m.p. 148°.

155° Isobutyric $(CH_3)_2CHCOOH$
Sol. in 5 vols. cold H_2O. Equiv. wt. 88. Heat (140°–155°) with conc. $H_2SO_4 \rightarrow CO$ (and SO_2). Amide m.p. 129°. Anilide m.p. 105°. *p*-Toluidide m.p. 104°. SBT m.p. 143°. *p*-Bromophenacyl ester m.p. 77°.

163° n-Butyric $CH_3(CH_2)_2COOH$
Miscible with H_2O. Rancid odour. Equiv. wt. 88. Amide m.p. 114°. Anilide m.p. 96°. *p*-Toluidide m.p. 72°. SBT m.p. 146°. *p*-Bromophenacyl ester m.p. 63°.

176° Isovaleric $(CH_3)_2CHCH_2COOH$
Sl. sol. in cold H_2O. Unpleasant odour. Equiv. wt. 102. Amide m.p. 135°. Anilide m.p. 110°. *p*-Toluidide m.p. 106°. SBT m.p. 153°. *p*-Bromophenacyl ester m.p. 68°.

186° n-Valeric $CH_3(CH_3)COOH$
Spar. sol. in cold H_2O. Unpleasant odour. Equiv. wt. 102. Amide m.p. 104° (114°). Anilide m.p. 63°. *p*-Toluidide m.p. 74°. *p*-Bromophenacyl ester m.p. 75°.

205° n-Caproic $CH_3(CH_2)_4COOH$
Sl. sol. in cold H_2O. Equiv. wt. 116. Amide m.p. 100°. Anilide m.p. 95°. *p*-Toluidide m.p. 73°. *p*-Bromophenacyl ester m.p. 72°.

223° n-Heptylic $CH_3(CH_2)_5COOH$, Oenanthic acid.
Very sl. sol. in cold H_2O. Equiv. wt. 130. Amide m.p. 94°. Anilide m.p. 70°. *p*-Toluidide m.p. 81°. *p*-Bromophenacyl ester m.p. 72°.

237° n-Caprylic $CH_3(CH_2)_6COOH$
M.p. 16°. Very sl. sol. in cold H_2O. Equiv. wt. 144. Amide m.p. 106°. Anilide m.p. 57°. *p*-Toluidide m.p. 70°. *p*-Bromophenacyl ester m.p. 67°.

254° Pelargonic $CH_3(CH_2)_7COOH$
M.p. 12°. Almost insol. in H_2O. Equiv. wt. 158. Amide m.p. 99°. Anilide m.p. 57°. *p*-Toluidide m.p. 84°. *p*-Bromophenacyl ester m.p. 69°.

SOLID

M.P.°C

31° n-Capric $CH_3(CH_2)_8COOH$
B.p. 269°. Almost insol. in H_2O. Equiv. wt. 172. Steam volatile. Amide m.p. 108° (100°). Anilide m.p. 70°. *p*-Toluidide m.p. 78°. *p*-Bromophenacyl ester m.p. 67°.

35° Pivalic $(CH_3)_3CCOOH$, Trimethylacetic acid.
B.p. 163°. Sl. sol. in H_2O. Equiv. wt. 102. Amide m.p. 155°. Anilide m.p. 128°. *p*-Toluidide m.p. 120°. *p*-Bromophenacyl ester m.p. 76°.

M.P.°C

44° Lauric $CH_3(CH_2)_{10}COOH$
B.p. 299°. Insol. in H_2O. Equiv. wt. 200. Steam volatile. Amide m.p. 99°. Anilide m.p. 77°. p-Toluidide m.p. 87°. p-Bromophenacyl ester m.p. 76°. SBT m.p. 141°.

48° Hydrocinnamic $C_6H_5CH_2CH_2COOH$
B.p. 280°. Insol. in H_2O. Equiv. wt. 150. CrO_3 oxdtn. → benzoic acid. Amide m.p. 105°. Anilide m.p. 96°. p-Toluidide m.p. 136°. p-Bromophenacyl ester m.p. 104°.

52° 4-Phenylbutyric $C_6H_5(CH_2)_3COOH$
B.p. 290°. Insol. in H_2O. Equiv. wt. 164. CrO_3 oxdtn. → benzoic acid. Amide m.p. 84°.

54° Myristic $CH_3(CH_2)_{12}COOH$
B.p. 192°/10 mmHg. Insol. in H_2O. Equiv. wt. 228. Amide m.p. 103°. Anilide m.p. 84°. p-Toluidide m.p. 93°. SBT m.p. 139°. p-Bromophenacyl ester m.p. 81°.

62° Palmitic $CH_3(CH_2)_{14}COOH$
B.p. 211°/10 mmHg. Insol. in H_2O. Alc. soln. reacts acid to litmus. Equiv. wt. 256. Amide m.p. 106°. Anilide m.p. 90°. p-Toluidide m.p. 98°. SBT m.p. 141°. p-Bromophenacyl ester m.p. 84°.

69° Stearic $CH_3(CH_2)_{16}COOH$
B.p. 227°/10 mmHg. Like palmitic (above) but less sol. in alcohol. Equiv. wt. 284. Amide m.p. 108°. Anilide m.p. 94°. p-Toluidide m.p. 102°. SBT m.p. 143°. p-Bromophenacyl ester m.p. 90°.

76° Phenylacetic $C_6H_5CH_2COOH$
B.p. 256°. Sol. in hot H_2O. Odour like urine. Equiv. wt. 136. Alk. $KMnO_4$ oxdtn. → benzoic acid. Amide m.p. 154°. Anilide m.p. 118°. p-Toluidide m.p. 136°. SBT m.p. 165°. p-Bromophenacyl ester m.p. 89°.

98° Glutaric $COOH(CH_2)_3COOH$
B.p. 303°. Very sol. in H_2O. Sl. sol. in benzene. Equiv. wt. 66. Gives the fluorescein reaction. Heat with aniline → anil m.p. 144°. Amide m.p. 174° gives imide m.p. 154° on heating above m.p. Anilide m.p. 224°. p-Toluidide m.p. 217°. p-Bromophenacyl ester m.p. 137°.

98° Phenoxyacetic. See Alkyl and Acyl Hydroxy Acids.

101° Oxalic $(COOH)_2 \cdot 2H_2O$
Usually as dihydrate. Loses H_2O on fusion and sublimes above 150°. Anhydrous m.p. 189° (sealed tube). Fairly sol. in cold H_2O. Heat with conc. H_2SO_4 → $CO + CO_2$. Acid $KMnO_4$ oxdtn., on warming → CO_2. Ca salt insol. in H_2O or dil. AcOH. Equiv. wt. 63. Amide (di) m.p. 419°d. Anilide (di) m.p. 246°. p-Nitrobenzyl ester (di) m.p. 204°. SBT m.p. 195°d. p-Bromophenacyl ester m.p. 242°d.

104° o-Toluic $CH_3C_6H_4COOH$
B.p. 259°. Spar. sol. in cold H_2O. Equiv. wt. 136. Alk. $KMnO_4$ oxdtn. → phthalic acid. Heat with soda-lime → toluene. Amide

M.P.°C

m.p. 142°. Anilide m.p. 125°. *p*-Toluidide m.p. 144°. SBT m.p. 145°. *p*-Nitrobenzyl ester m.p. 91°.

105°   Pimelic   COOH(CH$_2$)$_5$COOH
B.p. 223°/15 mmHg. Sublimes. Equiv. wt. 80. Amide (di) m.p. 175°. Anilide (di) m.p. 155°. *p*-Toluidide (di) m.p. 206°. *p*-Bromophenacyl ester (di) m.p. 136°.

106°   Azelaic   COOH(CH$_2$)$_7$COOH
B.p. 237°/15 mmHg. Very sl. sol. in cold H$_2$O. Miscible hot H$_2$O. Sl. sol. in ether. Equiv. wt. 94. Amide (di) m.p. 175°. Anilide (di) m.p. 186°. *p*-Toluidide (di) m.p. 201°. *p*-Bromophenacyl ester m.p. 130°.

110°   *m*-Toluic   CH$_3$C$_6$H$_4$COOH
B.p. 263°. Spar. sol. in cold H$_2$O. Equiv. wt. 136. Alk. KMnO$_4$ oxdtn. → isophthalic acid. Heat with soda-lime → toluene. Amide m.p. 94°. Anilide m.p. 126°. *p*-Toluidide m.p. 118°. SBT m.p. 164°. *p*-Bromophenacyl ester m.p. 108°.

111°   Ethyl malonic   C$_2$H$_5$CH(COOH)$_2$
Sol. in H$_2$O. Heat above m.p. → n-butyric acid + CO$_2$. Equiv. wt. 66. Amide (di) m.p. 214°. Anilide m.p. 150°. Diethyl ester b.p. 200°

115°   Methyl succinic   COOHCH(CH$_3$)CH$_2$COOH, Pyrotartaric acid.
Sol. in H$_2$O. Equiv. wt. 66. Amide (di) m.p. 225°. Anilide (di) m.p. 200°. *p*-Toluidide (di) m.p. 164°.

117°d   Benzyl malonic   C$_6$H$_5$CH$_2$CH(COOH)$_2$
Sol. in H$_2$O. Heat above m.p. → hydrocinnamic acid + CO$_2$. Equiv. wt. 97. Amide (di) m.p. 225°. Anilide (di) m.p. 217°. *p*-Nitrobenzyl ester m.p. 119°.

122°   Benzoic   C$_6$H$_5$COOH
B.p. 249°. Spar. sol. in cold H$_2$O, sol. in hot. Heat with soda-lime → benzene. Equiv. wt. 122. Buff ppt. with neutral aq. FeCl$_3$ soln. Amide m.p. 128°. Anilide m.p. 164°. *p*-Toluidide m.p. 158°. SBT m.p. 167°. *p*-Bromophenacyl ester m.p. 119°. *p*-Nitrobenzyl ester m.p. 89°.

133°   Sebacic   COOH(CH$_2$)$_8$COOH
Sl. sol. in hot H$_2$O. Equiv. wt. 101. Amide (di) m.p. 210°. Anilide (di) m.p. 201°. *p*-Toluidide (di) m.p. 207°. *p*-Bromophenacyl ester (di) m.p. 147°.

133°d   Malonic   COOHCH$_2$COOH
Very sol. in H$_2$O, sol. in ether. Ca, Ba, Ag salts insol. Heat above m.p. → acetic acid + CO$_2$. Warm with acetic anhydride → orange colour, green fluorescence. Br$_2$/H$_2$O → tribromoacetic acid m.p. 135° + CO$_2$. Warm with C$_6$H$_5$NCO → anilide (di) m.p. 224°. Equiv. wt. 52. Amide (di) m.p. 170°. *p*-Toluidide (di) m.p. 252°. SBT m.p. 147°. *p*-Nitrobenzyl ester (di) m.p. 85°. Diethyl ester b.p. 198°.

133°   Furoic   (C$_4$H$_4$O)COOH, Pyromucic.
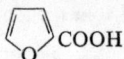
Mod. sol. in cold H$_2$O, easily sol. in hot. Sol. in benzene and ether. Heat with NH$_4$OH → pyrrole (vapour turns pine splint, moist with conc. HCl, red colour). Equiv. wt. 112. Amide m.p.

M.P.°C

142°. Anilide m.p. 123°. *p*-Toluidide m.p. 107°. SBT m.p. 211°. *p*-Bromophenacyl ester m.p. 139°. *p*-Nitrobenzyl ester m.p. 133°.

137°d   Methyl malonic $CH_3CH(COOH)_2$, Isosuccinic.
Sol. in H_2O. Ca, Ba, Pb, Ag salts insol. Heat above m.p. → propionic acid + CO_2. Equiv. wt. 59. Amide m.p. 206°(217°). Diethyl ester b.p. 196°.

144°   Suberic $COOH(CH_2)_6COOH$
Sl. sol. in H_2O. Equiv. wt. 87. Amide (di) m.p. 216°. Anilide (di) m.p. 186°. *p*-Toluidide (di) m.p. 218°. *p*-Bromophenacyl ester m.p. 144°. *p*-Nitrobenzyl ester m.p. 85°.

151°   Adipic $COOH(CH_2)_4COOH$
Spar. sol. in cold H_2O. Readily sol. in alcohol. Gives the fluorescein reaction. Equiv. wt. 73. Amide (di) m.p. 220°. Anilide (di) m.p. 240°. *p*-Toluidide m.p. 241°. SBT m.p. 159°. *p*-Bromophenacyl ester m.p. 155°. *p*-Nitrobenzyl ester m.p. 106°. Diethyl ester b.p. 245°.

153°   Phenyl malonic $C_6H_5CH(COOH)_2$
Sol. in H_2O. Heat above m.p. → phenylacetic acid + CO_2. Equiv. wt. 90. Amide m.p. 233°.

162°   1-Naphthoic $C_{10}H_7COOH$
Spar. sol. in H_2O. Heat with soda-lime → naphthalene. Equiv. wt. 172. Amide m.p. 205°. Anilide m.p. 160°. SBT m.p. 151°. *p*-Bromophenacyl ester m.p. 135°.

180°   *p*-Toluic $CH_3C_6H_4COOH$
Sol. in hot H_2O. Alk. $KMnO_4$ oxdtn. → terephthalic acid. Heat with soda-lime → toluene. Equiv. wt. 136. Amide m.p. 158°. Anilide m.p. 144°. *p*-Toluidide m.p. 161°. SBT m.p. 190°. *p*-Bromophenacyl ester m.p. 153°. *p*-Nitrobenzyl ester m.p. 104°.

185°   2-Naphthoic $C_{10}H_7COOH$
Insol. in H_2O. Heat with soda-lime → naphthalene. Equiv. wt. 172. Amide m.p. 192°. Anilide m.p. 170°. *p*-Toluidide m.p. 192°. *p*-Nitrophenyl ester m.p. 142°.

185°   Succinic $COOH(CH_2)_2COOH$
B.p. 235°(d) giving anhydride m.p. 119°. Equiv. wt. 59. Gives the fluorescein reaction. Buff ppt. (sol. in dil. HCl) with neutral aq. $FeCl_3$ soln. Amide (di) m.p. 260°(d) giving the imide m.p. 126°. Amilide (di) m.p. 228°, (mono) m.p. 148° (succinanilic acid). *p*-Toluidide (di) m.p. 255°. Anil m.p. 156°. SBT m.p. 154°. *p*-Bromophenacyl ester m.p. 211°. *p*-Nitrobenzyl ester m.p. 88°.

195°d   Phthalic $C_6H_4(COOH)_2$
M.p. rises with more rapid heating 195° → 230°. Loses H_2O at m.p. → anhydride m.p. 131°. Sol. in hot H_2O. Heat with soda-lime → benzene. Gives fluorescein reaction. NH_4 salt gives imide m.p. 233° on prolonged fusion. Equiv. wt. 83. Amide (di) m.p. 220°. Anilide (di) m.p. 250°. *p*-Toluidide (di) m.p. 201°. SBT m.p. 157°. *p*-Bromophenacyl ester m.p. 153°. *p*-Nitrobenzyl ester m.p. 155°. Dimethyl ester B.p. 282°. Anil m.p. 210°.*

* Reported at various temps. 203°–210°

M.P.°C

228° Diphenic $COOHC_6H_4C_6H_4COOH$ 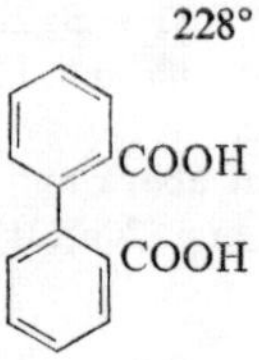
Sl. sol. in H_2O. Equiv. wt. 121. Amide (di) m.p. 212°. Anilide (di) m.p. 229°. p-Nitrobenzyl ester m.p. 187°.

>300° Isophthalic $C_6H_4(COOH)_2$ 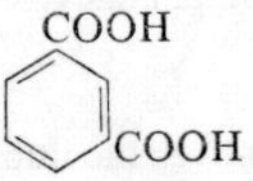
Sublimes. Insol. in H_2O. Heat with soda-lime → benzene. Ba salt very sol. in H_2O (differentiate from terephthalic). Equiv. wt. 83. Amide m.p. 280°. Anilide m.p. 250°. SBT m.p. 215°. p-Bromophenacyl ester m.p. 179°. p-Nitrobenzyl ester m.p. 202°.

>300° Terephthalic $C_6H_4(COOH)_2$ 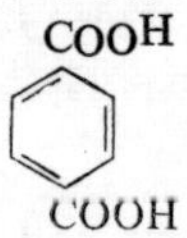
Sublimes. Insol. in H_2O. Heat with soda-lime → benzene. Ba salt almost insol. H_2O. Equiv. wt. 83. SBT m.p. 202°. p-Bromophenacyl ester m.p. 225°. p-Nitrobenzyl ester m.p. 263°.

UNSATURATED ACIDS

LIQUID

B.P.°C

140° Acrylic $CH_2:CHCOOH$
M.p. 13°. Miscible with H_2O. Br_2 → dibromo m.p. 67°. Equiv. wt. 72. Amide m.p. 84°. Anilide m.p. 104°. p-Toluidide m.p. 141°. Ethyl ester b.p. 101°.

144°d Propiolic $CH:CCOOH$
M.p. 18°. Miscible with H_2O. Br_2 → dibromo m.p. 85°. Equiv. wt. 70. Amide m.p. 61°. Anilide m.p. 87°.

165°d Isocrotonic $CH_3CH:CHCOOH$
M.p. 15°. Cis isomer. Miscible with H_2O. Heat at 150° with trace of I_2 → trans isomer (α-crotonic, see below). Br_2 → dibromo (oil). Equiv. wt. 86. Amide m.p. 102°. Anilide m.p. 102°. p-Toluidide m.p. 132°. p-Bromophenacyl ester m.p. 81°. p-Nitrobenzyl ester m.p. 67°.

285°d Oleic $CH_3(CH_2)_7CH:CH(CH_2)_7COOH$
B.p. 223°/10 mmHg. M.p. 14°. Cis isomer. Insol. in H_2O. Acid reaction to litmus in alcohol soln. only. $KMnO_4$ oxdtn. → pelargonic, azelaic, dihydroxystearic acids. Equiv. wt. 282. Amide m.p. 76°.

SOLID

M.P.°C

44° Elaidic $CH_3(CH_2)_7CH:CH(CH_2)_7COOH$
(51°) B.p. 226°/10 mmHg. Trans isomer of oleic acid. Insol. in H_2O. Equiv. wt. 282. Acid reaction to litmus in alcohol soln. only. $KMnO_4$ oxdtn. → pelargonic and azelaic acids. Amide m.p. 90°. p-Bromophenacyl ester m.p. 65°.

M.P.°C

72°   α-Crotonic $CH_3CH:CHCOOH$
Trans isomer. Sol. in H_2O. Reduces Tollen's reagent on warming.
Yields n-butyric acid with Zn dust/dil. H_2SO_4. $Br_2 \rightarrow$ dibromo
m.p. 87°. Equiv. wt. 86. Amide m.p. 158°. Anilide m.p. 116°.
p-Toluidide m.p. 132°. p-Bromophenacyl ester m.p. 96°. p-
Nitrobenzyl ester m.p. 67°.

92°d   Citraconic $COOHCH:C(CH_3)COOH$, Methyl maleic.
Cis isomer. Sol. in H_2O. Equiv. wt. 65. Heat → anhydride b.p.
213°. Warm Br_2/H_2O → dibromo m.p. 150°. Anilide (di) m.p.
175°. p-Nitrobenzyl ester m.p. 70°.

130°   Maleic $COOHCH:CHCOOH$
M.p. rises with more rapid heating 130° → 143°. Heat at 160°
(ca.) → anhydride m.p. 56°. Heat above 200° → fumaric acid.
Cis isomer. Sol. in H_2O. Equiv. wt. 58. Br_2 → dibromo m.p. 166°.
Reflux with aniline (2 parts) for 1 hour → phenylaspartic anil
m.p. 210°. Anilide (di) m.p. 187°. p-Toluidide (di) m.p. 142°.
SBT m.p. 175°(163°). p-Nitrobenzyl ester m.p. 89°.

133°   Cinnamic $C_6H_5CH:CHCOOH$
Trans isomer. Spar. sol. in H_2O. Alk. $KMnO_4$ oxdtn. → benzoic
acid. Warm Br_2/H_2O → dibromo m.p. 203°.* Equiv. wt. 148.
Amide m.p. 147°. Anilide m.p. 151°. p-Toluidide m.p. 168°.
SBT m.p. 175°. p-Bromophenacyl ester m.p. 145°. p-Nitrobenzyl
ester m.p. 116°.

136°   Phenyl propiolic $C_6H_5C:CCOOH$
Sublimes. Insol. in cold H_2O. Heat with Zn dust/AcOH →
cinnamic acid. Equiv. wt. 146. Amide m.p. 100°. Anilide m.p.
126°. p-Toluidide m.p. 142°. p-Nitrobenzyl ester m.p. 83°.

165°   Itaconic $COOHC(:CH_2)CH_2COOH$, Methylene succinic.
Sol. in H_2O. Warm with CH_3COCl → anhydride m.p. 68°. Heat
with aniline → anilic acid deriv. m.p. 190°. Equiv. wt. 65. Amide
m.p. 191°. p-Bromophenacyl ester m.p. 117°. p-Nitrobenzyl ester
m.p. 90°.

179°   Acetylene dicarboxylic $COOHC:CCOOH$
Crystalline dihydrate from H_2O. Equiv. wt. 57. Amide (di) m.p.
294°d. Diethyl ester b.p. 107°/13 mmHg.

191°d   Aconitic $COOHCH:C(COOH)CH_2COOH$
Trans isomer. Sol. in H_2O. Heat at 200° → itaconic anhydride +
CO_2. Decolourises warm Br_2/H_2O. Equiv. wt. 58. p-Bromo-
phenacyl ester (tri) m.p. 186°.

200°   Fumaric $COOHCH:CHCOOH$
Sublimes. M.p. 287° (sealed tube). Trans isomer. Spar. sol. in
H_2O. Equiv. wt. 58. Reflux with aniline (2 parts) for 1 hour →
phenylaspartic anil m.p. 210°. SBT m.p. 178°. p-Nitrobenzyl
ester m.p. 151°.

204°   Mesaconic $COOHC(CH_3):CHCOOH$, Methyl fumaric.
Sublimes. Trans isomer. Sl. sol. in cold H_2O. Equiv. wt. 65.

* Another form has m.p. 90°.

M.P.°C

Amide (di) m.p. 176°. Anilide (di) m.p. 185°. *p*-Toluidide (di) m.p. 212°. *p*-Nitrobenzyl ester (di) m.p. 134°.

ALIPHATIC HYDROXY ACIDS

SOLID

M.P.°C

18° (±)-Lactic $CH_3CHOHCOOH$
Usually a liquid, b.p. 122°/15 mmHg. Miscible with H_2O. Chars rapidly with warm conc. H_2SO_4. Decolourises acid $KMnO_4$ on warming. Gives iodoform test. Warm gently, until yellow, 2 or 3 drops of 0.1 % soln. with 2 cm³ conc. H_2SO_4, cool, add 2 drops 5% alcoholic guaiacol → red colour. CrO_3 oxdtn. → $CH_3CHO +$ CH_3COOH. Equiv. wt. 90. Amide m.p. 74°. *p*-Toluidide m.p. 107°. SBT m.p. 153°. *p*-Bromophenacyl ester m.p. 112°. Benzoyl m.p. 112°.

80° Glycollic $CH_2OHCOOH$
Sol. in H_2O. Deliquescent. Heat → HCHO. In guaiacol test (above) in presence of 1 cm³ CH_3COOH→ violet colour. Ag,Ca, Cu salts spar. sol. in cold H_2O. HNO_3 oxdtn. → oxalic acid. Equiv. wt. 76. Amide m.p. 120°. Anilide m.p. 96°. *p*-Toluidide m.p. 143°. SBT m.p. 142°. *p*-Bromophenacyl ester m.p. 138° *p*-Nitrobenzyl ester m.p. 107°.

100° (−)-Malic $COOHCH_2CH(OH)COOH$, Hydroxy succinic.
Opt. act. Conc. solns. (−)-rotatory, dil. solns. (+)-rotatory. Hygroscopic. Ba salt very sol., Ca salt insol. in H_2O. Neutral soln, $(+NH_4Cl)$ on boiling with $CaCl_2$ → no ppt. but ppt. forms on addtn. of 1–2 vols. alcohol. Cobalt nitrate + excess NaOH → blue colour. 2-Naphthol and conc. H_2SO_4 → green-yellow colour, clear yellow on warming, pale orange on dilution. Equiv. wt. 67. Amide (di) m.p. 157°. Anilide (di) m.p. 197°. *p*-Toluidide (di) m.p. 207°. *p*-Bromophenacyl ester (di) m.p. 179°. *p*-Nitrobenzyl ester (di) m.p. 124°.

100° Citric $COOHC(OH)(CH_2COOH)_2$
Monohydrate. Heat at 130° → anhydrous acid m.p. 153°. Sol. in H_2O. Gives iodoform test. With $CaCl_2$ soln. slowly forms ppt. on boiling or on addtn. of NaOH. Does not char with warm conc. H_2SO_4 → CO,CO_2 and yellow soln. containing acetone dicarboxylic acid (*q.v.*), on neutralisation → red colour with Na nitroprusside changed to violet by CH_3COOH. Insol. Ba salt separates slowly. 2-Naphthol and conc. H_2SO_4 → blue colour, unchanged on warming. Acid or neutral $KMnO_4$ → acetone. Equiv. wt. (hydrated form) 70. Amide (tri) m.p. 210°(d). Anilide (tri) m.p. 192°. *p*-Toluidide (tri) m.p. 189°. *p*-Bromophenacyl ester (tri) m.p. 148°. *p*-Nitrobenzyl ester (tri) m.p. 102°.

118° (±)-Mandelic $C_6H_5CHOHCOOH$
Sol. in H_2O. Warm acid $KMnO_4$ → C_6H_5CHO. Equiv. wt. 152.

M.P.°C

Amide m.p. 133°. Anilide m.p. 151°. p-Toluidide m.p. 172°. SBT m.p. 166°. p-Nitrobenzyl ester m.p. 124°. Nitrile b.p. 244°.

125° ($\pm$)-Saccharic $COOH(CHOH)_4COOH$
Opt. act. $[\alpha]_D^{20} +7° \rightarrow +20°(H_2O)$. Sol. in H_2O, alcohol. HNO_3 oxdtn. product from glucose. Amide (di) m.p. 172°.

133° ($+$) or ($-$)-Mandelic. See ($\pm$)-Mandelic acid.

140° Mesotartaric acid.
Anhydrous m.p.

150° Benzilic $(C_6H_5)_2C(OH)COOH$
Sol. in hot H_2O. CrO_3 oxdtn. $\rightarrow$ benzophenone. Soln. in conc. $H_2SO_4 \rightarrow$ red colour. Gives Molisch's test (for carbohydrates). Equiv. wt. 228. Amide m.p. 154°. Anilide m.p. 175°. p-Toluidide m.p. 189°. SBT m.p. 125°. p-Bromophenacyl ester m.p. 152°. p-Nitrobenzyl ester m.p. 99°. Acetate m.p. 98°.

153° Citric (anhydrous). See Citric m.p. 100°.

170° ($+$)-Tartaric $COOH(CHOH)_2COOH$
$[\alpha]_D +12°$ (20% H_2O soln.). Ca salt insol. in H_2O but sol. in cold conc. NaOH soln. Aq. soln. $+ CaCl_2$ gives no ppt. Salts char on heating. Acid chars rapidly and effervesces with conc. H_2SO_4. Cobalt nitrate and excess NaOH $\rightarrow$ colourless soln. in cold, turns blue on warming, colourless on cooling. 2-Naphthol and conc. $H_2SO_4 \rightarrow$ green colour, unchanged on warming, orange on diltn. Add in turn $FeSO_4$, H_2O_2, NaOH $\rightarrow$ violet colour (Fenton's test). Add 5% soln. NH_4 vanadate, acidify with dil. $CH_3COOH \rightarrow$ orange colour. Equiv. wt. 75. Amide (di) m.p. 195°d. Anilide (di) m.p. 264°d. p-Bromophenacyl ester (di) m.p. 216°. p-Nitrobenzyl ester (di) m.p. 163°.

206° ($\pm$)-Tartaric. See above.
Anhydrous. Monohydrate m.p. 203°. Amide (di) m.p. 226°. Anilide (di) m.p. 235°. p-Nitrobenzyl ester (di) m.p. 147°.

214°d Mucic $COOH(CHOH)_4COOH$
M.p. varies with heating rate. Meso form. HNO_3 oxdtn. product from lactose or galactose. Spar. sol. in cold H_2O, insol. in alcohol. Heat the spar. sol. NH_4 salt $\rightarrow$ pyrrole. Equiv. wt. 105. Amide (di) m.p. 220°. p-Bromophenacyl ester (di) m.p. 225°. SBT m.p. 194°. Acetate (tetra) m.p. 266°.

AROMATIC HYDROXY ACIDS

SOLID

M.P.°C

118° Mandelic. See Aliphatic Hydroxy Acids.

150° Benzilic. See Aliphatic Hydroxy Acids.

158° Salicylic HOC_6H_4COOH
Difficultly sol. in H_2O. Purple colour with neutral aq. $FeCl_3$ soln. Heat with soda-lime $\rightarrow$ phenol. $Me_2SO_4 + NaOH \rightarrow o$-methoxy-benzoic acid m.p. 100°. Equiv. wt. 138. Amide m.p. 139°. Anilide m.p. 134°. p-Toluidide m.p. 156°. SBT m.p. 146°.

M.P.°C

p-Bromophenacyl ester m.p. 140°. p-Nitrobenzyl ester m.p. 96°. Methyl ester (oil of wintergreen) b.p. 223°. Acetyl ('aspirin') q.v.

200°d   Protocatechuic $(HO)_2C_6H_3COOH$, 3,4-Dihydroxybenzoic.
Sol. in H_2O. Blue-green colour with neutral aq. $FeCl_3$ soln. Heat → catechol + CO_2. Reduces Tollen's reagent. Me_2SO_4 + NaOH → veratric acid m.p. 181° (dimethyl ether deriv.). Equiv. wt. 154. Amide m.p. 212°. Anilide m.p. 166°. p-Nitrobenzyl ester m.p. 188°. Acetyl (di) m.p. 162°.

200°   m-Hydroxy benzoic HOC_6H_4COOH
Sl. sol. in H_2O. No colour with neutral aq. $FeCl_3$ soln. Heat with soda-lime → phenol. Me_2SO_4 + NaOH → m-methoxy benzoic acid m.p. 110°. Equiv. wt. 138. Amide m.p. 170°. Anilide m.p. 155°. p-Toluidide m.p. 163°. p-Bromophenacyl ester m.p. 176° (168°). p-Nitrobenzyl ester m.p. 106°. Acetyl m.p. 131°.

207°d   o-Coumaric $HOC_6H_4CH:CHCOOH$
Unsaturated. Trans isomer. Spar. sol. in cold H_2O. Buff ppt. with neutral aq. $FeCl_2$ soln. KOH fusion → salicylic acid. Equiv. wt. 164. Amide m.p. 209°(d). p-Nitrobenzyl ester m.p. 152°. Acetyl m.p. 149°.

213°d   β-Resorcylic $(HO)_2C_6H_3COOH$, 2,4-Dihydroxybenzoic.
Loses H_2O of crystallisation at 100°. Loses CO_2 easily giving variable m.p. 194° → 236°. Equiv. wt. 154. Amide m.p. 222°. Anilide m.p. 126°. p-Nitrobenzyl ester m.p. 188°.

213°   p-Hydroxy benzoic HOC_6H_4COOH
Spar. sol. in H_2O. Red colour with neutral aq. $FeCl_3$ soln. Heat with soda-lime → phenol. Me_2SO_4 + NaOH → anisic acid (q.v.). Equiv. wt. 138. Amide m.p. 162°. Anilide m.p. 197°. p-Toluidide m.p. 204°. SBT m.p. 143°. p-Bromophenacyl ester m.p. 184°. p-Nitrobenzyl ester m.p. 198°. Acetyl m.p. 190°.

222°   2-Hydroxy 3-naphthoic $HOC_{10}H_6COOH$
Crystallised from H_2O, m.p. 239°. Sl. sol. in H_2O. Blue colour with neutral aq. $FeCl_3$ soln. Equiv. wt. 188. Amide m.p. 217°. Anilide m.p. 243°. p-Toluidide m.p. 222°. Acetyl m.p. 184°.

263°d* Gallic $(HO)_3C_6H_2COOH$, 3,4,5-Trihydroxybenzoic.
Readily sol. in hot H_2O. Readily oxidised. Blue-black ppt. with neutral aq. $FeCl_3$ soln. Heat → pyrogallol + CO_2. Me_2SO_4 + NaOH → trimethyl ether m.p. 170°. Equiv. wt. 170. Amide m.p. 244°. Anilide m.p. 207°. p-Bromophenacyl ester m.p. 134°. p-Nitrobenzyl ester m.p. 141°.

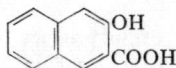

ALKYL AND ACYL HYDROXY ACIDS

LIQUID

B.P.°C

203°   Methoxyacetic CH_3OCH_2COOH, Glycollic acid methyl ether. Equiv. wt. 90. Amide m.p. 96°. Anilide m.p. 58°.

* Various temps. reported from 225° upwards.

B.P.°C

206° Ethoxyacetic $C_2H_5OCH_2COOH$, Glycollic acid ethyl ether.
Equiv. wt. 104. Amide m.p. 80°. Anilide m.p. 92°. p-Bromo-
phenacyl ester m.p. 105°.

SOLID

M.P.°C

25° o-Ethoxybenzoic $C_2H_5OC_6H_4COOH$, Salicyclic acid ethyl
ether.
B.p. 300°(d). Heat with soda-lime → phenetole. Equiv. wt. 166.
Amide m.p. 132°.

98° Phenoxyacetic $C_6H_5OCH_2COOH$, Glycollic acid phenyl ether.
Sol. in H_2O. Yellow ppt. with neutral aq. $FeCl_3$ soln. Warm with
dil. HNO_3 → 2,4-dinitrophenol m.p. 114°. Equiv. wt. 152.
Amide m.p. 101°. Anilide m.p. 99°. p-Bromophenacyl ester m.p.
148°.

100° o-Methoxybenzoic $CH_3OC_6H_4COOH$, Salicyclic acid methyl
ether.
B.p. 200°. Spar. sol. in H_2O. No colour with neutral aq. $FeCl_3$
soln. Heat with soda-lime → anisole. Equiv. wt. 152. Amide
m.p. 128°. Anilide m.p. 131°. p-Bromophenacyl ester m.p. 113°.

135° Acetyl salicylic $CH_3COOC_6H_4COOH$, 'Aspirin'.
Spar. sol. in H_2O. No colour with neutral aq. $FeCl_3$ soln. but
hydrolyses easily to give salicylic acid which gives a purple
colour. Equiv. wt. 180. Amide m.p. 138°. Anilide m.p. 136°.
SBT m.p. 144°. p-Nitrobenzyl ester m.p. 90°.

183° Anisic $CH_3OC_6H_4COOH$, p-Methoxybenzoic.
Spar. sol. in H_2O. Heat with soda-lime → anisole. Equiv. wt. 152.
Amide m.p. 162°. Anilide m.p. 168°. p-Toluidide m.p. 186°.
SBT m.p. 185°. p-Bromophenacyl ester m.p. 152°. p-Nitrobenzyl
ester m.p. 132°.

KETONIC ACIDS

SOLID

M.P.°C

13° Pyruvic $CH_3COCOOH$
B.p. 165°(d), 80°/25 mmHg. Miscible with H_2O. N_2H_4 →
hydrazone m.p. 116°. Decolourises Br_2/H_2O, $KMnO_4$; reduces
Tollen's reagent. Gives iodoform reaction. 2-Naphthol + conc.
H_2SO_4 → red colour, blue on warming. Aq. NH_4OH soln. with
Na nitroprusside → violet colour, changed to red by KOH, and
to blue by acetic acid. Equiv. wt. 88. Amide m.p. 124°. Anilide
m.p. 104°. p-Toluidide m.p. 109°. SBT m.p. 158°d. Phenylhydrazone m.p. 192°d. 2,4-Dinitrophenylhydrazone m.p. 218°.

33° Levulinic $CH_3CO(CH_2)_2COOH$, Laevulic.
B.p. 245°(d). Deliquescent. Sol. in H_2O. Gives iodoform reaction.
Equiv. wt. 116. Amide m.p. 108°. Anilide m.p. 102°. p-Toluidide
m.p. 109°. p-Bromophenacyl ester m.p. 84°. Phenylhydrazone

M.P.°C

m.p. 108°. *p*-Nitrophenylhydrazone m.p. 175°. Oxime m.p. 95°. Semicarbazone m.p. 184°.

66° Phenylglyoxylic $C_6H_5COCOOH$, Benzoylformic.
Sol. in H_2O. Distil → $C_6H_5COOH + CO_2$. Heat with aniline → benzylidene aniline m.p. 54° + CO_2. Equiv. wt. 158. Amide m.p. 91° (α-form). 2,4-Dinitrophenylhydrazone m.p. 196°d.

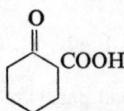

81°d Cyclohexanone 2-carboxylic $C_7H_{10}O_3$
Heat, or dil. acid → cyclohexanone + CO_2. Blue colour with neutral aq. $FeCl_3$ soln. Equiv. wt. 142. Ethyl ester b.p. 107°/11 mmHg., 159°/100 mmHg.

128° *o*-Benzoyl benzoic $C_6H_5COC_6H_4COOH$
Crystallises from hot H_2O m.p. 91°. Heat with conc. H_2SO_4 → anthraquinone (*q.v.*) Equiv. wt. 226. Amide m.p. 165°. Anilide m.p. 195°. *p*-Nitrobenzyl ester m.p. 100°.

135°d Acetone dicarboxylic $COOHCH_2COCH_2COOH$
Sol. in H_2O. Heat, or H_2O, or acid, or alkali → acetone + CO_2. Violet colour with neutral aq. $FeCl_3$ soln. Equiv. wt. 73. Anilide (di) m.p. 155°. Oxime m.p. 54° (89° anhydrous). Diethyl ester b.p. 250°, 138°/12 mmHg.

190°d (−)-Ascorbic acid. Vitamin C. See Lactones.

ACID ANHYDRIDES

LIQUID

B.P.°C

138° Acetic $(CH_3CO)_2O$
Sl. sol. in H_2O, hydrolysis → acetic acid. Irritating odour. Warm gently with aniline in benzene soln. → acetanilide m.p. 114°. Equiv. wt. 51.

168° Propionic $(C_2H_5CO)_2O$
Hydrolysis → propionic acid. Warm gently with aniline in benzene soln. → propionanilide m.p. 104°. Equiv. wt. 65.

SOLID

M.P.°C

42° Benzoic $(C_6H_5CO)_2O$
Slowly decomp. by boiling H_2O → benzoic acid. Warm gently with aniline in benzene soln. → benzanilide m.p. 164°. Equiv. wt. 113.

56° Glutaric $C_5H_6O_3$
Equiv. wt. 57. Anil m.p. 144°. Amide (di) of acid m.p. 174°. Anilide (di) of acid m.p. 224°.

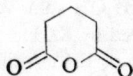

56° Maleic $C_4H_2O_3$
Readily hydrolysed by warm H_2O → maleic acid. Reflux with aniline (2 parts) for one hour → phenylaspartic anil m.p. 210°. Warm gently with aniline in benzene soln. → anilic acid m.p. 187°. Equiv. wt. 49. Anil m.p. 91°.

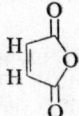

M.P.°C

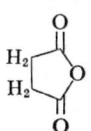

119°   Succinic $C_4H_4O_3$

Readily hydrolysed by warm H_2O → succinic acid. Warm gently with aniline in benzene soln. → anilic acid m.p. 148°. Equiv. wt. 50. Anil m.p. 156°.

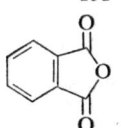

131°   Phthalic $C_8H_4O_2$

Insol. in cold H_2O, readily sol. in hot (with hydrolysis). Gives fluorescein reaction. Warm gently with aniline in benzene soln. → anilic acid m.p. 170°. Equiv. wt. 74. Anil m.p. 210°.* Zn dust + acetic acid → phthalide m.p. 73°.

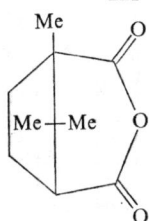

221°   (+)-Camphoric $C_{10}H_{14}O_3$

Boiling H_2O → cis-camphoric acid m.p. 187°, $[\alpha]_D^{20}$ +48°. Equiv. wt. 91. Anilic acid m.p. 204°. With conc. NH_4OH → camphoraminic acid m.p. 174°.

SIMPLE ALKYL AND ARYL ESTERS OF CARBOXYLIC ACIDS

B.P.°C	LIQUID		d_4^{20}	n_D^{20}
32°	Methyl formate	$HCOOCH_3$	0.974	1.344
53°	Ethyl formate	$HCOOC_2H_5$	0.923	1.360
56°	Methyl acetate	CH_3COOCH_3	0.929	1.362
71°	Isopropyl formate	$HCOOCH(CH_3)_2$	0.873	1.368
77°	Ethyl acetate	$CH_3COOC_2H_5$	0.901	1.372
79°	Methyl propionate	$C_2H_5COOCH_3$	0.915	1.377
81°	n-Propyl formate	$HCOOCH_2C_2H_5$	0.904	1.377
88°	Isopropyl acetate	$CH_3COOCH(CH_3)_2$	0.872	1.377
90°	Dimethyl carbonate	$CH_3OCOOCH_3$	1.071	1.369
92°	Methyl isobutyrate	$(CH_3)_2CHCOOCH_3$	0.888	1.383
97°	t-Butyl acetate	$CH_3COOC(CH_3)_3$	0.867	1.386
98°	Isobutyl formate	$HCOOCH_2CH(CH_3)_2$	0.876	1.386
98°	Ethyl propionate	$C_2H_5COOC_2H_5$	0.892	1.384
101°	n-Propyl acetate	$CH_3COOCH_2C_2H_5$	0.887	1.384
102°	Methyl n-butyrate	$C_2H_5CH_2COOCH_3$	0.898	1.387
105°	Methyl orthoformate	$HC(OCH_3)_3$	0.968	1.379
106°	n-Butyl formate	$HCOO(CH_2)_3CH_3$	0.892	1.389
110°	Ethyl isobutyrate	$(CH_3)_2CHCOOC_2H_5$	0.869	1.387
111°	Isopropyl propionate	$C_2H_5COOCH(CH_3)_2$	0.893(%)	—
112°	s-Butyl acetate	$CH_3COOCH(CH_3)C_2H_5$	0.872	1.389
116°	Isobutyl acetate	$CH_3COOCH_2CH(CH_3)_2$	0.871	1.390
116°	Methyl isovalerate	$(CH_3)_2CHCH_2COOCH_3$	0.881	1.393
118°	Ethyl pivalate	$(CH_3)_3CCOOC_2H_5$	0.855	1.391
121°	Ethyl n-butyrate	$C_2H_5CH_2COOC_2H_5$	0.879	1.400
122°	n-Propyl propionate	$C_2H_5COOCH_2C_2H_5$	0.882	1.393

* Reported at various temps. 203°–210°.

B.P.°C			d_4^{20}	n_D^{20}
124°	Isoamyl formate	$HCOO(CH_2)_3CH(CH_3)_2$	0.882	1.398
126°	n-Butyl acetate	$CH_3COO(CH_2)_3CH_3$	0.881	1.394
126°	Diethyl carbonate	$C_2H_5OCOOC_2H_5$	0.976	1.384
127°	Methyl n-valerate	$CH_3(CH_2)_3COOCH_3$	0.885	1.397
128°	Isopropyl n-butyrate	$C_2H_5CH_2COOCH(CH_3)_2$	0.879(°¼)	—
133°	Ethyl isovalerate	$(CH_3)_2CHCH_2COOC_2H_5$	0.865	1.401
137°	Isobutyl propionate	$C_2H_5COOCH_2CH(CH_3)_2$	0.888(°¼)	1.398
141°	Isoamyl acetate	$CH_3COO(CH_2)_2CH(CH_3)_2$	0.872	1.400
143°	n-Propyl n-butyrate	$C_2H_5CH_2COOCH_2C_2H_5$	0.872	1.400
144°	Ethyl orthoformate	$HC(OC_2H_5)_3$	0.891	1.392
145°	Ethyl n-valerate	$CH_3(CH_2)_3COOC_2H_5$	0.874(°¼)	1.400
146°	n-Butyl propionate	$C_2H_5COO(CH_2)_3CH_3$	0.875	1.401
148°	n-Amyl acetate	$CH_3COO(CH_2)_4CH_3$	0.875	1.402
151°	Methyl n-caproate	$CH_3(CH_2)_4COOCH_3$	0.885	1.405
160°	Isoamyl propionate	$C_2H_5COO(CH_2)_2CH(CH_3)_2$	0.870	1.406
161°	Cyclohexyl formate	$HCOOC_6H_{11}$	0.994	1.443
165°	n-Butyl n-butyrate	$C_2H_5CH_2COO(CH_2)_3CH_3$	0.869	1.406
165°	n-Propyl carbonate	$C_2H_5CH_2OCOOCH_2C_2H_5$	0.943	1.400
167°	Ethyl n-caproate	$CH_3(CH_2)_4COOC_2H_5$	0.871	1.407
169°	n-Hexyl acetate	$CH_3COO(CH_2)_5CH_3$	0.872	1.409
172°	Methyl n-heptoate	$CH_3(CH_2)_5COOCH_3$	0.882	1.412
175°	Cyclohexyl acetate	$CH_3COOC_6H_{11}$	0.970	1.442
181°	Dimethyl malonate	$CH_2(COOCH_3)_2$	1.154	1.414
185°	Diethyl oxalate	$(COOC_2H_5)_2$	1.078	1.410
188°	Ethyl n-heptoate	$CH_3(CH_2)_5COOC_2H_5$	0.870	1.413
190°	Ethylene glycol diacetate	$(CH_2OCOCH_3)_2$	1.104	1.415
193°	Methyl n-caprylate	$CH_3(CH_2)_6COOCH_3$	0.878	1.417
195°	Dimethyl succinate	$(CH_2COOCH_3)_2$	1.120	1.420
196°	Phenyl acetate	$CH_3COOC_6H_5$	1.078	1.503
198°	Diethyl malonate	$CH_2(COOC_2H_5)_2$	1.055	1.416
199°	Methyl benzoate	$C_6H_5COOCH_3$	1.089	1.517
203°	Benzyl formate	$HCOOCH_2C_6H_5$	1.082	1.515
205°	n-Butyl carbonate	$C_4H_9OCOOC_4H_9$	0.925	1.412
208°	Ethyl n-caprylate	$CH_3(CH_2)_6COOC_2H_5$	0.869	1.418
208°	o-Cresyl acetate	$CH_3COOC_6H_4CH_3$	1.045	—
211°	Phenyl propionate	$C_2H_5COOC_6H_5$	1.050	—
212°	n-Propyl oxalate	$(COOCH_2C_2H_5)_2$	1.019	1.416
212°	m-Cresyl acetate	$CH_3COOC_6H_4CH_3$	1.046	1.498
212°	p-Cresyl acetate	$CH_3COOC_6H_4CH_3$	1.050	1.500
212°	Ethyl benzoate	$C_6H_5COOC_2H_5$	1.047	1.505
213°	Methyl o-toluate	$CH_3C_6H_4COOCH_3$	1.068	—
214°	Methyl pelargonate	$CH_3(CH_2)_7COOCH_3$	0.892	—
215°	Benzyl acetate	$CH_3COOCH_2C_6H_5$	1.057	1.523
215°	Methyl phenylacetate	$C_6H_5CH_2COOCH_3$	1.068	1.507
215°	Methyl m-toluate	$CH_3C_6H_4COOCH_3$	1.061	—
217°	Diethyl succinate	$(CH_2COOC_2H_5)_2$	1.042	1.420
218°	Isopropyl benzoate	$C_6H_5COOCH(CH_3)_2$	1.015	1.491
222°	Benzyl propionate	$C_2H_5COOCH_2C_6H_5$	—	—
227°	Ethyl pelargonate	$CH_3(CH_2)_7COOC_2H_5$	0.866	1.422
227°	Ethyl o-toluate	$CH_3C_6H_4COOC_2H_5$	1.034	1.508
227°	Ethyl m-toluate	$CH_3C_6H_4COOC_2H_5$	1.028	1.506
228°	Methyl n-caprate	$CH_3(CH_2)_8COOCH_3$	0.873	1.426
228°	Ethyl phenylacetate	$C_6H_5CH_2COOC_2H_5$	1.033	1.497
228°	Ethyl p-toluate	$CH_3C_6H_4COOC_2H_5$	1.025	1.507
230°	n-Propyl benzoate	$C_6H_5COOCH_2C_2H_5$	1.023	1.500
233°	Diethyl glutarate	$CH_2(CH_2COOC_2H_5)_2$	1.023	1.424
238°	Benzyl n-butyrate	$C_2H_5CH_2COOCH_2C_6H_5$	1.033(¹⁸¼)	—
240°	Guaiacol acetate	$CH_3COOC_6H_4OCH_3$	1.133	1.512

B.P.°C			d_4^{20}	n_D^{20}
242°	n-Butyl oxalate	$(COOC_4H_9)_2$	0.987	1.423
242°	Isobutyl benzoate	$C_6H_5COOCH_2CH(CH_3)_2$	0.999	—
245°	Ethyl n-caprate	$CH_3(CH_2)_8COOC_2H_5$	0.865	1.426
245°	Diethyl adipate	$(CH_2CH_2COOC_2H_5)_2$	1.009	1.428
246°	n-Propyl succinate	$(CH_2COOCH_2C_2H_5)_2$	1.006	1.425
248°	n-Butyl benzoate	$C_6H_5COO(CH_2)_3CH_3$	1.005	1.497
248°	Ethyl hydrocinnamate	$C_6H_5CH_2CH_2COOC_2H_5$	1.016	1.495
251°	Ethyl phenoxyacetate	$C_6H_5OCH_2COOC_2H_5$	1.101	—
255°	Diethyl pimelate	$CH_2(CH_2CH_2COOC_2H_5)_2$	0.993	1.430
258°	Glyceryl triacetate (Triacetin)	$C_3H_5(OCOCH_3)_3$	1.161	—
260°	Glyceryl diacetate (Diacetin)	$C_3H_5(OH)(OCOCH_3)_2$	1.180	—
262°	Methyl laurate	$CH_3(CH_2)_{10}COOCH_3$	0.870	1.432
262°	Isoamyl benzoate	$C_6H_5COO(CH_2)_2CH(CH_3)_2$	1.004	1.495
269°	Ethyl anisate	$CH_3OC_6H_4COOC_2H_5$	1.103	1.524
273°	Ethyl laurate	$CH_3(CH_2)_{10}COOC_2H_5$	0.862	1.431
278°	Resorcinol diacetate	$C_6H_4(OCOCH_3)_2$	1.180	—
282°	Diethyl suberate	$[(CH_2)_3COOC_2H_5]_2$	0.981	1.432
282°	Dimethyl phthalate	$C_6H_4(COOCH_3)_2$	1.191	1.516
291°	Diethyl azelate	$CH_2[(CH_2)_3COOC_2H_5]_2$	0.973	1.435
298°	Diethyl phthalate	$C_6H_4(COOC_2H_5)_2$	1.118	1.502
302°	Diethyl isophthalate	$C_6H_4(COOC_2H_5)_2$	1.121	1.507
306°(295°)	Ethyl myristate	$CH_3(CH_2)_{12}COOC_2H_5$	0.865	1.436
307°	Diethyl sebacate	$[(CH_2)_4COOC_2H_5]_2$	0.964	1.437
307°	o-Cresyl benzoate	$C_6H_4COOC_6H_4CH_3$	1.114	—
309°	Ethyl 1-naphthoate	$C_{10}H_7COOC_2H_5$	1.122	—
317°	Benzyl phenylacetate	$C_6H_5CH_2COOCH_2C_6H_5$	—	—
323°	Benzyl benzoate	$C_6H_5COOCH_2C_6H_5$	1.114	1.568
339°	Di n-butyl phthalate	$C_6H_4(COOC_4H_9)_2$	1.050	1.490

(Reduced pressure b.p. (temperature order)

B.P.°C			d_4^{20}	n_D^{20}
109°/21 mmHg	Dimethyl glutarate	$CH_2(CH_2COOCH_3)_2$	1.087	1.424
121°/17 mmHg	Dimethyl adipate	$(CH_2CH_2COOCH_3)_2$	1.063	1.428
128°/16 mmHg	Dimethyl pimelate	$CH_2(CH_2CH_2COOCH_3)_2$	1.038	1.431
158°/15 mmHg	Glycerol monoacetate	$C_3H_7O_2(OCOCH_3)$	1.206	1.416

SOLID

M.P.°C			B.P.°C
21°	Benzyl benzoate	See Liquids	
24°	Ethyl palmitate	$CH_3(CH_2)_{14}COOC_2H_5$	185°/10 mmHg
32°	Ethyl 2-naphthoate	$C_{10}H_7COOC_2H_5$	304°
33°	Ethyl stearate	$CH_3(CH_2)_{16}COOC_2H_5$	200°/10 mmHg
34°	Ethyl furoate	$(C_4H_3O)COOC_2H_5$	197°
34°	Methyl p-toluate	$CH_3C_6H_4COOCH_3$	217°
42°	Dibenzyl phthalate	$C_6H_4(COOCH_2C_6H_5)_2$	
44°	Diethyl terephthalate	$C_6H_4(COOC_2H_5)_2$	302°
45°	Dibenzyl succinate	$(CH_2COOCH_2C_6H_5)_2$	238°/14 mmHg
45°	Methyl anisate	$CH_3OC_6H_4COOCH_3$	255°
49°	1-Naphthyl acetate	$CH_3COOC_{10}H_7$	
53°	Dimethyl oxalate	$(COOCH_3)_2$	163°
54°	m-Cresyl benzoate	$C_6H_4COOC_6H_4CH_3$	314°
56°	1-Naphthyl benzoate	$C_6H_5COOC_{10}H_7$	
63°	Catechol diacetate	$C_6H_4(OCOCH_3)_2$	
65°	Glyceryl tripalmitate (Tripalmitin)	$C_3H_5(OCOC_{15}H_{31})_3$	
68°	Dimethyl isophthalate	$C_6H_4(COOCH_3)_2$	

M.P.°C			B.P.°C
68°	Phenyl benzoate	$C_6H_4COOC_6H_5$	299°(314°)
70°	Diphenyl phthalate	$C_6H_4(COOC_6H_5)_2$	
70°	2-Naphthyl acetate	$CH_3COOC_{10}H_7$	
71°	Glyceryl tristearate (Tristearin)	$C_3H_5(OCOC_{17}H_{35})_3$	
71°	p-Cresyl benzoate	$C_6H_5COOC_6H_4CH_3$	316°
71°	Glyceryl tribenzoate	$C_3H_5(OCOC_6H_5)_3$	
73°	Ethylene glycol dibenzoate	$(CH_2OCOC_6H_5)_2$	
77°	Methyl 2-naphthoate	$C_{10}H_7COOCH_3$	290°
78°	Diphenyl carbonate	$C_6H_5OCOOC_6H_5$	306°
80°	Dibenzyl oxalate	$(COOCH_2C_6H_5)_2$	235°/14 mmHg
83°	Benzoin acetate	$C_6H_5COCH(OCOCH_3)C_6H_5$	
84°	Catechol dibenzoate	$C_6H_4(OCOC_6H_5)_2$	
84°	Pentaerythritol tetraacetate	$C_5H_8(OCOCH_3)_4$	
90°	Pyrogallol tribenzoate	$C_6H_3(OCOC_6H_5)_3$	
95°	Pentaerythritol tetrabenzoate	$C_5H_8(OCOC_6H_5)_4$	
96°	Hydroxyhydroquinone triacetate	$C_6H_3(OCOCH_3)_3$	
104°	Phloroglucinol triacetate	$C_6H_3(OCOCH_3)_3$	
107°	2-Naphthyl benzoate	$C_6H_5COOC_{10}H_7$	
117°	Resorcinol dibenzoate	$C_6H_4(OCOC_6H_5)_2$	
120°	Hydroxyhydroquinone tribenzoate	$C_6H_3(OCOCH_3)_3$	
121°	Diphenyl succinate	$(CH_2COOC_6H_5)_2$	330°
123°	Quinol diacetate	$C_6H_4(OCOCH_3)_2$	
136°	Diphenyl oxalate	$(COOC_6H_5)_2$	191°
141°	Dimethyl terephthalate	$C_6H_4(COOCH_3)_2$	
165°	Pyrogallol triacetate	$C_6H_3(OCOCH_3)_2$	
185°	Phloroglucinol tribenzoate	$C_6H_3(OCOC_6H_5)_3$	
199°	Quinol dibenzoate	$C_6H_4(OCOC_6H_5)_2$	

UNSATURATED ESTERS

LIQUID

B.P.°C			d_4^{20}	n_D^{20}
72°	Vinyl acetate	$CH_3COOCH:CH_2$	—	—
83°	Allyl formate	$HCOOCH_2CH:CH_2$	0.946	—
85°	Methyl acrylate	$CH_2:CHCOOCH_3$	0.961	1.398
99°	Methyl methacrylate	$CH_2:C(CH_3)COOCH_3$	0.936	1.413
101°	Ethyl acrylate	$CH_2:CHCOOC_2H_5$	0.925(0/4)	1.405
104°	Allyl acetate	$CH_3COOCH_2CH:CH_2$	0.928	1.404
104°	Allyl succinate	$(CH_2COOCH_2CH:CH_2)_2$	1.051	1.452
119°	Methyl crotonate	$CH_3CH:CHCOOCH_3$	0.946	1.425
123°	Allyl propionate	$C_2H_5COOCH_2CH:CH_2$	0.914	1.410
137°	Ethyl crotonate	$CH_3CH:CHCOOC_2H_5$	0.918	1.425
142°	Allyl n-butyrate	$C_2H_5CH_2COOCH_2CH:CH_2$	0.902	1.416
204°	Dimethyl maleate	$(:CHCOOCH_3)_2$	1.150	1.442
214°	Diethyl fumarate	$(:CHCOOC_2H_5)_2$	1.052	1.441
222°	Diethyl maleate	$(:CHCOOC_2H_5)_2$	1.066	1.440
230°	Allyl benzoate	$C_6H_5COOCH_2CH:CH_2$	1.067(¼)	1.058(15/16)
271°	Ethyl cinnamate	$C_6H_5CH:CHCOOC_2H_5$	1.049	1.560

B.P.°C

Reduced pressure b.p. (*temperature order*)

102°/20 mmHg Acetylene dicarbomethoxylate (:̤CCOOCH$_3$)$_2$* Adduct m.p. 160° with anthracene
107°/13 mmHg Acetylene dicarboethoxylate (:̤CCOOC$_2$H$_5$)$_2$* Adduct m.p. 88° with 1,4-diphenyl-
butadiene

			SOLID	
M.P.°C				**B.P.°C**
36°	Methyl cinnamate	C$_6$H$_5$CH:CHCOOCH$_3$		262° dibromo m.p. 117°
39°	Benzyl cinnamate	C$_6$H$_5$CH:CHCOOCH$_2$C$_6$H$_5$		
72°	Phenyl cinnamate	C$_6$H$_5$CH:CHCOOC$_6$H$_5$		
102°	Dimethyl fumarate	(:CHCOOCH$_3$)$_2$		193°

ESTERS OF HYDROXY ACIDS

B.P.°C	LIQUID		d_4^2	n_D^{20}
145°	Methyl lactate	CH$_3$CHOHCOOCH$_3$	1.089	1.414
151°	Methyl glycollate	HOCH$_2$COOCH$_3$	1.166	—
154°	Ethyl lactate	CH$_3$CHOHCOOC$_2$H$_5$	1.030	1.415
160°	Ethyl glycollate	HOCH$_2$COOC$_2$H$_5$	1.082	—
223°	Methyl salicylate	HOC$_6$H$_4$COOCH$_3$	1.184	1.537
234°	Ethyl salicylate	HOC$_6$H$_4$COOC$_2$H$_5$	1.131	1.522
237°	Isopropyl salicylate	HOC$_6$H$_4$COOCH(CH$_3$)$_2$	1.073	1.506
240°	n-Propyl salicylate	HOC$_6$H$_4$COOCH$_2$C$_2$H$_5$	1.098	1.516
242°	Dimethyl (−)malate	COOCH$_3$CH(OH)CH$_2$-COOCH$_3$	1.233	1.442
253°	Diethyl (−)malate	COOC$_2$H$_5$CH(OH)CH$_2$-COOC$_2$H$_5$	1.129	1.436
268°	n-Butyl salicylate	HOC$_6$H$_4$COO(CH$_2$)$_3$CH$_3$	1.073	1.512
280°	Diethyl (+)tartrate	(CHOHCOOC$_2$H$_5$)$_2$	1.203	1.447
294°	Triethyl citrate	COOC$_2$H$_5$CHOH-(CH$_2$COOC$_2$H$_5$)$_2$	1.137	1.446
320°	Benzyl salicylate	HOC$_6$H$_4$COOCH$_2$C$_6$H$_5$	—	—

M.P.°C	SOLID		**B.P.°C**	Benzoyl M.P.°C
37°	Ethyl (±)mandelate	C$_6$H$_5$CHOHCOOC$_2$H$_5$	254°	73°
42°	Phenyl salicylate	HOC$_6$H$_4$COOC$_6$H$_5$	172°/12 mmHg	80°
58°	Methyl (±)mandelate	C$_6$H$_5$CHOHCOOCH$_3$	144°/20 mmHg	—
61°	Dimethyl (+)tartrate	(CHOHCOOCH$_3$)$_2$	280°	132°(di)
70°	Methyl m-hydroxy-benzoate	HOC$_6$H$_4$COOCH$_3$	—	—
73°	Ethyl m-hydroxy-benzoate	HOC$_6$H$_4$COOC$_2$H$_5$	282°	58°
95°	2-Naphthyl salicylate	HOC$_6$H$_4$COOC$_{10}$H$_7$	—	—
116°	Ethyl p-hydroxy-benzoate	HOC$_6$H$_4$COOCH$_3$	—	89°
131°	Methyl p-hydroxy-benzoate	HOC$_6$H$_4$COOCH$_3$	—	135°

* Powerful lachrymator and vesicant.

ESTERS OF KETONIC ACIDS

LIQUID

B.P.°C

130° Ethyl glyoxylate $CHOCOOC_2H_5$
Easily polymerised. Gives bright colours with NH_3 in air. Phenylhydrazone m.p. 130°.

135° Methyl pyruvate $CH_3COCOOCH_3$
Readily hydrolysed. Heat $\rightarrow CH_3COOCH_3 + CO$. Oxime m.p. 68°. Hydrazone m.p. 82°. Semicarbazone m.p. 208°. 2,4-Dinitrophenylhydrazone m.p. 187°.

155° Ethyl pyruvate $CH_3COCOOC_2H_5$
Readily hydrolysed. Heat $\rightarrow CH_3COOC_2H_5 + CO$. Oxime m.p. 94°. 2,4-Dinitrophenylhydrazone m.p. 154°. Semicarbazone m.p. 206°d.

169° Methyl acetoacetate $CH_3COCH_2COOCH_3$
Sl. sol. in H_2O, sol. in dil. NaOH. Keto-enolic. Red colour with neutral aq. $FeCl_3$ soln. NH_3/ether $\rightarrow$ methyl 2-aminocrotonate m.p. 85°. Semicarbazone m.p. 152°. *Cf.* ethyl acetoacetate.

181° Ethyl acetoacetate $CH_3COCH_2COOC_2H_5$, Acetoacetic ester.
d_4^{20} 1.028. Sl. sol. in H_2O, sol. in dil. NaOH. Keto-enolic. Steam-volatile. Red colour with neutral aq. $FeCl_3$ soln. Decolourises Br_2 water. Cold dil. alc. KOH or mineral acid $\rightarrow$ acetone, $C_2H_5OH + CO_2$. Cold conc. alc. KOH $\rightarrow$ K acetate. Forms Na deriv. with C_2H_5ONa in abs. alc. Phenylhydrazine $\rightarrow$ methyl phenylpyrazolone m.p. 127°. Warm aq. $NH_2OH \rightarrow$ methyl isoxazolone m.p. 169°d. Semicarbazone m.p. 129°d.

187° Ethyl methylacetoacetate $CH_3COCH(CH_3)COOC_2H_5$
Sol. in dil. NaOH. Keto-enolic. Violet colour with neutral aq. $FeCl_3$ soln. Dil. alkali $\rightarrow$ methyl ethyl ketone. Aq. $NH_3 \rightarrow$ crotonate deriv. m.p. 73°.

198° Ethyl ethylacetoacetate $CH_3COCH(C_2H_5)COOC_2H_5$
Sol. in dil. NaOH. Keto-enolic. Blue colour with neutral aq. $FeCl_3$ soln. Dil. alkali $\rightarrow$ methyl n-propyl ketone. Aq. $NH_3 \rightarrow$ crotonate deriv. m.p. 96°.

205° Ethyl levulinate $CH_3COCH_2CH_2COOC_2H_5$
Fairly sol. in H_2O and in dil. NaOH. Boiling dil. NaOH $\rightarrow$ Na levulinate. $NH_3 \rightarrow$ levulinamide m.p. 107°. Phenylhydrazone m.p. 103°. Semicarbazone m.p. 147°.

250° Acetone dicarbethoxylate $CO(CH_2COOC_2H_5)_2$
Spar. sol. in H_2O, sol. in dil. Na_2CO_3 or NaOH soln. Keto-enolic. Red colour with neutral aq. $FeCl_3$ soln. Aniline (cold) $\rightarrow$ ethyl 2-anilino glutaconate m.p. 97°. Phenylhydrazine $\rightarrow$ methyl phenylpyrazolone carbethoxylate m.p. 85° (free acid m.p. 134°). Semicarbazone m.p. 94°.

257° Ethyl benzoylformate $C_6H_5COCOOC_2H_5$, Ethyl phenylglyoxylate.
Hydrolysis $\rightarrow$ benzoic acid. Oxime m.p. 122°. Phenylhydrazone m.p. 156° (rapid heat).

B.P.°C

269° Ethylbenzoylacetate $C_6H_5COCH_2COOC_2H_5$
Sol. in dil. NaOH. Keto-enolic. Steam-volatile. Boiling dil. $H_2SO_4 \rightarrow$ acetophenone. Cold. conc. $H_2SO_4 \rightarrow$ diphenyl coumalin m.p. 138°. Phenylhydrazine $\rightarrow$ diphenylpyrazolone m.p. 137°.

Reduced pressure b.p. (temperature order)

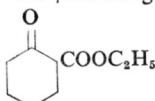

107°/11 mmHg Cyclohexanone 2-carbethoxylate $C_9H_{14}O_3$
Sol. in dil. NaOH. Keto-enolic. Hot alc. KOH $\rightarrow$ pimelic acid.

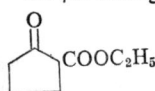

117°/26 mmHg Cyclopentanone 2-carbethoxylate $C_8H_{12}O_3$
Sol. in dil. NaOH. Keto-enolic. Strongly acidic. Acid hydrolysis $\rightarrow$ adipic acid.

LACTONES

LIQUID

B.P.°C

51°/10 mmHg β-Propiolactone $C_3H_4O_2$
Very reactive. Heat with acids or bases $\rightarrow$ linear polyester. Aq. NaCl soln. $\rightarrow$ Na 2-chloropropionate. Alc. alkali $\rightarrow$ glycollic esters.

204° γ-Butyrolactone $C_4H_6O_2$
Stable. Miscible with H_2O. Steam-volatile. Prolonged boiling in $H_2O \rightarrow$ butyric acid (partially). Alkali $\rightarrow$ salt of hydroxy acid. Cold conc. $NH_3 \rightarrow$ amide of hydroxy acid. Reduces Tollen's reagent. CrO_3 oxdtn. $\rightarrow$ succinic acid.

207° δ-Valerolactone $C_5H_8O_2$
Miscible with H_2O. More readily hydrolysed than γ-lactones. Polymerises on standing $\rightarrow$ polyester m.p. 48° (approx). Alk. hydrolysis $\rightarrow$ Na δ-hydroxyvalerate.

SOLID

M.P.°C

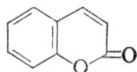

67° Coumarin $C_9H_6O_2$
Characteristic odour. Unsaturated. V. sol. in boiling H_2O. Steam volatile. Yellow soln. in KOH, coumarin repptd. by CO_2. Long boiling with conc. KOH soln. $\rightarrow$ green fluorescence. $Br_2/CS_2 \rightarrow$ dibromo m.p. 105°. Methyl sulphate and alkali $\rightarrow$ o-methoxycinnamic acid m.p. 185°. Nitration $\rightarrow$ mononitro m.p. 183°. Hydrazine $\rightarrow$ deriv. m.p. 128°.

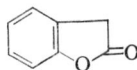

73° Phthalide $C_8H_6O_2$
B.p. 290°. Almost insol. in H_2O. Bromine or N-bromosuccinimide $\rightarrow$ bromo m.p. 85°. $KMnO_4$ oxdtn. $\rightarrow$ phthalic acid. Benzene + $AlCl_3 \rightarrow$ o-benzyl benzoic acid m.p. 144°. Boiling alkali $\rightarrow$ o-hydroxymethyl benzoic acid m.p. 120°.

M.P.°C

84°   Glycollide $C_4H_4O_4$
Sol. in hot H_2O. Heat → polyglycollide $(OCH_2CO)_x$ m.p. 223°.
NH_3 → glycolamide m.p. 120°. Warm with aniline → glycolanilide
m.p. 91°.

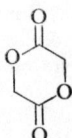

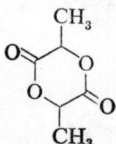

124°   Lactide $C_6H_8O_4$
B.p. 255°. Almost insol. in H_2O. Heat in H_2O → lactic acid.
NH_3 → lactamide m.p. 74°. Warm with aniline → lactanilide
m.p. 58°.

190°d   L-Ascorbic acid $C_6H_8O_6$, Vitamin C.
Easily sol. in H_2O, sl. sol. in alc., solns. unstable (especially
alkaline). $[\alpha]_D^{20}$ +23° (in H_2O). Gives rapid furfural test with
HCl. Powerful reducing agent: reduces Fehling's soln. (cold),
Tollen's reagent, neutral $KMnO_4$; decolourises Br_2 and I_2/H_2O.
No colour with Schiff's reagent. Intense violet colour with
neutral aq. $FeCl_3$ soln. Phenylhydrazine → diphenylhydrazone
m.p. 187°d. λ_{max} (in H_2O) 265 nm, $\varepsilon = 7000$; λ_{max} (in acid)
245 nm, $\varepsilon = 7500$.

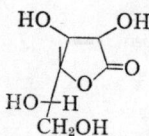

250°   Phenolphthalein. See Phenols (Solid).

CARBOHYDRATES

N.B. The m.p. of most carbohydrates can only be regarded as
approximate since variation with the rate of heating is common
as also is decomposition at or near the m.p.

M.p.

86°   D-Ribose $C_5H_{10}O_5$
$[\alpha]_D^{20}$ −23° → −24°* (add trace of NH_3; $c = 4$ in H_2O). Sol.
in H_2O, v. sl. sol. in alc. Reduces Fehling's soln. and Barfoed's
reagent. Phenylosazone m.p. 163° (rapid heat).

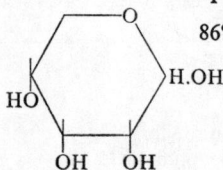

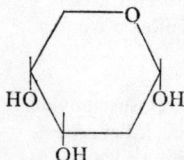

90°   2-Deoxy D-ribose $C_5H_{10}O_4$,
$[\alpha]_D^{20}$ −56° (final; $c = 1$ in H_2O). Sol. in H_2O, crystallises from
isopropyl alcohol. Reduces Fehling's soln. and Barfoed's reagent.
Gives Molisch's test. p-Nitrophenylhydrazone m.p. 160°.

* The rotation rises and then falls to the equilibrium value (−24°).

M.P.°C

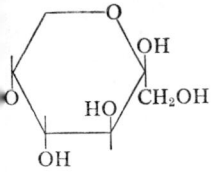

102° D-Fructose $C_6H_{12}O_6$, Laevulose. Fruit sugar.

$[\alpha]_D^{20}$ $-92°$ (30 mins. after dissolving; $c = 4$ in H_2O). Extremely sol. in H_2O, fairly sol. in alc. Reduces Fehling's soln. (cold) and Barfoed's reagent (warm). Red colour on heating with resorcinol + conc. HCl (Seliwanoff test). Blue colour on warming with NH_4 molybdate (specific). Not oxidised by cold Br_2/H_2O. Gives rapid furfural test–boiling dil. HCl → furfural (vapours redden aniline acetate paper). Phenylosazone m.p. 204° (rapid heat; separates in 2–5 mins.). p-Nitrophenylhydrazone m.p. 180°. Pentaacetate, α form m.p. 70°, β form m.p. 108°.

102° Maltose $C_{12}H_{22}O_{11}$ $(+H_2O)$

Monohydrate. Anhydrous m.p. 160–165°. $[\alpha]_D^{20}$ $+111°$ → $+130°$ ($c = 4$, hydrate in H_2O). Sol. in H_2O, v. sl. sol. in alc. Reduces Fehling's soln. but not Barfoed's reagent. Warm v. dil. HCl → soln. of $[\alpha]_D^{20}$ $+53°$. Phenylosazone m.p. 206° (rapid heat; formed after 1 hour warming). Octaacetate α form m.p. 125°, β form m.p. 159°.

118° Raffinose $C_{18}H_{32}O_{16}$ $(+5H_2O)$

Anhydrous. Pentahydrate (from dil. alc.) m.p. 78°. $[\alpha]_D^{20}$ $+105°$ (no mutarotn., hydrate in H_2O). Fairly sol. in H_2O, insol. in alc. Does not reduce Fehling's soln. or react with phenylhydrazine. Gives ppt. with Pb acetate, also with sat'd. $Ca(OH)_2$ soln. Hydrolysis → galactose, glucose, fructose. Hendecaacetate m.p. 99°.

132° D-Mannose $C_6H_{12}O_6$

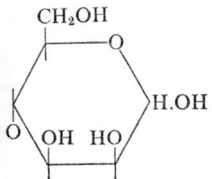

β-Form: $[\alpha]_D^{20}$ $-17°$ → $+14°$ (add trace of NH_3; $c = 4$ in H_2O). Extremely sol. in H_2O, spar. sol. in alc. Reduces Fehling's soln. and Barfoed's reagent. Phenylhydrazone m.p. 199° (rapid heat; colourless, insol. ppt. produced almost immediately). Pentaacetate α form m.p. 74°(64°), β form m.p. 117°. Mannose. $CaCl_2 \cdot 4H_2O$ m.p. 101°.

145° D-Xylose $C_5H_{10}O_5$, Wood sugar.

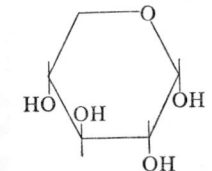

$[\alpha]_D^{20}$ $+93°$ → $+18°$ (add trace of NH_3; $c = 4$ in H_2O). Very sol. in H_2O, sol. in alc. Reduces Fehling's soln. and Barfoed's reagent. Gives rapid furfural test. Red colour on warming with phloroglucinol + dil. HCl. Phenylosazone m.p. 163° (rapid heat; separates in 10–15 mins.). Tetraacetate α form m.p. 59°, β-form m.p. 126°.

146° D-Glucose $C_6H_{12}O_6$, Dextrose. Grape sugar.

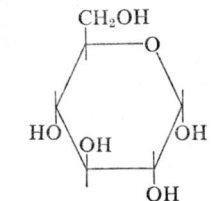

Usually α form. Anhydrous. Monohydrate m.p. 85°. $[\alpha]_D^{20}$ $+112°$ → $+52.8°$ (add trace of NH_3; $c = 3.9$ anhydrous in H_2O). Easily sol. in H_2O, sl. sol. in alc. Reduces Fehling's soln. and Barfoed's reagent. HNO_3 oxdtn. → saccharic acid. Phenylosazone m.p. 204° (rapid heat; separates in 5–10 mins.). Pentaacetate α form m.p. 114°, $[\alpha]_D^{20}$ $+102°$ (CHCl$_3$ soln.); β form m.p. 135°, $[\alpha]_D^{20}$ $+4°$ (CHCl$_3$ soln.).

M.P.°C

159° L-Arabinose $C_5H_{10}O_5$

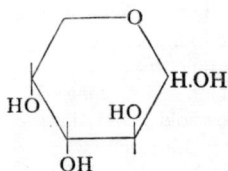

$[\alpha]_D^{20}$ +180° (approx.) → +105° (add trace of NH_3; $c = 1$ in H_2O). Sol. in H_2O, very sl. sol. in alc. Reduces Fehling's soln. and Barfoed's reagent. Gives rapid furfural test. Red colour on warming with phloroglucinol + dil. HCl. Phenylosazone m.p. 163° (rapid heat; appears as a turbidity after 15–30 mins.). Tetraacetate α form m.p. 96°, β form m.p. 86°.

168° D-Galactose $C_6H_{12}O_6$

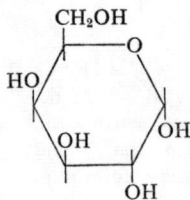

Anhydrous. Monohydrate m.p. 119°. $[\alpha]_D^{20}$ +150° → +80° (add trace of NH_3; $c = 4$ in H_2O). Readily sol. in hot H_2O, crystallises as hydrate on cooling. Almost insol. in cold alc. Reduces Fehling's soln. and Barfoed's reagent. Phenylosazone m.p. 196° (rapid heat; separates in 20–30 mins.). Pentaacetate α form m.p. 96°, β form m.p. 142°.

169° Sucrose $C_{12}H_{22}O_{11}$, Cane sugar.

B form crystallises from methyl alcohol. A form m.p. 185° crystallises from alcohol and most solvents. $[\alpha]_D^{20}$ +66.5° (no mutarotn., $c = 26$ in H_2O). Extremely sol. in H_2O, sl. sol. in alc., insol. ether, benzene, chloroform. Non-reducing sugar. No reaction with Fehling's soln. or phenylhydrazine. Forms blue soln. with $CuSO_4$ + NaOH. Violet colour on warming with $Co(NO_3)_2$ + NaOH. HNO_3 oxdtn. → saccharic acid. Warming with very dil. HCl → invert sugar (mixture of glucose + fructose) $[\alpha]_D^{20}$ −37°, which reduces Fehling's soln. and reacts with phenylhydrazine. Octaacetate m.p. 69°, $[\alpha]_D^{20}$ +60° ($CHCl_3$ soln.).

202° Lactose $C_{12}H_{22}O_{11}$ (+H_2O), Milk sugar.

Monohydrate. Anhydrous m.p. 223°. Loses H_2O at 130° (approx.) and turns yellow at 160° (approx.). $[\alpha]_D^{20}$ +85° → +52.5° (add trace of NH_3; $c = 7.6$, hydrate in H_2O). Readily sol. in hot H_2O, insol. in alc. Reduces Fehling's soln. but not Barfoed's reagent. Warm dil. HCl → glucose + galactose. Phenylosazone m.p. 200° (rapid heat; separates in 90 mins.). Octaacetate α form m.p. 152°, β form m.p. 100° (90°).

d. Inulin $(C_6H_{10}O_5)_n$ + H_2O

Mol. Wt. = 5,000 (approx.). $[\alpha]_D^{15}$ −40°. Sol. in warm H_2O, separating very slowly on cooling, more readily on addtn. of alc. Does not reduce Fehling's soln. No colour with iodine. Boiling HCl (or better, oxalic acid) → fructose.

d. Glycogen $(C_6H_{10}O_5)_n$ + H_2O, Animal starch.

Mol. Wt. = 1 to 2 × 10^6. $[\alpha]_D^{20}$ +196° (approx.) in H_2O. Very sol. in H_2O giving opalescent soln., separates on adding alc. Does not reduce Fehling's soln., weakly reduces Barfoed's reagent. Red-brown colour with iodine, fading at 60° and returning on cooling. Aq. soln. gives ppt. with basic Pb acetate soln. Boiling v. dil. HCl → glucose.

M.P.°C
    d.   Starch $(C_6H_{10}O_5)_n + H_2O$
       Mol. Wt. depends on source. Mixture of amylose and amylopectin. Insol. in cold H_2O, colloidal soln. in boiling H_2O, does not separate on cooling. Blue colour with iodine, stable on heating. Gives ppt. with tannin and with basic Pb acetate soln. Boiling dil. HCl → glucose.
    d.   Cellulose $(C_6H_{10}O_5)_n + H_2O$
       Mol. Wt. = 2 to 4 × 10^4. Insol. in H_2O, alc., ether, chloroform. Sol. in ammoniacal Cu soln., and in $ZnCl_2$/conc. HCl soln.

FLUORO AROMATIC HYDROCARBONS

LIQUID

B.P.°C
   82°   *m*-Difluorobenzene $C_6H_4F_2$
   85°   Pentafluorobenzene C_6HF_5
   86°   Fluorobenzene C_6H_5F
       d_4^{20} 1.024. Sulphonamide m.p. 125°. 2,4-Dinitro m.p. 25°.
   86°   Hexafluorobenzene C_6F_6
       d_4^{25} 1.612.
   88°   *p*-Difluorobenzene $C_6H_4F_2$
   92°   *o*-Difluorobenzene $C_6H_4F_2$
  100°   Trifluoromethylbenzene $C_6H_5CF_3$
  114°   *o*-Fluorotoluene $CH_3C_6H_4F$
       d_4^{13} 1.004. Oxdtn. → *o*-fluorobenzoic acid m.p. 127°. Sulphonamide m.p. 105°.
  116°   *m*-Fluorotoluene $CH_3C_6H_4F$
       d_4^{13} 0.997. Oxdtn. → *m*-fluorobenzoic acid m.p. 124°. Sulphonamide m.p. 174°.
  117°   *p*-Fluorotoluene $CH_3C_6H_4F$
       d_4^{15} 1.001. Oxdtn. → *p*-fluorobenzoic acid m.p. 186°. Sulphonamide m.p. 141°.
  117°   Pentafluorotoluene $CH_3C_6F_5$
  214°   1-Fluoronaphthalene $C_{10}H_7F$
       d_4^{20} 1.134. Picrate m.p. 113°.

SOLID

M.P.°C
   60°   2-Fluoronaphthalene $C_{10}H_7F$
       Picrate m.p. 101°.

CHLORO HYDROCARBONS

GAS

Methyl chloride CH_3Cl
Alkyl 2-naphthyl ether m.p. 72°. SATP* m.p. 224°.

* *S*-Alkyl thiourea picrate derivative.

LIQUID

B.P.°C

12° Ethyl chloride C_2H_5Cl
SATP m.p. 188°.

35° Isopropyl chloride $(CH_3)_2CHCl$
d_4^{20} 0.859. SATP m.p. 196° (formed on prolonged heating).

40° Methylene dichloride CH_2Cl_2, Dichloromethane.
d_4^{20} 1.336. Boiling $NaOH/CH_3OH \rightarrow$ formaldehyde. Alkyl 2-naphthyl ether (di) m.p. 133°.

45° Allyl chloride $CH_2:CHCH_2Cl$
d_4^{20} 0.938. Unsaturated. $Cl_2 \rightarrow$ glycerol trichlorohydrin b.p. 156°. Dibromo b.p. 195°. SATP m.p. 154°.

46° n-Propyl chloride $C_2H_5CH_2Cl$
d_4^{20} 0.890. SATP m.p. 177° (small yield).

48° *Trans*-dichloroethylene CHCl:CHCl
d_4^{20} 1.256. n_D^{20} 1.446. Unsaturated. Dibromo m.p. 192°.

51° t-Butyl chloride $(CH_3)_3CCl$
d_4^{20} 0.846. Hydrolysed by cold $H_2O \rightarrow$ t-butyl alcohol. SATP m.p. 151°.

57° Ethylidene dichloride CH_3CHCl_2, 1,1-Dichloroethane.
d_4^{20} 1.176. Reduces Fehling's soln. (diff. from ethylene dichloride). Hydrolysis $\rightarrow CH_3CHO$.

59° Chloroprene $CH_2:CHCCl:CH_2$, 2-chlorobuta-1,3-diene.
d_4^{20} 0.958. Maleic anhydride adduct, with boiling $H_2O \rightarrow$ adduct (acid) m.p. 173°.

60° *Cis*-dichloroethylene CHCl:CHCl
d_4^{20} 1.284. n_D^{20} 1.449. Unsaturated. Dibromo m.p. 192°.

61° Chloroform $CHCl_3$
d_4^{20} 1.489. n_D^{20} 1.446. Characteristic odour. Warm with aniline + alc. $KOH \rightarrow$ carbylamine C_6H_5NC (foul odour; destroy with HCl). Warm with resorcinol + dil. $NaOH \rightarrow$ red colour. Reduces Fehling's soln. on heating (diff. from CCl_4).

67° s-Butyl chloride $C_2H_5CHClCH_3$
Racemate. d_4^{20} 0.873. SATP m.p. 166° (190°).

68° Isobutyl chloride $(CH_3)_2CHCH_2Cl$
d_4^{20} 0.877. SATP m.p. 167° (174°).

77° n-Butyl chloride $CH_3(CH_2)_3Cl$
d_4^{20} 0.885. SATP m.p. 177°.

77° Carbon tetrachloride CCl_4
d_4^{20} 1.594. n_D^{20} 1.463. Gives carbylamine test more slowly than $CHCl_3$. Boil (30 secs.) with 1-naphthol + NaOH pellet in cyclohexanol, cool $\rightarrow$ blue colour.

84° Ethylene dichloride CH_2ClCH_2Cl, 1,2-Dichloroethane.
d_4^{20} 1.253. Does not reduce Fehling's soln. (diff. from ethylidene dichloride). Alkyl 2-naphthyl ether (di) m.p. 217°. SATP m.p. 260°.

86° t-Amyl chloride $(CH_3)_2CClC_2H_5$
d_4^{20} 0.865. Hydrolysed by cold $H_2O \rightarrow$ t-amyl alcohol. With boiling H_2O rapidly loses $HCl \rightarrow$ olefines.

B.P.°C

87° Trichloroethylene   CHCl:CCl₂
d_4^{20} 1.466. Warm with pyridine + aq. alkali → red colour, changing to orange (Fujiwara reaction). Adds Br₂ readily. Alc. 1-naphthol + conc. H₂SO₄ → red colour, on dilution with H₂O.

97° Propylene dichloride   CH₃CHClCH₂Cl, 1,2-Dichloropropane.
d_4^{20} 1.156. Alkyl 2-naphthyl ether (di) m.p. 152°. SATP m.p. 232°.

100° Isoamyl chloride   (CH₃)₂CHCH₂CH₂Cl
d_4^{20} 0.872. SATP m.p. 173°.

107° n-Amyl chloride   CH₃(CH₂)₄Cl
d_4^{20} 0.882. SATP m.p. 154°.

114° 1,1,2-Trichloroethane   CH₂ClCHCl₂
d_4^{20} 1.441. No colour in Fujiwara reaction (see above).

120° Trimethylene dichloride   CH₂ClCH₂CH₂Cl, 1,3-Dichloropropane.
d_4^{20} 1.180. Alc. KOH → allyl chloride. Alkyl-2-naphthyl ether (di) m.p. 148°.

121° Tetrachloroethylene   CCl₂:CCl₂
d_4^{20} 1.623. Adds Cl₂ or Br₂ in sunlight.

132° Chlorobenzene   C₆H₅Cl
d_4^{20} 1.106. Conc. HNO₃ + H₂SO₄(90°) → 2,4-dinitro m.p. 52°. Add to large excess ClSO₃H → p-sulphonylchloride, +NH₃ → sulphonamide m.p. 143°.

142° Chlorocyclohexane   C₆H₁₁Cl, Cyclohexyl chloride.
d_4^{20} 0.989. Boiling alc. KOH → cyclohexene. Alkyl 2-naphthyl ether m.p. 116°.

147° Acetylene tetrachloride   CHCl₂CHCl₂, 1,1,2,2-Tetrachloroethane.
d_4^{20} 1.598. Alc. KOH → trichloroethylene. Alc. + Zn dust → acetylene.

155° 1,4-Dichlorobutane   Cl(CH₂)₄Cl
d_4^{20} 1.159. Slowly hydrolysed to tetramethylene glycol.

156° Glycerol trichlorohydrin   CH₂ClCHClCH₂Cl, 1,2,3-Trichloropropane.

159° o-Chlorotoluene   CH₃C₆H₄Cl
d_4^{20} 1.082. Alk. KMnO₄ oxdtn. → o-chlorobenzoic acid m.p. 140°. Sulphonamide m.p. 126°.

162° Pentachloroethane   CCl₃CHCl₂
d_4^{20} 1.680. Boiling alc. KOH → tetrachloroethylene.

162° m-Chlorotoluene   CH₃C₆H₄Cl
d_4^{20}1.072. Alk. KMnO₄ oxdtn. → m-chlorobenzoic acid m.p. 158° Sulphonamide m.p. 185°.

162° p-Chlorotoluene   CH₃C₆H₄Cl
M.p. 7°. d_4^{20} 1.070. Alk. KMnO₄ oxdtn. → p-chlorobenzoic acid m.p. 240°. Sulphonamide m.p. 143°.

172° m-Dichlorobenzene   C₆H₄Cl₂
d_4^{20} 1.288. Sulphonamide has same m.p. 180°as p-dichlorobenzene (distinguish by mixed m.p. or m.ps. of sulphonyl chlorides, m–53°, p–38°).

B.P.°C

179° Benzyl chloride $C_6H_5CH_2Cl$
d_4^{20} 1.100. Lachrymator and vesicant. Boiling $Pb(NO_3)_2$ soln. →
C_6H_5CHO. $K_2Cr_2O_7/H_2SO_4$ oxdtn. → benzoic acid. Thiourea →
S-benzyl thiuronium chloride m.p. 150°, giving picrate m.p. 188°
(SATP).

179° o-Dichlorobenzene $C_6H_4Cl_2$
d_4^{20} 1.306. Sulphonamide m.p. 135°.

190° 2-Phenylethyl chloride $C_6H_5CH_2CH_2Cl$
Sl. decomp. at b.p. d_4^{25} 1.069. CrO_3/H_2SO_4 oxdtn. → benzoic acid
(more slowly than benzyl chloride).

195°d 1-Phenylethyl chloride $C_6H_5CHClCH_3$
Racemate. Tends to lose HCl, even on standing → styrene.
Boiling Na_2CO_3 soln. → phenylmethyl carbinol. $Cu(NO_3)_2$
oxdtn. → acetophenone.

205° Benzal chloride $C_6H_5CHCl_2$
Boiling Na_2CO_3 soln. → C_6H_5CHO. Excess phenylhydrazine
(reflux in alc.) → benzaldehyde phenylhydrazone m.p. 156°.
Oxdtn. → benzoic acid.

215° Hexachlorobuta-1,3-diene $Cl_2C:CCl\cdot CCl:CCl_2$
d_4^{20} 1.682. Stable to acids and alkalis. Does not add halogens or
maleic anhydride. Mulling agent in infrared spectroscopy
(detection of C—H stretching vibrations).

220° Benzotrichloride $C_6H_5CCl_3$
Warm with dimethylaniline + $ZnCl_2$ → intense green colour
(Malachite Green). Boiling Na_2CO_3 soln. → benzoic acid.
Boiling alc. + $ZnCl_2$ → ethyl benzoate.

259° 1-Chloronaphthalene $C_{10}H_7Cl$
d_4^{20} 1.194. Picrate m.p. 137°. Sulphonamide m.p. 186°.

SOLID

M.P.°C

32° 1-Chloromethyl naphthalene $C_{10}H_7CH_2Cl$
Lachrymator and vesicant. Picrate m.p. 82°. Thiourea →
thiuronium chloride m.p. 238°.

53° p-Dichlorobenzene $C_6H_4Cl_2$
B.p. 173°. Anti-moth agent and fumigant. Sublimes. Steam-
volatile. Sulphonamide m.p. 180°.

56° 2-Chloronaphthalene $C_{10}H_7Cl$
(60°) B.p. 265°. Picrate m.p. 81°. Sulphonamide m.p. 232°.

77° p-Chlorobiphenyl $C_6H_5C_6H_4Cl$
B.p. 291°. $CrO_3/AcOH$ oxdtn. → p-chlorobenzoic acid m.p. 240°.

99° p-Xylylene dichloride $ClCH_2C_6H_4CH_2Cl$
Lachrymator and vesicant. Boiling $Pb(NO_3)_2$ soln. → terephthal-
aldehyde m.p. 115°.

M.P.°C

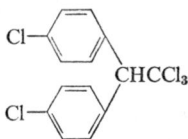

108° 1,1,1-Trichloro-2,2-bis (*p*-chlorophenyl) ethane, DDT.
B.p. 260°. Powerful insecticide. Benzene + AlCl₃ → *sym*-tetraphenylethane m.p. 211°. Nitro m.p. 148°.

112° γ-Benzene hexachloride $C_6H_6Cl_6$, Gammexane.
Powerful insecticide. One of 9 stereoisomers.

113° Triphenylmethyl chloride $(C_6H_5)_3CCl$, Trityl chloride.
Cold conc. H_2SO_4 → yellow soln. Boiling H_2O (5 mins.) → triphenylcarbinol m.p. 162°. Ethanol → trityl ethyl ether m.p. 83°. Anhydrous HCOOH (reflux) → triphenylmethane m.p. 93°.

185° Hexachloroethane CCl_3CCl_3
Sublimes. Camphor-like odour. Zn/dil. H_2SO_4 → tetrachloroethylene (slowly).

226° Hexachlorobenzene C_6Cl_6
B.p. 326°. Boiling fuming HNO_3 → chloranil m.p. 290°.

BROMO HYDROCARBONS

LIQUID

B.P.°C

4°(gas) Methyl bromide CH_3Br

38° Ethyl bromide C_2H_5Br
d_4^{20} 1.460. *cf.* Chloro Hydrocarbons for derivatives *et seq.*

60° Isopropyl bromide $(CH_3)_2CHBr$
d_4^{20} 1.314. SATP m.p. 196°.

70° Allyl bromide $CH_2:CHCH_2Br$
d_4^{20} 1.398. Unsaturated. Br_2/CCl_4 → glycerol tribromohydrin b.p. 219°.

71° n-Propyl bromide $C_2H_5CH_2Br$
d_4^{20} 1.353.

72° t-Butyl bromide $(CH_3)_3CBr$
d_4^{20} 1.222. Hydrolysed by cold H_2O → t-butyl alcohol.

90° s-Butyl bromide $C_2H_5CHBrCH_3$
Racemate. d_4^{20} 1.258.

91° Isobutyl bromide $(CH_3)_2CHCH_2Br$
d_4^{20} 1.264.

98° Methylene dibromide CH_2Br_2, Dibromomethane.
d_4^{20} 2.495. Alkyl 2-naphthyl ether (di) m.p. 133°.

100° n-Butyl bromide $CH_3(CH_2)_3Br$
d_4^{20} 1.299.

108° *Trans*-dibromoethylene CHBr:CHBr

110° *Cis*-dibromoethylene CHBr:CHBr
d_4^{20} 2.271. n_D^{20} 1.5428.

131° Ethylene dibromide CH_2BrCH_2Br, 1,2-Dibromoethane.
M.p. 10°. d_4^{20} 2.179.

B.P.°C

142° Propylene dibromide $CH_3CHBrCH_2Br$, 1,2-Dibromopropane. d_4^{20} 1.933.

150° Bromoform $CHBr_3$
M.p. 9°. d_4^{20} 2.890. Gives the carbylamine reaction.

156° Bromobenzene C_6H_5Br
d_4^{20} 1.494. Conc. $HNO_3 + H_2SO_4$ (90°) → 2,4-dinitro m.p. 75°. Add to excess $ClSO_3H$ → p-sulphonyl chloride m.p. 75°, which with NH_3 → sulphonamide m.p. 160°.

165° Bromocyclohexane $C_6H_{11}Br$, Cyclohexyl bromide. d_4^{20} 1.336.

167° Trimethylene dibromide $CH_2BrCH_2CH_2Br$, 1,3-Dibromopropane.
d_4^{20} 1.979. Alkyl 2-naphthyl ether (di) m.p. 148°.

181° o-Bromotoluene $CH_3C_6H_4Br$
d_4^{20} 1.222. Alk. $KMnO_4$ or dil. HNO_3 oxdtn. → o-bromobenzoic acid m.p. 150°. Br_2 (2 moles) → 2,4,5-tribromotoluene m.p. 117°. Sulphonamide m.p. 168°.

197° 1,4-Dibromobutane $Br(CH_2)_4Br$

198°d Benzyl bromide $C_6H_5CH_2Br$
d_4^{20} 1.438. Very irritating odour.

200°d sym-Tetrabromoethane $CHBr_2CHBr_2$

205° 1-Phenylethyl bromide $C_6H_5CHBrCH_3$

218° 2-Phenylethyl bromide $C_6H_5CH_2CH_2Br$

219° Glycerol tribromohydrin $CH_2BrCHBrCH_2Br$, 1,2,3-Tribromopropane.
M.p. 16°. B.p. 92°/10 mmHg. d_4^{20} 2.402.

219° m-Dibromobenzene $C_6H_4Br_2$
d_4^{20} 1.952. Excess $ClSO_3H$ → sulphonyl chloride m.p. 79°, which with NH_3 → sulphonamide m.p. 190°.

220°d β-Bromostyrene $C_6H_5CH:CHBr$
M.p. 7°. Not attacked by boiling alc. KOH or Zn dust. Adds Br_2.

224° o-Dibromobenzene $C_6H_4Br_2$
M.p. 7°. d_4^{20} 1.956. Warm conc. $HNO_3 + H_2SO_4$ → dinitro m.p. 114°. Excess $ClSO_3H$ → sulphonyl chloride, which with NH_3 → sulphonamide m.p. 175°.

280° 1-Bromonaphthalene $C_{10}H_7Br$
M.p. 4°. d_4^{20} 1.488. Picrate m.p. 134°. Conc. HNO_3 → 4-nitro m.p. 85°. Sulphonamide m.p. 192°. CrO_3/gl. AcOH → phthalic acid.

SOLID

M.P.°C

27° p-Bromotoluene $CH_3C_6H_4Br$
B.p. 185°. Oxdtn. → p-bromobenzoic acid m.p. 251°. Sulphonamide m.p. 165°.

59° 2-Bromonaphthalene $C_{10}H_7Br$
B.p. 281°. Sulphonamide m.p. 208°. Picrate m.p. 79° (86°).

M.P.°C

63° *p*-Bromobenzyl bromide $BrC_6H_4CH_2Br$
Oxdtn. → *p*-bromobenzoic acid m.p. 251°.

65° 9-Bromophenanthrene $C_{14}H_9Br$

89° *p*-Dibromobenzene $C_6H_4Br_2$
B.p. 219°. Sulphonamide m.p. 195°. Nitro m.p. 84°.

89° *p*-Bromobiphenyl $C_6H_5C_6H_4Br$
B.p. 310°. CrO_3 oxdtn. → *p*-bromobenzoic acid m.p. 251°.

92° Carbon tetrabromide CBr_4
B.p. 189°d.

94° *o*-Xylylene dibromide $C_6H_4(CH_2Br)_2$
Lachrymator and vesicant.

120° 1,3,5-Tribromobenzene $C_6H_3Br_3$
B.p. 271°. Sulphonamide m.p. 222°d.

145° *p*-Xylylene dibromide $BrCH_2C_6H_4CH_2Br$

163° Pentaerythrityl tetrabromide $C(CH_2Br)_4$

IODO HYDROCARBONS

LIQUID

B.P.°C

43° Methyl iodide CH_3I
d_4^{20} 2.282. Cold alc. $AgNO_3$ → ppt. (AgI). Quaternary salt with pyridine m.p. 117°; with isoquinoline m.p. 159°; with quinaldine m.p. 194°.

72° Ethyl iodide C_2H_5I
d_4^{20} 1.933. Cold. alc. $AgNO_3$ → ppt. Quaternary salt with pyridine m.p. 90°; with quinoline m.p. 159°; with dimethylaniline m.p. 136°.

89° Isopropyl iodide $(CH_3)_2CHI$
d_4^{20} 1.703. Quaternary salt with pyridine m.p. 114°; with dimethylaniline m.p. 163° (slowly).

98° t-Butyl iodide $(CH_3)_3CI$
d_4^0 1.571. Readily hydrolysed by cold H_2O. Usually contains free iodine. Boiling K_2CO_3 soln. → isobutylene + KI.

101° Allyl iodide $CH_2{:}CHCH_2I$
d_4^{20} 1.777. Cold alc. $AgNO_3$ → ppt. Quaternary salt with quinoline m.p. 177°; with dimethylaniline m.p. 86°. Br_2 → glycerol tribromohydrin b.p. 219° + I_2.

102° n-Propyl iodide $C_2H_5CH_2I$
d_4^{20} 1.743. Quaternary salt with quinoline m.p. 145°; with dimethylaniline m.p. 68° (slowly).

119° s-Butyl iodide $C_2H_5CHICH_3$
Racemate. d_4^{20} 1.592.

120° Isobutyl iodide $(CH_3)_2CHCH_2I$
d_4^{20} 1.605. Quaternary salt with dimethylaniline m.p. 155° (very slowly).

B.P.°C

130° n-Butyl iodide $CH_3(CH_2)_3I$
d_4^{20} 1.616. Quaternary salt with quinoline m.p. 174°.

180° Methylene iodide CH_2I_2, Diiodomethane.
B.p. 80°/25 mmHg. m.p. 6°. d_4^{20} 3.325. Quaternary salt with pyridine m.p. 220°(d); with quinoline m.p. 132°.

188° Iodobenzene C_6H_5I
d_4^{20} 1.831. Na/Hg in moist ether → benzene. Warm conc. HNO_3 + H_2SO_4 → p-nitro m.p. 171°. Br_2 → p-bromo m.p. 92°. Cl_2/$CHCl_3$ → dichloro $C_6H_5ICl_2$ (yellow needles, decomp. 80°).

204° m-Iodotoluene $CH_3C_6H_4I$
d_4^{20} 1.698. Oxdtn. → m-iodobenzoic acid m.p. 186°. Cl_2/$CHCl_3$ → dichloro ($RICl_2$) decomp. 88°. Nitro m.p. 108°.

211° o-Iodotoluene $CH_3C_6H_4I$
d_4^{20} 1.697. Oxdtn. → o-iodobenzoic acid m.p. 162°. Cl_2/$CHCl_3$ → dichloro ($RICl_2$) decomp. 85°. Cold fuming HNO_3 → nitro m.p. 103°.

305° 1-Iodonaphthalene $C_{10}H_7I$
Picrate m.p. 128°

SOLID

M.P.°C

24° Benzyl iodide $C_6H_5CH_2I$
Very irritating odour. Decomp. on heating. Quaternary salt with dimethylaniline m.p. 165°.

35° p-Iodotoluene $CH_3C_6H_4I$
B.p. 211°. CrO_3 oxdtn. → p-iodobenzoic acid m.p. 265°. Heat with Cu powder (210°–260°) → di p-tolyl m.p.121° Cl_2/$CHCl_3$ → dichloro, 2 forms–85°d and 110°d.

55° 2-Iodonaphthalene $C_{10}H_7I$
B.p. 309°. Picrate m.p. 95°.

81° Ethylene iodide CH_2ICH_2I, 1,2-Diiodoethane.
Br_2 → ethylene bromide.

113° p-Iodobiphenyl $C_6H_5C_6H_4I$
Cl_2/$CHCl_3$ → dichloro ($RICl_2$) m.p. 102°.

119° Iodoform CHI_3
Yellow. Characteristic odour. Warm with phenol in dil. alc. NaOH → red colour. Quinoline/dry ether → adduct m.p. 65°.

MIXED HALOGENO HYDROCARBONS

LIQUID

B.P.°C

67° Bromochloromethane CH_2BrCl

106° 1-Bromo 2-chloroethane CH_2BrCH_2Cl
d_4^{20} 1.689. Alkyl 2-naphthyl ether (di) m.p. 217°.

130° p-Chlorofluorobenzene ClC_6H_4F

136° o-Chlorofluorobenzene ClC_6H_4F

150° p-Bromofluorobenzene BrC_6H_4F

B.P.°C
 156° *o*-Bromofluorobenzene BrC_6H_4F
 195° *p*-Bromochlorobenzene BrC_6H_4Cl, See Solid.
 197° *m*-Bromochlorobenzene BrC_6H_4Cl
 201° *o*-Bromochlorobenzene BrC_6H_4Cl
 251° *p*-Bromoiodobenzene BrC_6H_4I, See Solid.
 257° *o*-Bromoiodobenzene BrC_6H_4I

SOLID

M.P.°C
  26° 1-Bromo 2,4-dichlorobenzene $BrC_6H_3Cl_2$
  56° *p*-Chloroiodobenzene ClC_6H_4I
  59° 1-Bromo 2,3-dichlorobenzene $BrC_6H_3Cl_2$
  67° *p*-Bromochlorobenzene BrC_6H_4Cl
       B.p. 195°. Nitro m.p. 72°.
  92° *p*-Bromoiodobenzene BrC_6H_4I
       B.p. 251°.

ALIPHATIC HALOGEN SUBSTITUTED ETHERS

LIQUID

B.P.°C
  59° Chloromethyl ether CH_3OCH_2Cl
       Cold H_2O → methyl alc. + formaldehyde. NH_4OH → hexamethylene tetramine. Picrate m.p. 163°.
  80° Chloromethyl ethyl ether $C_2H_5OCH_2Cl$
       Cold H_2O → ethyl alc. + formaldehyde. NH_4OH → hexamethylene tetramine.
  98° α-Chloroethyl ether $CH_3CHClOC_2H_5$
       Cold H_2O → ethyl alc. + acetaldehyde.
 105° αα'-Dichloromethyl ether $(CH_2Cl)_2O$
       Cold H_2O → formaldehyde. NH_4OH → hexamethylene tetramine.
 107° β-Chloroethyl ether $C_2H_5OCH_2CH_2Cl$
 116° αα'-Dichloroethyl ether $(CH_3CHCl)_2O$
       Cold H_2O → acetaldehyde (slowly).
 117° Epichlorhydrin $CH_2Cl(CHOCH_2)$, 1-Chloro 2,3-epoxypropane
       d_4^{20} 1.181. Chloroform-like odour. Insol. in H_2O, miscible with alc. Phenol + NaOH in boiling alc. → glycerol diphenyl ether m.p. 81°.

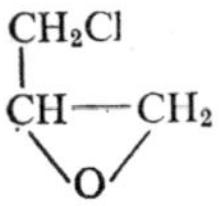

CH_2Cl
 |
$CH{-}{-}CH_2$
  \\O/

 127° β-Bromoethyl ether $C_2H_5OCH_2CH_2Br$
 150° Dibromomethyl ether $(CH_2Br)_2O$
       Cold H_2O → formaldehyde. NH_4OH → hexamethylene tetramine.
 177° ββ'-Dichloroethyl ether $(CH_2CH_2Cl)_2O$
       Heat with excess aniline + NaOH → *N*-phenyl morpholine m.p. 57°.

AROMATIC HALOGEN SUBSTITUTED ETHERS

LIQUID

B.P.°C

157° *p*-Fluoroanisole $CH_3OC_6H_4F$

194° *m*-Chloroanisole $CH_3OC_6H_4Cl$
Sulphonamide m.p. 131°.

195° *o*-Chloroanisole $CH_3OC_6H_4Cl$
Fuming HNO_3 → nitro m.p. 95°. Sulphonamide m.p. 131°.

200° *p*-Chloroanisole $CH_3OC_6H_4Cl$
Fuming HNO_3 → nitro m.p. 98°. Sulphonamide m.p. 151°.

208° *o*-Chlorophenetole $C_2H_5OC_6H_4Cl$
M.p. 17°. Fuming HNO_3 → nitro m.p. 82°. Sulphonamide m.p. 133°.

212° *p*-Chlorophenetole $C_2H_5OC_6H_4Cl$
M.p. 21°. Sulphonamide m.p. 134°.

218° *o*-Bromoanisole $CH_3OC_6H_4Br$
(210°) Fuming HNO_3/glacial AcOH (1:1) → nitro m.p. 106°. Sulphonamide m.p. 140°.

222° *p*-Bromoanisole $CH_3OC_6H_4Br$
(215°) M.p. 12°. Fuming HNO_3/glacial AcOH (1:1) → nitro m.p. 88°. Sulphonamide m.p. 148°.

224° *o*-Bromophenetole $C_2H_5OC_6H_4Br$
(218°) Fuming HNO_3 → nitro m.p. 98°. Sulphonamide m.p. 135°.

229° *p*-Bromophenetole $C_2H_5OC_6H_4Br$
M.p. 12°. Sulphonamide m.p. 144°.

240° *o*-Iodoanisole $CH_3OC_6H_4I$

245° *o*-Iodophenetole $C_2H_5OC_6H_4I$

SOLID

M.P.°C

27° *p*-Iodophenetole $C_2H_5OC_6H_4I$
B.p. 252°.

43° *s*-Trichlorophenetole $C_2H_5OC_6H_2Cl_3$
B.p. 246°. Warm conc. HNO_3/H_2SO_4 (slowly) → nitro (di) m.p. 100°.

52° *p*-Iodoanisole $CH_3OC_6H_4I$

60° *s*-Trichloroanisole $CH_3OC_6H_2Cl_3$
B.p. 240°. Warm conc. HNO_3/H_2SO_4 (slowly) → nitro (di) m.p. 95°.

72° *s*-Tribromophenetole $C_2H_5OC_6H_2Br_3$

87° *s*-Tribromoanisole $CH_3OC_6H_2Br_3$

HALOGEN SUBSTITUTED ALCOHOLS

LIQUID

B.P.°C

105° 2-Fluoroethanol CH_2FCH_2OH
1-Naphthyl carbamate m.p. 128°.

B.P.°C

129°   Ethylene chlorohydrin CH_2ClCH_2OH, 2-Chloroethanol.
d_4^{20} 1.202. H_2O miscible. Reduced by aq. Na/Hg to ethyl alcohol.
$K_2Cr_2O_7$/conc. H_2SO_4 oxdtn. → chloracetic acid. Solid NaOH →
ethylene oxide. Boiling dil. NaOH → ethylene glycol. 1-Naphthyl
carbamate m.p. 101°. 3,5-Dinitrobenzoate m.p. 88°(95°). Heat
(<135°) with thiourea → thiuronium salt m.p. 111°.

150°d   Ethylene bromohydrin CH_2BrCH_2OH, 2-Bromoethanol.
d_4^{20} 1.772. H_2O miscible. 1-Naphthyl carbamate m.p. 86°. Properties like ethylene chlorohydrin.

161°d   Trimethylene chlorohydrin $CH_2ClCH_2CH_2OH$, 3-Chloropro-
panol.
Sol. in 2–3 vols. cold H_2O. Solid NaOH → trimethylene oxide
b.p. 48°. Conc. HNO_3 oxdtn. → 3-chloropropionic acid m.p. 41°.
1-Naphthyl carbamate m.p. 76°. 3,5-Dinitrobenzoate m.p. 77°.

176°   Glycerol αα'-dichlorohydrin $(CH_2Cl)_2CHOH$, 1,3-Dichloro-
propan-2-ol.
Sol. in H_2O. Solid NaOH → epichlorhydrin. $K_2Cr_2O_7$/dil.
H_2SO_4 oxdtn. → *sym*-dichloroacetone m.p. 44°. Phenol + NaOH
in boiling alc. → glycerol-1,3-diphenyl ether m.p. 81°. 1-Naphthyl
carbamate m.p. 115°.

176°d   Trimethylene bromohydrin $CH_2BrCH_2CH_2OH$, 3-Bromopro-
panol.
Sol. in 9 vols. cold H_2O. Solid NaOH → trimethylene oxide.
HNO_3 oxdtn. → 3-bromopropionic acid m.p. 62°. 1-Naphthyl
carbamate m.p. 73°.

182°   Glycerol αβ-dichlorohydrin $CH_2ClCHClCH_2OH$, 2,3-Dichloro-
propanol.
Solid NaOH → epichlorhydrin. 1-Naphthyl carbamate m.p. 93°.

213°d   Glycerol α-chlorohydrin $CH_2ClCHOHCH_2OH$
Miscible with H_2O, alc., ether. Reduced by Na/Hg in H_2O →
propylene glycol. Phenol + NaOH in boiling alc. → glycerol
phenyl ether m.p. 69°.

219°d   Glycerol αα'-dibromohydrin $(CH_2Br)_2CHOH$
$K_2Cr_2O_7$/dil. H_2SO_4 oxdtn. → *sym*-dibromoacetone m.p. 24°.

219°   Glycerol αβ-dibromohydrin $CH_2BrCHBrCH_2OH$, 2,3-Di-
bromopropanol.
Phenylurethane m.p. 84°.

SOLID

M.P.°C

19°   2,2,2-Trichloroethyl alcohol Cl_3CCH_2OH
B.p. 151°. Reduces Fehling's soln. on warming. 3,5-Dinitro-
benzoate m.p. 142°. 1-Naphthyl carbamate m.p. 120°.

80°   2,2,2-Tribromoethyl alcohol Br_3CCH_2OH

96°   Trichloro-t-butyl alcohol $Cl_3C(CH_3)_2COH$, Chloretone.
Anhydrous m.p. Hydrate m.p. 80°. B.p. 167°. Camphor-like
odour. Gives iodoform reaction. *p*-Nitrobenzoate m.p. 145°

HALOGEN SUBSTITUTED PHENOLS

LIQUID

B.P.°C

175° *o*-Chlorophenol ClC$_6$H$_4$OH
M.p. 7°. Violet colour with aq. FeCl$_3$ soln. Methyl ether b.p.
195°. Ethyl ether b.p. 208°. Aryloxyacetic acid m.p. 144°.
p-Tosyl m.p. 74°. 3,5-Dinitrobenzoate m.p. 143°. 1-Naphthyl
carbamate m.p. 120°.

195° *o*-Bromophenol BrC$_6$H$_4$OH
M.p. 5°. Violet colour with aq. FeCl$_3$ soln. Methyl ether b.p.
218°. Ethyl ether b.p. 224°. Br$_2$/H$_2$O → tribromophenol m.p. 95°.
Aryloxyacetic acid m.p. 142°. 1-Naphthyl carbamate m.p. 129°.

SOLID

M.P.°C

32° *m*-Chlorophenol ClC$_6$H$_4$OH
B.p. 214°. Aryloxyacetic acid m.p. 109°. Benzoate m.p. 71°.
3,5-Dinitrobenzoate m.p. 156°. 1-Naphthyl carbamate m.p. 158°.

32° *m*-Bromophenol BrC$_6$H$_4$OH
B.p. 236°. Aryloxyacetic acid m.p. 108°. Benzoate m.p. 86°.
1-Naphthyl carbamate m.p. 108°.

36° 2,4-Dibromophenol Br$_2$C$_6$H$_3$OH
B.p. 238°. Violet colour with aq. FeCl$_3$ soln. Br$_2$/H$_2$O → tri-
bromophenol. Aryloxyacetic acid m.p. 153°. *p*-Tosyl m.p. 120°.
Benzoate m.p. 97°.

38° *p*-Chlorophenol ClC$_6$H$_4$OH
(42°) B.p. 217°. Blue-violet colour with aq. FeCl$_3$ soln. Aryloxyacetic
acid m.p. 156°. Benzoate m.p. 87°. 3,5-Dinitrobenzoate m.p.
186°. 1-Naphthyl carbamate m.p. 166°.

40° *m*-Iodophenol IC$_6$H$_4$OH
Aryloxyacetic acid m.p. 115°. Benzoate m.p. 73°. 3,5-Dinitro-
benzoate m.p. 183°.

43° 2,4-Dichlorophenol Cl$_2$C$_6$H$_3$OH
B.p. 209°. Blue-violet colour with aq. FeCl$_3$ soln. Br$_2$/H$_2$O →
bromo m.p. 68°. Aryloxyacetic acid m.p. 138°. Benzoate m.p.
97°. *p*-Tosyl m.p. 125°.

43° *o*-Iodophenol IC$_6$H$_4$OH
Aryloxyacetic acid m.p. 135°. *p*-Tosyl m.p. 80°.

64° *p*-Bromophenol BrC$_6$H$_4$OH
Violet colour with aq. FeCl$_3$ soln. Br$_2$/H$_2$O → tribromophenol.
Aryloxyacetic acid m.p. 157°. Benzoate m.p. 102°. *p*-Tosyl m.p.
94°. 3,5-Dinitrobenzoate m.p. 191°. 1-Naphthyl carbamate m.p.
169°.

68° 2,4,6-Trichlorophenol Cl$_3$C$_6$H$_2$OH
B.p. 245°. Strongly acidic, NaHCO$_3$ → CO$_2$. CrO$_3$/AcOH →
Chloranil m.p. 290°. Aryloxyacetic acid m.p. 182°(177°). Benzo-
ate m.p. 74°. 3,5-Dinitrobenzoate m.p. 136°.

M.P.°C

94° p-Iodophenol IC_6H_4OH
Aryloxyacetic acid m.p. 156°. Benzoate m.p. 119°. p-Tosyl m.p. 99°.

95° 2,4,6-Tribromophenol $Br_3C_6H_2OH$
Strongly acidic, $NaHCO_3 \rightarrow CO_2$. Aryloxyacetic acid m.p. 200°. Benzoate m.p. 81°. p-Tosyl m.p. 113°. 3,5-Dinitrobenzoate m.p. 174°.

106° Chlorohydroquinone $ClC_6H_3(OH)_2$
B.p. 263°. Very sol. in H_2O and alc. Reduces Tollen's reagent rapidly on warming. $K_2Cr_2O_7$/dil. H_2SO_4 oxdtn. → chloroquinone. m.p. 57°. Benzoate (di) m.p. 130°. Acetate (di) m.p. 72°.

158° 2,4,6-Triiodophenol $I_3C_6H_2OH$
Acetate m.p. 156°. Benzoate m.p. 137°.

190° Pentachlorophenol Cl_5C_6OH
Anhydrous. Hydrate m.p. 174°. Strongly acidic, titrates quant. in alc. soln. Methyl ether m.p. 107°. Acetate m.p. 150°. Benzoate m.p. 159°(164°). p-Tosyl m.p. 145°. Aryloxyacetic acid m.p. 196°.

236° Tetrachlorohydroquinone $Cl_4C_6(OH)_2$
Boiling conc. HNO_3 oxdtn. → chloranil m.p. 290° (sealed tube). Acetate (di) m.p. 245°. Benzoate (di) m.p. 233°.

HALOGEN SUBSTITUTED ALDEHYDES AND ACETALS

LIQUID

B.P.°C

85° Chloroacetaldehyde. See Chloroacetal.

90° Dichloroacetaldehyde $CHCl_2CHO$

98° Chloral CCl_3CHO
Usually as chloral hydrate $CCl_3CH(OH)_2$ m.p. 57°. Conc. H_2SO_4 → polymers. Cold conc. NaOH → $CHCl_3$ + Na formate. Zn/dil. H_2SO_4 → acetaldehyde. Reduces Fehling's soln. and Tollen's reagent on warming. KI → $CHCl_3$ + free I_2. Forms bisulphite cpd. 2,4-Dinitrophenylhydrazone m.p. 131°. Semicarbazone m.p. 90°d.

104° Bromoacetaldehyde. See Bromoacetal.

157° Chloroacetal $CH_2ClCH(OC_2H_5)_2$
Stable to alkali. Warm dil. H_2SO_4 → chloroacetaldehyde b.p. 85° (very sharp odour), semicarbazone m.p. 134°(d).

170°d Bromoacetal $CH_2BrCH(OC_2H_5)_2$
Stable to alkali. Warm dil. H_2SO_4 → bromoacetaldehyde b.p. 104° (very sharp odour), semicarbazone m.p. 130°.

174° Bromal CBr_3CHO
Yellow. Usually as bromal hydrate $CBr_3CH(OH)_2$ m.p. 53°. NaOH → $CHBr_3$ (b.p. 150°) + Na formate. Acetic anhydride → acetate (di) m.p. 76°. Oxime m.p. 115°.

213° o-Chlorobenzaldehyde ClC_6H_4CHO
M.p. 11°. $KMnO_4$ oxdtn. → o-chlorobenzoic acid m.p. 140°.

B.P.°C

2,4-Dinitrophenylhydrazone m.p. 209°. Semicarbazone m.p. 226°. Dimethone m.p. 205°.

214°   *m*-Chlorobenzaldehyde ClC$_6$H$_4$CHO
M.p. 18°. KMnO$_4$ oxdtn. → *m*-chlorobenzoic acid m.p. 158°. 2,4-Dinitrophenylhydrazone m.p. 255°. Semicarbazone m.p. 228°.

234°   *m*-Bromobenzaldehyde BrC$_6$H$_4$CHO
KMnO$_4$ oxdtn. → *m*-bromobenzoic acid m.p. 155°. 2,4-Dinitrophenylhydrazone m.p. 257°. *p*-Nitrophenylhydrazone m.p. 220°. Semicarbazone m.p. 205°.

SOLID

M.P.°C

22°   *o*-Bromobenzaldehyde BrC$_6$H$_4$CHO
B.p. 230°. KMnO$_4$ oxdtn. → *o*-bromobenzoic acid m.p. 150°. Semicarbazone m.p. 214°. *p*-Nitrophenylhydrazone m.p. 240°.

47°   *p*-Chlorobenzaldehyde ClC$_6$H$_4$CHO
B.p. 214°. KMnO$_4$ oxdtn. → *p*-chlorobenzoic acid m.p. 243° (240°). Boiling acetic anhydride → acetate (di) m.p. 82°. Warm with NH$_2$OH·HCl + Na$_2$CO$_3$ → α-oxime m.p. 106°. 2,4-Dinitrophenylhydrazone m.p. 265°. Semicarbazone m.p. 231°.

53°   Bromal hydrate. See Bromal.

57°   Chloral hydrate. See Chloral.

67°   *p*-Bromobenzaldehyde BrC$_6$H$_4$CHO
(57°)   KMnO$_4$ oxdtn. → *p*-bromobenzoic acid m.p. 251°. Semicarbazone m.p. 228°. *p*-Nitrophenylhydrazone m.p. 208°.

78°   *p*-Iodobenzaldehyde IC$_6$H$_4$CHO
KMnO$_4$ oxdtn. → *p*-iodobenzoic acid m.p. 270°. 2,4-Dinitrophenylhydrazone m.p. 257°. Semicarbazone m.p. 224°.

HALOGEN SUBSTITUTED KETONES

LIQUID

119°   Chloroacetone CH$_3$COCH$_2$Cl
Lachrymator. Sol. in H$_2$O. Very easily hydrolysed by alkali. K$_2$CO$_3$ soln. → red colour. Add slowly to excess NH$_2$OH soln. → methylglyoxime m.p. 156°. 2,4-Dinitrophenylhydrazone m.p. 125°.

120°   1,1-Dichloroacetone CH$_3$COCHCl$_2$
Spar. sol. in H$_2$O. Boiling dil. K$_2$CO$_3$ soln. → K acrylate. NH$_2$OH → methylglyoxime m.p. 156°. Semicarbazone m.p. 163°.

136°   Bromoacetone CH$_3$COCH$_2$Br
Properties like chloroacetone.

229°   *o*-Chloroacetophenone ClC$_6$H$_4$COCH$_3$
2,4-Dinitrophenylhydrazone m.p. 206°. Semicarbazone m.p. 160°. Oxime m.p. 113°.

236°   *p*-Chloroacetophenone ClC$_6$H$_4$COCH$_3$
M.p. 20°. 2,4-Dinitrophenylhydrazone m.p. 231°. Oxime m.p.

M.P.°C

95°. Does not form NaHSO$_3$ addtn. cmpd. Salicylidene m.p. 151°.

SOLID

M.P.°C

42° *o*-Bromobenzophenone BrC$_6$H$_4$COC$_6$H$_5$
Oxime m.p. 133°.

45° *sym*-Dichloroacetone CH$_2$ClCOCH$_2$Cl
B.p. 173°. Lachrymator. Spar. sol. in cold H$_2$O. Readily hydroly-sed → dihydroxyacetone m.p. 72°. Readily reduces Fehling's soln. KI → *sym*-diiodoacetone m.p. 61°.

50° ω-Bromoacetophenone C$_6$H$_5$COCH$_2$Br, Phenacyl bromide.
Lachrymator. CrO$_3$ oxdtn. → benzoic acid. Aniline (cold) → acetophenone anilide m.p. 93°. 2,4-Dinitrophenylhydrazone m.p. 220°. Widely used for characterisation of acyl cpds. e.g. benzoate m.p. 118°.

51° *p*-Bromoacetophenone BrC$_6$H$_4$COCH$_3$
B.p. 256°. 2,4-Dinitrophenylhydrazone m.p. 230°. Semicarbazone m.p. 208°. Oxime m.p. 128°.

59° ω-Chloroacetophenone C$_6$H$_5$COCH$_2$Cl, Phenacyl chloride.
B.p. 244°. Lachrymator. 2,4-Dinitrophenylhydrazone m.p. 212°. Oxime m.p. 89°. Yields acyl derivatives as for phenacyl bromide.

76° α-Bromocamphor C$_8$H$_{14}$$\begin{cases} \text{CHBr} \\ \text{CO} \end{cases}$
Opt. act. [α]$_D$ +140° (10% alc. soln.). KMnO$_4$ oxdtn. → camphoric acid m.p. 187°.

78° *p*-Chlorobenzophenone ClC$_6$H$_4$COC$_6$H$_5$
2,4-Dinitrophenylhydrazone m.p. 185°.

82° *p*-Bromobenzophenone BrC$_6$H$_4$COC$_6$H$_5$
2,4-Dinitrophenylhydrazone m.p. 230°.

108° *p*-Bromophenacyl bromide BrC$_6$H$_4$COCH$_2$Br
Alk. KMnO$_4$ oxdtn. → *p*-bromobenzoic acid m.p. 251°. Oxime m.p. 115°. Widely used for characterisation of acyl cpds. e.g. benzoate m.p. 118°.

125° *p*-Phenylphenacyl bromide C$_6$H$_5$C$_6$H$_4$COCH$_2$Br
Widely used for characterisation of acyl cpds. e.g. benzoate m.p. 167°.

290° Chloranil C$_6$Cl$_4$O$_2$
(sealed tube) Yellow-orange colour. SO$_2$ redtn. → tetrachlorohydroquinone m.p. 236°. Dioxan solns. give colours with amines e.g. aniline → violet, piperidine → red.

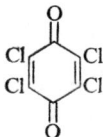

CARBOXYLIC ACID HALIDES

LIQUID

B.P.°C

52°	Acetyl chloride	CH_3COCl	
64°	Oxalyl chloride	$(COCl)_2$	
80°	Propionyl chloride	C_2H_5COCl	
81°	Acetyl bromide	CH_3COBr	
101°	n-Butyryl chloride	$C_2H_5CH_2COCl$	
105°	Chloracetyl chloride	$ClCH_2COCl$	
118°	Trichloroacetyl chloride	Cl_3CCOCl	
190°d	Succinoyl chloride	$(CH_2COCl)_2$	m.p. 18°
197°	Benzoyl chloride	C_6H_5COCl	
210°	Phenylacetyl chloride	$C_6H_5CH_2COCl$	95°/12 mmHg
221°	p-Chlorobenzoyl chloride	ClC_6H_4COCl	m.p. 14°
225°	p-Toluoyl chloride	$CH_3C_6H_4COCl$	
276°	Phthaloyl chloride	$C_6H_4(COCl)_2$	
	m.p. 12° (16° after distillation).		

SOLID

M.P.°C

20°	1-Naphthoyl chloride	$C_{10}H_7COCl$	b.p. 297°
22°	Anisoyl chloride	$CH_3OC_6H_4COCl$	b.p. 262°
35°	Cinnamoyl chloride	$C_6H_5CH:CHCOCl$	b.p. 257° (Trans isomer)
42°	p-Bromobenzoyl chloride	BrC_6H_4COCl	b.p. 245°
51° (43°)	2-Naphthoyl chloride	$C_{10}H_7COCl$	
83°	Terephthaloyl chloride	$C_6H_4(COCl)_2$	b.p. 263°

HALOGENOFORMATES (HALOCARBONATES)

LIQUID

B.P.°C

72°   Methyl chloroformate $ClCOOCH_3$
Lachrymator. Not readily decomp. by cold H_2O. $NH_3 \rightarrow$ methyl carbamate m.p. 57°. Aniline → methyl carbanilate m.p. 47°. Methanol → dimethyl carbonate b.p. 90°.

93°   Ethyl chloroformate $ClCOOC_2H_5$, Chloroformic ester.
Lachrymator. Not readily decomp. by cold H_2O. $NH_3 \rightarrow$ urethane m.p. 48°. Aniline → ethyl carbanilate m.p. 52°. Ethanol → diethyl carbonate b.p. 126°.

103°/20 mmHg   Benzyl chloroformate $ClCOOCH_2C_6H_5$, Carbobenzyloxy chloride.
Penetrating odour. Heat causes loss of $CO_2 \rightarrow$ benzyl chloride. Add to excess strong $NH_4OH \rightarrow$ benzyl carbamate m.p. 87°. Aniline → benzyl carbanilate m.p. 76°.

B.P.°C

127° Trichloromethyl chloroformate $ClCOOCCl_3$, Diphosgene.
Suffocating odour. Excess aniline → carbanilide m.p. 233° and
aniline hydrochloride.

150°d Cyclohexyl chloroformate $ClCOOC_6H_{11}$
Strong NH_4OH → cyclohexyl carbamate m.p. 110°.

HALOGEN SUBSTITUTED CARBOXYLIC ACIDS

LIQUID

B.P.°C

72° Trifluoroacetic CF_3COOH
Very strong acid. Equiv. wt. 114. Amide m.p. 74°. Anilide m.p.
91°.

186° 2-Chloropropionic $CH_3CHClCOOH$
Miscible with H_2O. Equiv. wt. 108.5. Zn/HCl redtn. → propionic
acid. Alk. hydrolysis → lactic acid. Amide m.p. 80°. Anilide m.p.
92°. *p*-Toluidide m.p. 124°. Phenol + NaOH → 2-phenoxy-
propionic acid m.p. 115°.

194°d Dichloroacetic $CHCl_2COOH$
M.p. 5°. Miscible with H_2O. Equiv. wt. 129. Amide m.p. 98°.
Anilide m.p. 118°. *p*-Toluidide m.p. 153°. SBT m.p. 178°.
p-Bromophenacyl ester m.p. 99°.

SOLID

M.P.°C

25° 2-Bromopropionic $CH_3CHBrCOOH$
B.p. 204°. Equiv. wt. 153. Amide m.p. 123°. Anilide m.p. 99°.
p-Toluidide m.p. 125°.

31° Fluoroacetic CH_2FCOOH
B.p. 168°. Equiv. wt. 78. Amide m.p. 108° (very toxic).

41° 3-Chloropropionic CH_2ClCH_2COOH
B.p. 204°d. Equiv. wt. 108.5. Alk. hydrolysis → acrylic acid
Amide m.p. 101°. Anilide m.p. 119°. *p*-Toluidide m.p. 121°.
Phenol + NaOH → 3-phenoxypropionic acid m.p. 97°.

48° Dibromoacetic $CHBr_2COOH$
B.p. 232–5°d. Equiv. wt. 218. Amide m.p. 156°.

50° Bromoacetic $CH_2BrCOOH$
B.p. 208°. Equiv. wt. 139. Amide m.p. 91°. Anilide m.p. 131°.
p-Toluidide m.p. 91°. *p*-Nitrobenzyl ester m.p. 89°.

57° Trichloroacetic CCl_3COOH
B.p. 197°. Equiv. wt. 163. Boiling H_2O, aq. alk. or organic
bases → $CHCl_3 + CO_2$. Amide m.p. 141°. Anilide m.p. 94°.
p-Toluidide m.p. 113°. SBT m.p. 149°.

62° 3-Bromopropionic CH_2BrCH_2COOH
Equiv. wt. 153. Amide m.p. 111°.

63° Chloroacetic $CH_2ClCOOH$
B.p. 189°. Equiv. wt. 94.5. Alk. hydrolysis → glycollic acid.

M.P.°C

Amide m.p. 120°. Anilide m.p. 137°. *p*-Toluidide m.p. 162°. SBT m.p. 160°. *p*-Bromophenacyl ester m.p. 105°. Phenol + NaOH → phenoxyacetic acid m.p. 98°. Similarly 2-naphthol → 2-naphthoxyacetic acid m.p. 156°. Boiling conc. soln. of Na salt + NaNO₂ → nitromethane b.p. 101°.

82° 3-Iodopropionic CH₂ICH₂COOH
Equiv. wt. 200. Br₂ water → 3-bromopropionic acid + I₂. Amide m.p. 101°.

83° Iodoacetic CH₂ICOOH
Spar. sol. in cold H₂O. Equiv. wt. 186. Amide m.p. 95°. Anilide m.p. 143°.

113°d Bromomalonic CHBr(COOH)₂
Diethyl ester b.p. 232–5°d.

126° *o*-Fluorobenzoic FC₆H₄COOH
Equiv. wt. 140. Amide m.p. 115°.

131° Tribromoacetic CBr₃COOH
D.p. 245°d. Equiv. wt. 297. Amide m.p. 122°.

140° *o*-Chorobenzoic ClC₆H₄COOH
Sol. in hot H₂O. Equiv. wt. 157. Amide m.p. 142°. Anilide m.p. 118°. *p*-Toluidide m.p. 131°. *p*-Nitrobenzyl ester m.p. 106°. *p*-Bromophenacyl ester m.p. 107°.

150° *o*-Bromobenzoic BrC₆H₄COOH
Sol. in hot H₂O. Equiv. wt. 201. Amide m.p. 155°. Anilide m.p. 141°. SBT m.p. 171°. *p*-Nitrobenzyl ester m.p. 110°. *p*-Bromophenacyl ester m.p. 102°.

155° *m*-Bromobenzoic BrC₆H₄COOH
Equiv. wt. 201. Amide m.p. 155°. Anilide m.p. 146°. SBT m.p. 168°. *p*-Nitrobenzyl ester m.p. 105°. *p*-Bromophenacyl ester m.p. 126°.

158° *m*-Chlorobenzoic ClC₆H₄COOH
Equiv. wt. 157. Amide m.p. 134°. Anilide m.p. 124°. SBT m.p. 155°. *p*-Nitrobenzyl ester m.p. 107°. *p*-Bromophenacyl ester m.p. 117°.

162° *o*-Iodobenzoic acid IC₆H₄COOH
Equiv. wt. 248. Amide m.p. 184°. Anilide m.p. 142°. *p*-Nitrobenzyl ester 111°. *p*-Bromophenacyl ester m.p. 110°.

167° (±)2,3-Dibromosuccinic (CHBrCOOH)₂
Equiv. wt. 138. *p*-Nitrobenzyl ester (di) m.p. 168°.

186° *p*-Fluorobenzoic FC₆H₄COOH
Equiv. wt. 140. Amide m.p. 154°.

187° *m*-Iodobenzoic IC₆H₄COOH
Equiv. wt. 248. Amide m.p. 186°. *p*-Nitrobenzyl ester m.p. 121°. *p*-Bromophenacyl ester m.p. 128°.

188° 3-Bromophthalic acid BrC₆H₃(COOH)₂

243° *p*-Chlorobenzoic ClC₆H₄COOH
(240°) Equiv. wt. 157. Amide m.p. 179°. Anilide m.p. 194°. SBT m.p. 186°. *p*-Nitrobenzyl ester m.p. 129°. *p*-Bromophenacyl ester m.p. 126°.

M.P.°C

251° p-Bromobenzoic BrC_6H_4COOH
Equiv. wt. 201. Amide m.p. 189°. Anilide m.p. 197°. p-Nitro-benzyl ester m.p. 140°. p-Bromophenacyl ester m.p. 134°.

270° p-Iodobenzoic IC_6H_4COOH
Equiv. wt. 248. Amide m.p. 217°. Anilide m.p. 210°. p-Nitro-benzyl ester m.p. 141°. p-Bromophenacyl ester m.p. 146°.

HALOGEN SUBSTITUTED CARBOXYLIC ESTERS

LIQUID

B.P.°C

72°   Methyl chloroformate. See Halogenoformates

93°   Ethyl chloroformate. See Halogenoformates.

115°   Chloromethyl acetate CH_3COOCH_2Cl
Sharp odour. Decomp. by $H_2O \rightarrow HCl$, CH_2O, acetic acid. Ethanol → ethylal + ethyl acetate.

121°   Ethyl fluoroacetate $CH_2FCOOC_2H_5$

130°   Methyl chloroacetate $CH_2ClCOOCH_3$
Cold conc. $NH_3 \rightarrow$ chloroacetamide m.p. 120°.

130°   Bromomethyl acetate CH_3COOCH_2Br

143°   Methyl dichloroacetate $CHCl_2COOCH_3$

144°   Ethyl chloroacetate $CH_2ClCOOC_2H_5$
Cold conc. $NH_3 \rightarrow$ chloroacetamide m.p. 120°.

144°d   Methyl bromoacetate $CH_2BrCOOCH_3$
Very irritating odour. Cold conc. $NH_3 \rightarrow$ bromoacetamide m.p. 91°.

147°   Ethyl 2-chloropropionate $CH_3CHClCOOC_2H_5$
Cold conc. $NH_3 \rightarrow$ amide m.p. 80°.

152°   Methyl trichloroacetate CCl_3COOCH_3

158°   Ethyl dichloroacetate $CHCl_2COOC_2H_5$
Alc. $NH_3 \rightarrow$ amide m.p. 98°.

159°   Ethyl bromoacetate $CH_2BrCOOC_2H_5$
Very irritating odour. Addtn. product with quinoline m.p. 180°.

162°   Ethyl 3-chloropropionate $CH_2ClCH_2COOC_2H_5$
Cold conc. alc. NaOH or heat (100°) with 10 parts. conc. $H_2SO_4 \rightarrow$ ethyl acrylate b.p. 101°.

162°   Ethyl 2-bromopropionate $CH_3CHBrCOOC_2H_5$

162°   n-Propyl chloroacetate $CH_2ClCOOCH_2C_3H_5$

167°   Ethyl trichloroacetate $CCl_3COOC_2H_5$
Cold conc. $NH_3 \rightarrow$ amide m.p. 141°. Yields $CHCl_3$ on long boiling with aq. alkali.

179°   Ethyl 3-bromopropionate $CH_2BrCH_2COOC_2H_5$
Boiling conc. NaOH → acrylic acid.

180°   Ethyl iodoacetate $CH_2ICOOC_2H_5$

184°   Ethyl 3-iodopropionate $CH_2ICH_2COOC_2H_5$

231°   Methyl m-chlorobenzoate $ClC_6H_4COOCH_3$

234°   Methyl o-chlorobenzoate $ClC_6H_4COOCH_3$

238°   Ethyl p-chlorobenzoate $ClC_6H_4COOC_2H_5$

B.P.°C
243° Ethyl o-chlorobenzoate $ClC_6H_4COOC_2H_5$
244° Methyl o-bromobenzoate $BrC_6H_4COOCH_3$
245° Ethyl m-chlorobenzoate $ClC_6H_4COOC_2H_5$
254° Ethyl o-bromobenzoate $BrC_6H_4COOC_2H_5$
259° Ethyl m-bromobenzoate $BrC_6H_4COOC_2H_5$
263° Ethyl p-bromobenzoate $BrC_6H_4COOC_2H_5$
275° Ethyl o-iodobenzoate $IC_6H_4COOC_2H_5$
278° Methyl o-iodobenzoate $IC_6H_4COOC_2H_5$

SOLID

M.P.°C
31° Methyl m-bromobenzoate $BrC_6H_4COOCH_3$
43° Methyl p-chlorobenzoate $ClC_6H_4COOCH_3$
54° Methyl m-iodobenzoate $IC_6H_4COOCH_3$
81° Methyl p-bromobenzoate $BrC_6H_4COOCH_3$
114° Methyl p-iodobenzoate $IC_6H_4COOCH_3$

ALIPHATIC PRIMARY AMINES

LIQUID

B.P.°C
−6°(Gas) Methylamine CH_3NH_2
Yellow ppt. with Nessler's soln. insol. in excess reagent. No ppt. with HgI_2/KI soln. Not attacked by boiling acid $KMnO_4$ soln. Hydrochloride m.p. 225° spar. sol. in alc., insol. in $CHCl_3$. $C_6H_5CHO \rightarrow$ benzylidene deriv. b.p. 180°. 2,4-Dinitrophenyl m.p. 178°. Benzoyl m.p. 80°. p-Tosyl m.p. 77°. Picrate m.p. 215° (207°).

17° Ethylamine $C_2H_5NH_2$
White ppt. with Nessler's soln. Slowly oxidised by $KMnO_4$ in boiling dil. H_2SO_4. $C_6H_5CHO \rightarrow$ benzylidene deriv. b.p. 195°. 2,4-Dinitrophenyl m.p. 114°. Benzoyl m.p. 69°. p-Tosyl m.p. 63°. Picrate m.p. 166°.

32° Isopropylamine $(CH_3)_2CHNH_2$
2,4-Dinitrophenyl m.p. 94°. Benzoyl m.p. 100°. p-Tosyl m.p. 51°. Picrate m.p. 150°.

46° t-Butylamine $(CH_3)_3CNH_2$
Benzoyl m.p. 134°. Picrate m.p. 198°.

49° n-Propylamine $C_2H_5CH_2NH_2$
Brown ppt. with Nessler's soln. Hydrochloride m.p. 157°. $C_6H_5CHO \rightarrow$ benzylidene deriv b.p. 209°. 2,4-Dinitrophenyl m.p. 96°. Benzoyl m.p. 85°. p-Tosyl m.p. 52°. Picrate m.p. 135°.

50° Cyclopropylamine $(CH_2)_2CHNH_2$
Hydrochloride m.p. 100°. Benzoyl m.p. 99°. Picrate m.p. 149°.

56° Allylamine $CH_2:CHCH_2NH_2$
Unsat'd. Dibromide an oil. Benzoyl of dibromide m.p. 135°. 2,4-Dinitrophenyl m.p. 76°. p-Tosyl m.p. 64°. Picrate m.p. 140°.

B.P.°C
 63° s-Butylamine $C_2H_5(CH_3)CHNH_2$
      Benzoyl m.p. 76°. *p*-Tosyl m.p. 55°. Picrate m.p. 140°.
 68° Isobutylamine $(CH_3)_2CHCH_2NH_2$
      CrO$_3$ oxdtn. → isobutyric acid. 2,4-Dinitrophenyl m.p. 80°.
      Benzoyl m.p. 57°. *p*-Tosyl m.p. 78°. Picrate m.p. 151°.
 77° n-Butylamine $CH_3(CH_2)_3NH_2$
      CrO$_3$ oxdtn. → n-butyric acid. Hydrochloride m.p. 195°. 2,4-
      Dinitrophenyl m.p. 58°. Picrate m.p. 151°.
 95° Isoamylamine $(CH_3)_2CHCH_2CH_2NH_2$
      CrO$_3$ oxdtn. → isovaleric acid. 2,4-Dinitrophenyl m.p. 91°.
      p-Tosyl m.p. 65°. Picrate m.p. 138°.
104° n-Amylamine $CH_3(CH_2)_4NH_2$
      CrO$_3$ oxdtn. → n-valeric acid. 2,4-Dinitrophenyl m.p. 81°.
      Picrate m.p. 139°.
116° Ethylenediamine $(CH_2NH_2)_2$
      M.p. 8°. H_2O miscible. Almost insol. ether, insol. in benzene.
      Forms hydrate b.p. 118° from which H_2O removed by Na. 2,4-
      Dinitrophenyl (di) m.p. 302°. Acetyl (di) m.p. 172°. Benzoyl (di)
      m.p. 249°. *p*-Tosyl (di) m.p. 160°. Picrate (di) m.p. 233°.
134° Cyclohexylamine $C_6H_{11}NH_2$
      2,4-Dinitrophenyl m.p. 156°. Acetyl m.p. 104°. Benzoyl m.p.
      148°. *p*-Tosyl m.p. 87°. Picrate m.p. 158°.
171° Ethanolamine $HOCH_2CH_2NH_2$
      Picrate m.p. 160°. 1-Naphthyl urea m.p. 186°.
178° Pentamethylene diamine $NH_2(CH_2)_5NH_2$, Cadaverine.
184° Benzylamine $C_6H_5CH_2NH_2$
      Hydrochloride m.p. 248°. 2,4-Dinitrophenyl m.p. 116°. Acetyl
      m.p. 60°. Benzoyl m.p. 106°. *p*-Tosyl m.p. 116°. Picrate m.p. 199°.
185° (±)1-Phenylethylamine $C_6H_5CH(CH_3)NH_2$
      Hydrochloride m.p. 158°. 2,4-Dinitrophenyl m.p. 118°. Acetyl
      m.p. 57°. Benzoyl m.p. 120°. Picrate m.p. 189°.
198° 2-Phenylethylamine $C_6H_5CH_2CH_2NH_2$
      Hydrochloride m.p. 217°. 2,4-Dinitrophenyl m.p. 154°. Acetyl
      m.p. 51°. Benzoyl m.p. 116°. *p*-Tosyl m.p. 64°. Picrate m.p. 167°
      (174°).
207° (−)Menthylamine $C_{10}H_{19}NH_2$
(212°) $[\alpha]_D^{19}$ − 34°. Acetyl m.p. 145°. Benzoyl m.p. 156°. Picrate m.p.
      215°.

SOLIDS

M.P.°C
 27° Tetramethylenediamine $(CH_2CH_2NH_2)_2$, Putrescine.
      B.p. 159°. Acetyl (di) m.p. 137°. Benzoyl (di) m.p. 177°. *p*-Tosyl
      (di) m.p. 224°. Picrate (di) m.p. 249°.
 42° Hexamethylenediamine $(CH_2CH_2CH_2NH_2)_2$
      B.p. 204°. Acetyl (di) m.p. 126°. Benzoyl (di) m.p. 155°. Picrate
      m.p. 220°.

AROMATIC PRIMARY AMINES

LIQUID

B.P.°C

184° Aniline $C_6H_5NH_2$
Transient purple colour with NaOCl soln. in very dil. aq. soln.
Warm with alc. KOH + $CHCl_3$ → C_6H_5NC (foul odour).* Br_2
water → white ppt. of tribromo deriv. m.p. 119°. Hydrochloride
m.p. 198°. Acetyl m.p. 114°. Benzoyl m.p. 163°. p-Tosyl m.p.
103°. 2,4-Dinitrophenyl m.p. 156°. Picrate m.p. 165°(180°).

200° o-Toluidine $CH_3C_6H_4NH_2$
Brown colour with NaOCl in ethereal soln. Soln. in strong
H_2SO_4 → blue colour with $K_2Cr_2O_7$ → purple on diltn. Alc.
KOH + $CHCl_3$ → foul odour. Br_2 water → dibromo deriv. m.p.
50°. Hydrochloride m.p. 215°. Acetyl m.p. 110°. Benzoyl m.p.
144°. p-Tosyl m.p. 109°. 2,4-Dinitrophenyl m.p. 120°. Picrate
m.p. 200°.

203° m-Toluidine $CH_3C_6H_4NH_2$
Ethereal soln. + NaOCl soln. → yellow-brown (H_2O layer) and
red (ether layer). Soln. in strong H_2SO_4 → yellow-brown colour
with $K_2Cr_2O_7$, red with HNO_3. Br_2 water → tribromo deriv. m.p.
97°. Hydrochloride m.p. 228°. Acetyl m.p. 65°. Benzoyl m.p.
125°. p-Tosyl m.p. 114°. 2,4-Dinitrophenyl m.p. 160°. Picrate
m.p. 200°.

212° 4-m-Xylidine $(CH_3)_2C_6H_3NH_2$, 2,4-Dimethylaniline.
Hydrochloride m.p. 235°. Acetyl m.p. 130°. Benzoyl m.p. 192°.
p-Tosyl m.p. 181°. 2,4-Dinitrophenyl m.p. 156°. Picrate m.p 209°.

214° 2-p-Xylidine $(CH_3)_2C_6H_3NH_2$, 2,5-Dimethylaniline.
M.p. 14°. Hydrochloride m.p. 228°. Acetyl m.p. 139°. Benzoyl
m.p. 140°. p-Tosyl m.p. 119°. 2,4-Dinitrophenyl m.p. 150°.
Picrate m.p. 171°.

215° 2-m-Xylidine $(CH_3)_2C_6H_3NH_2$, 2,6-Dimethylaniline.
M.p. 11°. Acetyl m.p. 177°. Benzoyl m.p. 168°. p-Tosyl m.p. 212°.
Picrate m.p. 180°.

216° p-Ethylaniline $C_2H_5C_6H_4NH_2$
Acetyl m.p. 94°. Benzoyl m.p. 151°. p-Tosyl m.p. 104°.

220° 5-m-Xylidine $(CH_3)_2C_6H_3NH_2$, 3,5-Dimethylaniline.
M.p. 10°. Acetyl m.p. 144°. Benzoyl m.p. 136°. Picrate m.p. 209°
(200°).

221° 3-o-Xylidine $(CH_3)_2C_6H_3NH_2$, 2,3-Dimethylaniline.
M.p. 3°. Acetyl m.p. 134°. Benzoyl m.p. 189°. Picrate m.p. 221°.

225° o-Anisidine $CH_3OC_6H_4NH_2$
(218°) M.p. 5°. Acetyl m.p. 85°. Benzoyl m.p. 60°. p-Tosyl m.p. 127°
2,4-Dinitrophenyl m.p. 151°. Picrate m.p. 200°.

225° p-Cumidine $(CH_3)_2CHC_6H_4NH_2$, p-Isopropylaniline.
Acetyl m.p. 102°. Benzoyl m.p. 162°.

* Readily destroyed by HCl.

B.P.°C
229°   *o*-Phenetidine $C_2H_5OC_6H_4NH_2$
Hydrochloride m.p. 214°. Acetyl m.p. 79°. Benzoyl m.p. 104°.
p-Tosyl m.p. 164°. 2,4-Dinitrophenyl m.p. 164°.

232°   Mesidine $(CH_3)_3C_6H_2NH_2$, 2,4,6-Trimethylaniline.
Acetyl m.p. 216°. Benzoyl m.p. 204°. *p*-Tosyl m.p. 167°. Picrate
m.p. 193°.

248°   *m*-Phenetidine $C_2H_5OC_6H_4NH_2$
Acetyl m.p. 97°. Benzoyl m.p. 103°. *p*-Tosyl m.p. 157°. Picrate
m.p. 158°.

251°   *m*-Anisidine $CH_3OC_6H_4NH_2$
Acetyl m.p. 81°. Benzoyl m.p. 110°. *p*-Tosyl m.p. 68°. 2,4-
Dinitrophenyl m.p. 138°. Picrate m.p. 169°.

254°   *p*-Phenetidine $C_2H_5OC_6H_4NH_2$
M.p. 2°. Red colour with NaOCl or $FeCl_3$ soln. Hydrochloride
m.p. 234°. Acetyl m.p. 134° (Phenacetin). Benzoyl m.p. 173°.
p-Tosyl m.p. 106°. 2,4-Dinitrophenyl m.p. 118°.

262°   *p*-Aminodiethylaniline $(C_2H_5)_2NC_6H_4NH_2$
Acetyl m.p. 104°. Benzoyl m.p. 172°.

267°   Ethyl anthranilate. See Amino acids.

294°   Ethyl *m*-aminobenzoate $NH_2C_6H_4COOC_2H_5$
Hydrochloride m.p. 185°. Acetyl m.p. 110°. Benzoyl m.p. 148°
(114°).

SOLID

M.P.°C
20°   *o*-Aminoacetophenone $NH_2C_6H_4COCH_3$
B.p. 252°. Acetyl m.p. 76°. Benzoyl m.p. 98°. *p*-Tosyl m.p. 148°.

25°   Methyl anthranilate. See Amino acids.

41°   *p*-Aminodimethylaniline $(CH_3)_2NC_6H_4NH_2$
(53°)   B.p. 262°. Darkens rapidly in air. Soln. treated successively with
H_2S and $FeCl_3 \rightarrow$ methylene blue. Acetyl m.p. 132°. Benzoyl
m.p. 228°. Picrate m.p. 186°.

43°   *p*-Toluidine $CH_3C_6H_4NH_2$
B.p. 200°. No colour with NaOCl soln. Soln. in strong $H_2SO_4 \rightarrow$
yellow colour with $K_2Cr_2O_7$, blue (changing to brown) with
HNO_3. Br_2 water $\rightarrow$ dibromo deriv. m.p. 73°. Hydrochloride
m.p. 240°. Acetyl m.p. 148°. Benzoyl m.p. 158°. *p*-Tosyl m.p.
117°. 2,4-Dinitrophenyl m.p. 136°. Picrate m.p. 169°(182°).

46°   *o*-Ethylaniline $C_2H_5C_6H_4NH_2$
B.p. 216° (210°). Acetyl m.p. 112°. Benzoyl m.p. 146°. Picrate
m.p. 194°.

48°   4-*o*-Xylidine $(CH_3)_2C_6H_3NH_2$, 3,4-Dimethylaniline.
B.p. 224°. Acetyl m.p. 99°. Benzoyl m.p. 118°. *p*-Tosyl m.p. 154°.
2,4-Dinitrophenyl m.p. 141°.

49°   2-Aminobiphenyl $C_6H_5C_6H_4NH_2$, *o*-Phenylaniline.
Acetyl m.p. 119°. Benzoyl m.p. 102° (86°).

50°   1-Naphthylamine $C_{10}H_7NH_2$
Carcinogenic. Blue ppt. with acid $FeCl_3$ soln. Hydrochloride

M.P.°C

m.p. 286°. Acetyl m.p. 160°. Benzoyl m.p. 161°. *p*-Tosyl m.p. 157°. 2,4-Dinitrophenyl m.p. 190°. Picrate m.p. 161°.

53°   4-Aminobiphenyl C₆H₅C₆H₄NH₂, *p*-Phenylaniline.
Acetyl m.p. 172°. Benzoyl m.p. 230°. *p*-Tosyl m.p. 255°.

57°   2-Aminopyridine. See Heterocyclic Bases.

57°   *p*-Anisidine CH₃OC₆H₄NH₂
B.p. 246°. Violet colour with acid FeCl₃ soln. Hydrochloride m.p. 216°. Acetyl m.p. 127°. Benzoyl m.p. 154°. *p*-Tosyl m.p. 114°. 2,4-Dinitrophenyl m.p. 141°. Picrate m.p. 117°.

63°   *m*-Phenylenediamine C₆H₄(NH₂)₂
Darkens rapidly in air. Sol. in hot H₂O. Brown ppt. with cold NaNO₂/HCl (Bismarck Brown). Bromine (dil. HCl) → tribromo m.p. 158°. Acetyl (di) m.p. 191°. Benzoyl (di) m.p. 240°. *p*-Tosyl (di) m.p. 172°. 2,4-Dinitrophenyl (di) m.p. 172°.

92°   Ethyl *p*-aminobenzoate. See Amino Acids.

99°   2,4-Diaminotoluene CH₃C₆H₄(NH₂)₂, *m*-Toluylenediamine.
Darkens rapidly in air. Sol. in hot H₂O. Like *m*-phenylenediamine. Hydrochloride m.p. 305°. Acetyl (di) m.p. 220°. Benzoyl (di) m.p. 224°. *p*-Tosyl (di) m.p. 192°. 2,4-Dinitrophenyl (di) m.p. 184°.

102°   *o*-Phenylenediamine C₆H₄(NH₂)₂
B.p. 256°. Darkens rapidly in air. Sol. in hot H₂O. Conc. soln. gives red ppt. with acid FeCl₃ soln. Na acetate + C₆H₅CHO → dibenzal deriv. m.p. 106°. Reflux with acetic acid → methylbenziminazole m.p. 175°. Acetic anhydride (cold H₂O soln.) → diacetyl deriv. m.p. 185°. Benzil in hot gl. AcOH → quinoxaline deriv. m.p. 126°. *p*-Tosyl (di) m.p. 201°.

106°   *p*-Aminoacetophenone NH₂C₆H₄COCH₃
Acetyl m.p. 167°. Benzoyl m.p. 205°. *p*-Tosyl m.p. 203°.

112°   2-Naphthylamine C₁₀H₇NH₂
Carcinogenic.* No colour with FeCl₃. Hydrochloride m.p. 260°. Acetyl m.p. 134°. Benzoyl m.p. 162°. *p*-Tosyl m.p. 133°. 2,4-Dinitrophenyl m.p. 179°. Picrate m.p. 179° (195°).

124°   *p*-Aminobenzophenone C₆H₅COC₆H₄NH₂
Acetyl m.p. 153°. Benzoyl m.p. 152°.

127°   Benzidine NH₂C₆H₄C₆H₄NH₂(4,4′)
Carcinogenic.* Sulphate almost insol. in H₂O. Boiling dil. H₂SO₄ + MnO₂ → *p*-benzoquinone m.p. 115°. C₆H₅CHO (reflux in alc.) → dibenzal deriv. m.p. 238°. *p*-Tosyl (di) m.p. 243°.

129°   *o*-Tolidine CH₃(NH₂)C₆H₃C₆H₃(NH₂)CH₃(3,4,4′,3′)
Carcinogenic. Like benzidine. Boiling dil. H₂SO₄ + MnO₂ → toluquinone m.p. 69°. C₆H₅CHO (reflux in alc.) → dibenzal deriv. m.p. 152°. Benzoyl (di) m.p. 265°. Picrate m.p. 185°.

137°   *o*-Dianisidine CH₃O(NH₂)C₆H₃C₆H₃(NH₂)OCH₃(3,4,4′,3′)
Like benzidine. Acetyl (di) m.p. 243°. Benzoyl (di) m.p. 236°.

* Skin contact with most amines should be avoided.

M.P.°C

140° p-Phenylenediamine $C_6H_4(NH_2)_2$
    Darkens rapidly in air. Sol. in hot H_2O. Boiling dil. $FeCl_3$ soln. → p-benzoquinone m.p. 115°. Acetyl (di) m.p. 303°. p-Tosyl (di) m.p. 266°. 2,4-Dinitrophenyl (di) m.p. 177°. Picrate m.p. 202°.

163° p-Aminoacetanilide. See Substituted Amides.

SECONDARY AMINES

LIQUID

B.P.°C

7°   Dimethylamine $(CH_3)_2NH$
    White ppt. with Nessler's soln. on diltn. with H_2O. 2,4-Dinitrophenyl m.p. 87°. Benzoyl m.p. 42°. p-Tosyl m.p. 79°. Picrate m.p. 158°.

55°   Diethylamine $(C_2H_5)_2NH$
    White ppt. with Nessler's soln. Yellow ppt. with HgI_2/KI soln. 2,4-Dinitrophenyl m.p. 80°. Benzoyl m.p. 42°. p-Tosyl m.p. 60°.

56°   Ethyleneimine $(CH_2)_2NH$, Aziridine.
    p-Tosyl m.p. 52°. Picrate m.p. 142°.

86°   Di-isopropylamine $[(CH_3)_2CH]_2NH$
    Picrate m.p. 140°. N-Nitroso m.p. 48°.

110°   Di-n-propylamine $(C_2H_5CH_2)_2NH$
    Picrate m.p. 75°. 2,4-Dinitrophenyl m.p. 40°.

138°   Di-isobutylamine $[(CH_3)_2CHCH_2]_2NH$
    Picrate m.p. 119°. 2,4-Dinitrophenyl m.p. 112°. Acetyl m.p. 86°.

159°   Di-n-butylamine $[CH_3(CH_2)_3]_2NH$
    Picrate m.p. 60°. N-Nitroso m.p. 61°.

181° N-Methylbenzylamine $C_6H_5CH_2NHCH_3$
    p-Tosyl m.p. 95°.

194° N-Methylaniline $C_6H_5NHCH_3$
    No colour with NaOCl soln. Hydrochloride m.p. 121°. Acetyl m.p. 102°. Benzoyl m.p. 63°. p-Tosyl m.p. 95°. 2,4-Dinitrophenyl m.p. 167°. Picrate m.p. 147°.

205° N-Ethylaniline $C_6H_5NHC_2H_5$
    No colour with NaOCl soln. Hydrochloride m.p. 176°. Acetyl m.p. 54°. Benzoyl m.p. 60°. p-Tosyl m.p. 88°. 2,4-Dinitrophenyl m.p. 95°. Picrate m.p. 132°.

206° N-Methyl-m-toluidine $CH_3C_6H_4NHCH_3$
    Acetyl m.p. 66°.

207° N-Methyl-o-toluidine $CH_3C_6H_4NHCH_3$
    Acetyl m.p. 55°. Benzoyl m.p. 66°. p-Tosyl m.p. 120°. 2,4-Dinitrophenyl m.p. 155°. Picrate m.p. 90°.

208° N-Methyl-p-toluidine $CH_3C_6H_4NHCH_3$
    Acetyl m.p. 83°. Benzoyl m.p. 53°. p-Tosyl m.p. 60°. Picrate m.p. 131°.

B.P.°C

217°   *N*-Ethyl *p*-toluidine $CH_3C_6H_4NHC_2H_5$
        p-Tosyl m.p. 71°. 2,4-Dinitrophenyl m.p. 120°.

218°   *N*-Ethyl *o*-toluidine $CH_3C_6H_4NHC_2H_5$
(214°) Benzoyl m.p. 72°. *p*-Tosyl m.p. 75°. 2,4-Dinitrophenyl m.p. 114°.

221°   *N*-Ethyl *m*-toluidine $CH_3C_6H_4NHC_2H_5$
        Hydrochloride m.p. 159°. Benzoyl m.p. 72°. Picrate m.p. 132°.

222°   n-Propylaniline $C_6H_5NHCH_2C_2H_5$
        Acetyl m.p. 48°. *p*-Tosyl m.p. 56°.

240°   n-Butylaniline $C_6H_5NH(CH_2)_3CH_3$
(235°) Benzoyl m.p. 56°. *p*-Tosyl m.p. 54°.

254°   Dicyclohexylamine $(C_6H_{11})_2NH$
        M.p. 20°. Acetyl m.p. 103°. Benzoyl m.p. 153°. *p*-Tosyl m.p. 119°.
        Picrate m.p. 173°.

300°d   Dibenzylamine $(C_6H_5CH_2)_2NH$
        Hydrochloride m.p. 256°. 2,4-Dinitrophenyl m.p. 105°. Benzoyl
        m.p. 112°.

SOLID

M.P.°C

28°   Diethanolamine $(HOCH_2CH_2)_2NH$
        B.p. 270°. *p*-Tosyl m.p. 99°. Picrate m.p. 110°.

37°   Benzylaniline $C_6H_5NHCH_2C_6H_5$
        Acetyl m.p. 58°. Benzoyl m.p. 107°. *p*-Tosyl m.p. 148°. 2,4-
        Dinitrophenyl m.p. 168°.

53°   Diphenylamine $(C_6H_5)_2NH$
        Weakly basic. Salts are hydrolysed by H_2O. Soln. in conc. H_2SO_4
        gives blue colour with trace of HNO_2 or HNO_3. Acetyl m.p.
        101°. Benzoyl m.p. 180°. *p*-Tosyl m.p. 142°. Picrate m.p. 182°.
        N-Nitroso m.p. 66°.

79°   Di-*p*-tolylamine $(CH_3C_6H_4)_2NH$
        Like diphenylamine, but yellow colour in conc. $H_2SO_4 + HNO_3$.
        Acetyl m.p. 85°. Benzoyl m.p. 125°. *N*-Nitroso m.p. 101°.

TERTIARY AMINES

LIQUID

B.P.°C

3°   Trimethylamine $(CH_3)_3N$
        Red-yellow ppt. with Nessler's soln.; yellow ppt. with HgI_2/KI
        soln. Hydrochloride m.p. 278°, readily sol. in alc. and in $CHCl_3$.
        Methiodide m.p. 230°d, spar. sol. in cold H_2O. Picrate m.p
        216°.

89°   Triethylamine $(C_2H_5)_3N$
        Hydrochloride m.p. 253°, sol. in alc. and in $CHCl_3$. Methiodide
        m.p. 280°. Picrate m.p. 173°.

123°   Dimethylaminoacetone $(CH_3)_2NCH_2COCH_3$
        Oxime m.p. 99°.

B.P.°C

156° Tri-n-propylamine $(C_2H_5CH_2)_3N$
Methiodide m.p. 208°. Picrate m.p. 116°.

185° Dimethyl o-toluidine $CH_3C_6H_4N(CH_3)_2$
Hydrochloride m.p. 156°. Methiodide m.p. 210°. Picrate m.p. 122°.

193° Dimethylaniline $C_6H_5N(CH_3)_2$
M.p. 2°. Hydrochloride m.p. ca. 90°, hygroscopic; sulphate m.p. 80°. $CH_2O \rightarrow$ Michler's hydride (q.v.). $Br_2/AcOH \rightarrow p$-bromo m.p. 55°. $HNO_3/AcOH \rightarrow$ dinitro m.p. 87°. Methiodide m.p. 220°(d). Ethiodide m.p. 136°. Picrate m.p. 162°. $NaNO_2/HCl \rightarrow p$-nitroso hydrochloride m.p. 177°, which with alkali $\rightarrow p$-nitroso m.p. 87°.

201° N-Ethyl-N-methylaniline $C_6H_5N(CH_3)C_2H_5$
Hydrochloride m.p. 114°. Methiodide m.p. 125°. Picrate m.p. 134°. p-Nitroso m.p. 66°.

209° Diethyl-o-toluidine $CH_3C_6H_4N(C_2H_5)_2$
Methiodide m.p. 224°. Picrate m.p. 180°.

210° Dimethyl-p-toluidine $CH_3C_6H_4N(CH_3)_2$
Methiodide m.p. 220°d. Picrate m.p. 129°.

211° Tri-n-butylamine $(C_2H_5CH_2CH_2)_3N$
(216°) Methiodide m.p. 180°. Picrate m.p. 106°.

212° Dimethyl m-toluidine $CH_3C_6H_4N(CH_3)_2$
Methiodide m.p. 177°. Picrate m.p. 131°.

216° Diethylaniline $C_6H_5N(C_2H_5)_2$
Methiodide m.p. 104°. Picrate m.p. 142°. p-Nitroso m.p. 84°.

229° Diethyl-p-toluidine $CH_3C_6H_5N(C_2H_5)_2$
Hydrochloride m.p. 157°. Methiodide m.p. 184°. Picrate m.p. 110°.

231° Diethyl-m-toluidine $CH_3C_6H_4N(C_2H_5)_2$
Picrate m.p. 97°.

245° Di-n-propylaniline $C_6H_5N(CH_2C_2H_5)_2$
Methiodide m.p. 156°. p-Nitroso m.p. 42°.

271° Di-n-butylaniline $C_6H_5N(CH_2CH_2C_2H_5)_2$
Picrate m.p. 125°.

296° N-Methyldiphenylamine $(C_6H_5)_2NCH_3$
Salts hydrolysed by H_2O. Violet colour with HNO_3. p-Nitroso m.p. 44°.

SOLID

M.P.°C

21° Triethanolamine $(HOCH_2CH_2)_3N$
B.p. 278°/150 mmHg. Alc. soln. + conc. HCl (0°) $\rightarrow$ hydrochloride m.p. 178°.

51° Tetramethyl p-phenylenediamine $(CH_3)_2NC_6H_4N(CH_3)_2$

70° Dibenzylaniline $C_6H_5N(CH_2C_6H_5)_2$
B.p. 300°. Salts hydrolysed by H_2O. $K_2Cr_2O_7/$dil. H_2SO_4 oxdtn. $\rightarrow$ C_6H_5CHO + benzalaniline. Methiodide m.p. 135°. Picrate m.p. 131°d. p-Nitroso m.p. 91°.

M.P.°C

73° *p*-Dimethylaminobenzaldehyde $(CH_3)_2NC_6H_4CHO$
Aniline acetate soln. → yellow anil m.p. 100°. Aniline hydrochloride soln. → yellow colour. With dimethylaniline, pass in HCl, then add $FeCl_3$ → violet dye. Oxime m.p. 148°. Phenylhydrazone m.p. 148°. *p*-Nitrophenylhydrazone m.p. 182°. Semicarbazone m.p. 222°.

90° 4,4′-Bis(dimethylamino) diphenylmethane $(CH_3)_2NC_6H_4CH_2C_6$-$H_4N(CH_3)_2$
Michler's hydride. Yields *p*-benzoquinone on oxdtn. Methiodide (di) m.p. 214°d. Picrate (mono) m.p. 185°; (di) m.p. 178°.

91° Tribenzylamine $(C_6H_5CH_2)_3N$
Very weak base. Hydrochloride m.p. 227° is hydrolysed by H_2O. Sulphate m.p. 106° is almost insol. in H_2O. Methiodide m.p. 184°. Picrate m.p. 190°.

98° 4,4′-Bis(dimethylamino)benzhydrol $(CH_3)_2NC_6H_4CHOHC_6$-$H_4N(CH_3)_2$
Michler's hydrol. Green. Oxidised to Michler's ketone and reduced to Michler's hydride. Methiodide (di) m.p. 195°.

127° Triphenylamine $(C_6H_5)_3N$
Does not form salts with acids. Soln. in conc. H_2SO_4 + HNO_3 (trace) → blue colour; soln. in acetic acid + HNO_3 (trace) → green colour. No picrate formed. Fuming $HNO_3/AcOH$ → trinitro m.p. 280°.

174° 4,4′-Bis(dimethylamino)benzophenone $(CH_3)_2NC_6H_4COC_6$-$H_4N(CH_3)_2$
Michler's ketone. Zn dust distltn. → Michler's hydride m.p. 90°. Methiodide (di) m.p. 105°. Picrate m.p. 156°. Oxime m.p. 233°. Phenylhydrazone m.p. 174°.

280° Hexamethylene tetramine. See Aldehyde-Ammonias.

SIMPLE AND SUBSTITUTED HETEROCYCLIC BASES

LIQUID

B.P.°C

89° Pyrrolidine C_4H_9N
Miscible with H_2O. Fumes in air. *p*-Tosyl m.p. 123°. Picrate (mono) m.p. 112°.

106° Piperidine $C_5H_{11}N$
Miscible with H_2O. Hydrochloride m.p. 237°. Benzoyl m.p. 48°. *p*-Tosyl m.p. 103° (96°). 2,4-Dinitrophenyl m.p. 92°. Picrate m.p. 152°.

115° Pyridine C_5H_5N
Characteristic odour. Miscible with H_2O. Salt with 1-chloro 2,4-dinitrobenzene m.p. 201°, gives purple colour with dil. NaOH. Addtn. product with chloroacetic acid m.p. 202°.

B.P.°C

Methiodide m.p. 117°. Ethiodide m.p. 91°. Picrate m.p. 167°. Perchlorate m.p. 209°.

129° 2-Methylpyridine C_6H_7N, α-Picoline.
Like pyridine. Boiling dil. $KMnO_4$ → picolinic acid m.p. 138°. Methiodide m.p. 227°. Ethiodide m.p. 123°. Picrate m.p. 169°.

130° Morpholine C_4H_9NO
Soluble in all common solvents. Steam-volatile. Hydrochloride m.p. 175°. Benzoyl m.p. 75°. p-Tosyl m.p. 147°. Picrate m.p. 146°.

131° Pyrrole C_4H_5N
Spar. sol. in H_2O. Polymerised by acids. Picrate m.p. 69°d. Phenylurethane m.p. 142°.

143° 3-Methylpyridine C_6H_7N, β-Picoline.
Like pyridine. Boiling dil. $KMnO_4$ → nicotinic acid m.p. 228°. Methiodide m.p. 92°. Picrate m.p. 150°.

143° 4-Methylpyridine C_6H_7N, γ-Picoline.
Like pyridine. Boiling dil. $KMnO_4$ → isonicotinic acid m.p. 308°. Methiodide m.p. 152°. Picrate m.p. 167°.

172° 2,4,6-Trimethylpyridine, γ-Collidine.
Picrate m.p. 156°.

202° N-Methyl-2-pyrrolidone C_5H_9NO
Very good solvent, especially for saccharides. Steam volatile. Hydrochloride m.p. 80°.

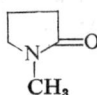

233° 1,2,3,4-Tetrahydroisoquinoline $C_9H_{11}N$
Reduces Tollen's reagent. Acetyl m.p. 46°. Benzoyl m.p. 129°. Picrate m.p. 200° (195°).

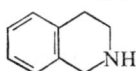

238° Quinoline C_9H_7N
Reduced by Sn + HCl → tetrahydroquinoline (q.v.). Methiodide m.p. 133° (anhydrous), 72° (hydrate). Ethiodide m.p. 159°. Picrate m.p. 203°.

247° 2-Methylquinoline $C_{10}H_9N$, Quinaldine.
Boiling conc. HNO_3 → nitroquinaldinic acid m.p. 219°. Methiodide m.p. 195° and ethiodide m.p. 233° formed more slowly than from quinoline. With phthalic anhydride (190°) → yellow quinophthalone m.p. 239°.

250° 1,2,3,4-Tetrahydroquinoline $C_9H_{11}N$
M.p. 20°. Hydrochloride m.p. 181°. Benzoyl m.p. 75°. Picrate m.p. 141°.

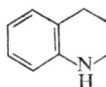

262° 4-Methylquinoline $C_{10}H_9N$, Lepidine.
Spar. sol. in H_2O. Methiodide m.p. 173°. Ethiodide m.p. 142°. Picrate m.p. 211°.

SOLID

M.P.°C

21°   Pyrimidine $C_4H_4N_2$
B.p. 124°. Miscible with H_2O → neutral soln. Picrate m.p. 156°.

26°   Isoquinoline C_9H_7N
B.p. 243°. Stronger base than quinoline. Methiodide m.p. 159°.
Ethiodide m.p. 148°. Picrate m.p. 222°.

44°   Piperazine hydrate. See piperazine (m.p. 104°).

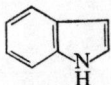

52°   Indole C_8H_7N
B.p. 253°. Sol. in hot H_2O. Acetyl m.p. 157°. Benzoyl m.p. 68°.
Picrate m.p. 187°. N-Nitroso m.p. 171°.

58°   2-Aminopyridine $C_5H_6N_2$
B.p. 204°. Acetyl m.p. 71°. Benzoyl m.p. 169°. Picrate m.p. 216°.

69°   2,2'-Bipyridyl $C_{10}H_8N_2$
Spar. sol. in H_2O. Ligand for many heavy metals. Red colour
with ferrous salts. Picrate m.p. 158°.

75°   8-Hydroxyquinoline C_9H_7NO, Oxine
B.p. 267°. Forms ppts. suitable for quantitative analysis with
many metals e.g. Al, Zn, Mg. Green colour with neutral $FeCl_3$
soln. Benzoyl m.p. 120°. p-Tosyl m.p. 115°. Methiodide m.p.
143°. Picrate m.p. 204°.

95°   3-Methylindole C_9H_9N, Skatole.
Foul odour. B.p. 267°. Hydrochloride m.p. 167°. Acetyl m.p.
68°. Picrate m.p. 170°.

104° Piperazine $C_4H_4N_2$
B.p. 140°. Readily sol. in H_2O, spar. sol. in ether. Odourless.
Acetyl (mono) m.p. 52°, (di) m.p. 134°. Benzoyl (mono) m.p.
75°, (di) m.p. 191°. Picrate m.p. 280°. N-Nitroso (di) m.p. 158°.

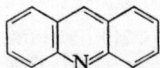

111° Acridine $C_{13}H_9N$
Sublimes at 100°. Solns. have blue fluorescence. Steam volatile.
Methiodide m.p. 224°. Picrate m.p. 208°.

113° Antipyrine. See Hydrazine derivatives.

121° 2,6-Diaminopyridine $C_5H_7N_3$
Acetyl (di) m.p. 203°. Benzoyl (di) m.p. 176°. Picrate m.p. 240°.

199° 2-Hydroxyquinoline C_9H_7NO, Carbostyril.
See Aminophenols.

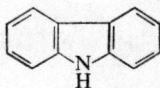

243° Carbazole $C_{12}H_9N$
Very weak base. Salts are hydrolysed by H_2O. Soln. in conc.
H_2SO_4 + $NaNO_2$ (trace) → green colour. Blue colour with dil.
soln. of isatin in conc. H_2SO_4. Forms K deriv. on fusion with
KOH. Yields NH_3 with soda lime. N-Nitroso m.p. 84°.

AMINOPHENOLS

SOLID

M.P.°C

75° *p*-Dimethylaminophenol $(CH_3)_2NC_6H_4OH$ (1,4).
Sol. cold H_2O and ether. Soln. in NaOH darkens giving odour of methyl isocyanide. No colour with $FeCl_3$ or $AgNO_3$ in cold, odour of quinone on warming. Acetyl m.p. 78°. Benzoyl (di) m.p. 158°. *p*-Tosyl m.p. 130°.

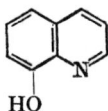

76° 8-Hydroxyquinoline C_9NH_7O, Oxine.
Forms insol. chelate coordination complexes with metallic ions. Sol. in dil. HCl and dil. NaOH. With aqu. $FeCl_3 \rightarrow$ blue-green colour (violet in MeOH). Benzoate, m.p. 118°. *O-p*-Tosyl, m.p 115°. Methiodide m.p. 143°. Picrate m.p. 204°.

79°d 2,4-Diaminophenol $HOC_6H_3(NH_2)_2$ (1,2,4).
Sol. cold H_2O. Soln. in NaOH rapidly darkens. With soln. of $FeCl_3$ or $AgNO_3 \rightarrow$ intense red colour. Acetyl (tri) m.p. 180°. Benzoyl (di) m.p. 253°, (tri) m.p. 231°.

85° *p*-Methylaminophenol $HOC_6H_4NHCH_3$ (1,4)
Mod. sol. cold H_2O, sol. in ether. Soln. in NaOH darkens giving odour of methyl isocyanide. With soln. of $FeCl_3$ or $AgNO_3 \rightarrow$ purple colour in cold, odour of quinone on warming. Acetyl (mono) m.p. 240°. Benzoyl (di) m.p. 175°.

96° *o*-Methylaminophenol $HOC_6H_4NHCH_3$ (1,2)
Sparingly sol. cold H_2O. Sol. in NaOH → dark green colour with odour of methyl isocyanide. With soln. of $FeCl_3 \rightarrow$ purple colour changing to red-brown on warming. Yellow-brown colour with $AgNO_3$ soln. Acetyl (mono) m.p. 150°. Benzoyl (di) m.p. 150°. Nitrosamine m.p. 130°d. (insol. cold H_2O).

122° *m*-Aminophenol $HOC_6H_4NH_2$ (1,3)
Sparingly sol. cold H_2O and benzene. Soln. in HCl gives no characteristic colour with $FeCl_3$ in cold, on warming soln. darkens. Acetyl, (mono) m.p. 148°, (di) m.p. 101°. Benzoyl (mono) m.p. 174°, (di) m.p. 153°. *p*-Tosyl m.p. 157°.

174°d *o*-Aminophenol $HOC_6H_4NH_2$ (1,2)
Sparingly sol. cold H_2O, readily in ether. Gives dark brown ppt. with $FeCl_3$ soln., no odour of quinone on warming. With $AgNO_3$ soln. → yellow-brown colour especially on warming. Acetyl (mono) m.p. 201°. Benzoyl (di) m.p. 184°. *p*-Tosyl m.p. 146°.

175° 4-Amino-2-methylphenol $HOC_6H_3(CH_3)NH_2$ (1,2,4).
Sparingly sol. cold H_2O, sol. in ether. Soln. in NaOH rapidly darkens. With soln. of $FeCl_3$ or $AgNO_3 \rightarrow$ purple colour in cold, and toluquinone (m.p. 68°) on warming. Acetyl, (mono) m.p. 179°, (di) m.p. 103°. Benzoyl (di) m.p. 194°. *p*-Tosyl m.p. 109°.

184°d *p*-Aminophenol $HOC_6H_4NH_2$ (1,4)
Sol. 100 parts cold H_2O. With soln. of $FeCl_3$ or $AgNO_3 \rightarrow$ purple colour in cold; on warming benzoquinone (m.p. 116°) is

M.P.°C

formed. Soln. in HCl, treated with NaOCl soln. → ppt. of quinonechlorimide, m.p. 85°. Acetyl, (mono) m.p. 168°, (di) m.p. 150°. Benzoyl, (mono) m.p. 216°, (di) m.p. 234°. Benzal, m.p. 183°.

4-Amino-1-naphthol $C_{10}H_6(OH)NH_2$

Turns blue in air. Oxdtn. → 1,4-naphthoquinone. Diacetyl, m.p. 158°. Dibenzoyl, m.p. 215°.

199°   Carbostyril C_9NH_7O, 2-Hydroxyquinoline.

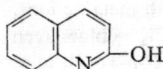

Almost insol. in cold H_2O. Sol. in dil. NaOH. No colour with $FeCl_3$ soln. Hot Zn dust → quinoline. Conc. $HNO_3 + H_2SO_4$ → 6-nitro deriv., m.p. 280°. PCl_5 at 130° → 2-chloroquinoline, m.p. 37° (b.p. 275°). 2-Methyl ether picrate m.p. 170°. p-Tosyl m.p. 82°.

SIMPLE AMIDES AND IMIDES, UREAS, GUANIDINES

SOLID

M.P.°C

3°   Formamide $HCONH_2$

B.p. 193°–decomposes yielding NH_3, CO and H_2O. d_4^{20} 1.1334. n_D^{20} 1.4472. Completely miscible with H_2O and with EtOH; spar. sol. in ether. Hot conc. H_2SO_4 → CO. With hot aniline → formanilide, m.p. 46° and NH_3; with 1-naphthylamine → formyl-1-naphthylamine m.p. 138°.

—   Guanidine $(NH_2)_2C:NH$

Alkaline deliquescent solid. Carbonate m.p. 197°. Nitrate m.p. 214°. Nitrite m.p. 77°. Acetate m.p. 229°. Hydrochloride m.p. 172°. Thiocyanate m.p. 118°. Picrate m.p. above 280° is almost insol. in cold water. Rapidly oxidised by $KMnO_4$. Boiling $Ba(OH)_2$ → urea and NH_3. With alk. NaOCl soln. → yellow colour changing to orange-red. With cold conc. $HNO_3 + H_2SO_4$ → nitroguanidine m.p. 230°d (long needles from hot water).

—   Methylguanidine $(NH_2)_2C:NCH_3$

Alkaline deliquescent solid. Nitrate m.p. 150°. Sulphate m.p. 240°. Platinichloride m.p. 194° is readily sol. in H_2O. Picrate m.p. 201°. Rapidly oxidised by $KMnO_4$. Boiling with HCl → NH_4Cl and $CH_3NH_2 \cdot HCl$.

49°   Ethyl carbamate $NH_2COOC_2H_5$, Ethyl urethane.

B.p. 184°. Sol. cold H_2O, EtOH and ether. Boiling dil. HCl → CO_2, EtOH and NH_4Cl. Cold alc. KOH → cryst. KOCN. Boiling aniline → NH_3, EtOH and carbanilide, m.p. 241°. Xanthyl m.p. 169°.

52°   Methyl carbamate NH_2COOCH_3, Methyl urethane.

B.p. 177°. Sol. cold H_2O, EtOH and ether. Boiling dil. HCl → CO_2, MeOH and NH_4Cl. Cold alc. KOH → cryst. KOCN. Boiling aniline → NH_3, MeOH and carbanilide m.p. 241°. Xanthyl m.p. 193°.

M.P.°C

54°   n-Butyl carbamate $NH_2COOC_4H_9$, n-Butyl urethane.
Sl. sol. in cold H_2O, readily in EtOH and ether. Boiling dil.
$HCl \rightarrow$ n-Butyl alc. and NH_4Cl. Boiling aniline $\rightarrow NH_3$, n-butyl
alc. and carbanilide m.p. 241°. Cold alc. KOH $\rightarrow$ cryst. KOCN
(slowly).

60°   n-Propyl carbamate $NH_2COOC_3H_7$, n-Propyl urethane.
B.p. 195°. Sol. in H_2O, EtOH and ether. Boiling dil. HCl $\rightarrow$
n-propyl alc. and NH_4Cl. Boiling aniline $\rightarrow NH_3$, n-propyl alc.
and carbanilide m.p. 241°. Cold. alc. KOH $\rightarrow$ cryst. KOCN
(slowly).

64°   Isoamyl carbamate $NH_2COOC_5H_{11}$, Isoamyl urethane.
B.p. 220°. Spar. sol. cold H_2O, readily in EtOH and ether.
Boiling dil. HCl $\rightarrow$ isoamyl alc. and NH_4Cl. Hot aniline $\rightarrow$
NH_3, isoamyl alc. and carbanilide m.p. 241°. Cold alc. KOH $\rightarrow$
cryst. KOCN (very slowly).

79°   Propionamide $CH_3CH_2CONH_2$
B.p. 213°. Readily sol. H_2O, EtOH and ether. At 120°, 75%
$H_2SO_4 \rightarrow$ propionic acid, b.p. 140°. Hot aniline $\rightarrow$ propionanilide
m.p. 103° and NH_3. Hot 1-naphthylamine $\rightarrow$ propionyl-1-
naphthylamine m.p. 116°. Xanthyl m.p. 214°.

82°   Acetamide CH_3CONH_2
B.p. 222°. Sol. H_2O and EtOH; insol. ether. Picrate m.p. 107°.
Distillation with conc. $H_2SO_4 \rightarrow$ acetic acid. Hot aniline $\rightarrow$
acetanilide m.p. 115° and NH_3. Hot 1-naphthylamine $\rightarrow$ acetyl-
1-naphthylamine m.p. 159°. Xanthyl m.p. 245° (240°).

96°   Semicarbazide. See Hydrazine derivatives.

97°   m-Toluamide $C_6H_4(CH_3)CONH_2$ (1,3)
Spar. sol. in benzene and ether. Hydrol. $\rightarrow$ m-toluic acid m.p.
112°. N-Di-Et deriv. is insect repellant b.p. 160°/19 mmHg.
N-Benzyl m.p. 75°.

105°   Dicyandiamidine $NH_2C(:NH)NHCONH_2$
Strong base. Sol. H_2O, spar. sol. EtOH, insol. ether. Hydrochloride m.p. 173°. Sulphate m.p. 194°. Picrate m.p. 265° is
almost insol. in H_2O. Heating to 160° or with boiling water $\rightarrow$
NH_3. Warm $Ba(OH)_2$ soln. $\rightarrow NH_3$, CO_2 and urea. Aqn.
$CuSO_4 + NaOH \rightarrow$ spar. sol. red Cu salt. Ni salt yellow and
insol. in H_2O.

106°   1,3-Dimethylurea $(CH_3)HNCONH(CH_3)$
B.p. 268°–270°. Sol. in H_2O, EtOH. Hot NaOH soln $\rightarrow CH_3NH_2$.
Boiling HCl $\rightarrow CH_3NH_2 \cdot HCl$ and CO_2. Hydrochloride m.p. 124°.
Nitrate m.p. 65°.

109°   Stearamide $CH_3(CH_2)_{16}CONH_2$, Octadecanamide.
B.p. 250°/12 mmHg. Sol. in ether, hot EtOH, hot H_2O. Refluxing
with aniline $\rightarrow$ stearanilide m.p. 94°. Hydrolysis $\rightarrow NH_3$ and
stearic acid m.p. 69°. Xanthyl m.p. 140°.

114°   Ethyl oxamate $NH_2COCOOC_2H_5$
Sol. hot H_2O, forming acid soln. Boiling dil. HCl $\rightarrow$ oxalic acid,

M.P.°C

EtOH and NH₄Cl. Hot aniline → NH₃, EtOH and oxanilide m.p. 246°. Alk. NaOCl soln. → N₂.

115°   n-Butyramide   CH₃(CH₂)₂CONH₂
B.p. 216°. Readily sol. in H₂O, EtOH and ether. At 120°, 75 % H₂SO₄ → n-butyric acid. Hot aniline → n-butyranilide m.p. 90°. Hot 1-naphthylamine → n-butyryl-1-naphthylamine m.p. 120°. Xanthyl m.p. 186°.

125°   Succinimide   C₄H₅O₂N

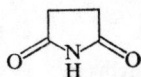

B.p. 287°. Boiling HCl → succinic acid m.p. 185° and NH₄Cl. Warm Ba(OH)₂ soln. → succinamic acid m.p. 157°. Hot Zn dust → pyrrole (vapours redden aniline acetate paper). Hot aniline → succinanil m.p. 156°, and NH₃. Xanthyl m.p. 246°.

128°   Benzamide   C₆H₅CONH₂
B.p. 290°. Spar. sol. in cold H₂O. At 120°, 75% H₂SO₄ → benzoic acid m.p. 121°. Hot aniline → benzanilide m.p. 164° and NH₃. Xanthyl m.p. 223°.

128°   Isobutyramide   (CH₃)₂CHCONH₂, 2-Methylpropionamide.
B.p. 218°. Sol. in H₂O, EtOH and ether. At 120°, 75% H₂SO₄ → isobutyric acid. Hot aniline → isobutyranilide m.p. 105° and NH₃.

132°   Urea   H₂NCONH₂, Carbamide.
Sol. in H₂O, EtOH. Insol. in ether. 50% HNO₃ → urea nitrate, m.p. 163°, spar. sol. in H₂O. Oxalate m.p. 171°. Heating to 150° → biuret m.p. 192° and NH₃. NaOBr soln. → N₂. Boiling HCl → CO₂ and NH₂Cl. Boiling aq. NaOH → NH₃ slowly. Warm aniline hydrochloride soln. → phenylurea m.p. 147 and carbanilide m.p. 241°. Hot aniline → carbanilide m.p. 241° and NH₃. Xanthyl m.p. 274°.

139°   Salicylamide   HOC₆H₄CONH₂ (1,2)
Spar. sol. in cold H₂O. FeCl₃ soln. → violet colour. Sol. in cold dil. NaOH, pptd unchanged by CO₂. Boiling NaOH soln. → salicylic acid, m.p. 158°, on acidification. Acetate m.p. 135°. Benzoate m.p. 143°.

147°   o-Toluamide   C₆H₄(CH₃)CONH₂ (1,2)
Sol. in hot H₂O, EtOH. Spar. sol. ether, benzene. Hydrol. → NH₃ and o-toluic acid m.p. 108°. Benzoyl m.p. 158–9°. Xanthyl m.p. 200°.

147°   N-Phenylurea   C₆H₅NHCONH₂
Sol. in hot H₂O. Heating above m.p. → NH₃, CO₂ and carbanilide, m.p. 241°. Hot aniline (or hot soln. aniline HCl) → carbanilide. Boiling HCl → aniline HCl and NH₄Cl. NaOH soln. → aniline and NH₃. Xanthyl m.p. 225°.

147°   Diphenylguanidine   (C₆H₅NH)₂C:NH
Insol. in H₂O. Sol. in HCl. Heat (170° for 1 hr) → NH₃, aniline and triphenyldicarbimide m.p. 74°. Boiling HCl → NH₄Cl and aniline HCl. Acetic anhydride at 100° → acetanilide m.p. 114° and sym-acetylphenylurea m.p. 183°.

M.P.°C

149°   *N*-Benzylurea $C_6H_5CH_2NHCONH_2$
Sol. in hot H_2O. Heating to 200° → CO_2, NH_3 and dibenzylurea.
m.p. 167°. Boiling HCl → benzylamine HCl and NH_4Cl.

157°   Phenylacetamide $C_6H_5CH_2CONH_2$
B.p. 262°d. Spar. sol. in cold H_2O. Boiling HCl → phenylacetic
acid m.p. 76° and NH_4Cl. Hot aniline → phenylacetanilide m.p.
117° and NH_3. Xanthyl m.p. 196°.

160°   *p*-Toluamide $C_6H_4(CH_3)CONH_2$ (1,4)
V. sol. in hot H_2O, EtOH, ether. Spar. sol. in cold H_2O, chloroform, benzene. Hydrol. → *p*-toluic acid m.p. 180°. *N*-Acetyl
m.p. 147°. *N*-Benzoyl m.p. 119°. Xanthyl m.p. 224°.

170°   Malonamide $CH_2(CONH_2)_2$
Sol. in 10 parts cold H_2O. Insol. abs. EtOH and ether. Aq
$CuSO_4$ + dil. NaOH → red colour. Boiling conc. HCl → acetic
acid and NH_4Cl. Boiling NaOH soln. → Na malonate and NH_3.
Hot aniline → malonanilide m.p. 224° and NH_3. Xanthyl m.p.
270°.

174°   *p*-Phenetylurea $C_2H_5OC_6H_4NHCONH_2$ (1,4), 'Dulcin'.
Spar. sol. in cold H_2O. Sweet taste. Heating above m.p. →
NH_3, CO_2 and di-*p*-phenetylurea m.p. 235°. Boiling HCl → *p*-
phenetidine·HCl and NH_4Cl. Boiling 48% HBr → *p*-amino-
phenol·HBr, NH_4Br and EtBr.

181°   *p*-Tolylurea $CH_3C_6H_4NHCONH_2$ (1,4)
Sol. in hot H_2O and EtOH. Heating above m.p. → CO_2, NH_3 and
di-*p*-tolylurea m.p. 268°. Boiling HCl → *p*-toluidine·HCl and
NH_4Cl. 3-*N*-Acetyl m.p. 200°. 3-*N*-Benzoyl m.p. 223°. 1-*N*-
Nitroso m.p. 83°.

182°   *asym*-Dimethylurea $(CH_3)_2NCONH_2$
Sol. in H_2O. Spar. sol. in cold EtOH. Insol. in ether. Boiling
NaOH soln. → $(CH_3)_2NH$ and NH_3. Boiling dil. HCl →
$(CH_3)_2NH·HCl$ and NH_4Cl. B·HNO_3 m.p. 103°. Picrate m.p.
130°d. Xanthyl m.p. 250°.

189°   *asym*-Diphenylurea $(C_6H_5)_2NCONH_2$
Boiling NaOH soln. → diphenylamine m.p. 54° and NH_3.
Boiling HCl → diphenylamine·HCl and NH_4Cl. Xanthyl m.p.
180°.

191°   *o*-Tolylurea $CH_3C_6H_4NHCONH_2$ (1,2)
Sol. in hot H_2O. Heating above m.p. → NH_3, CO_2 and di-*o*-
tolylurea m.p. 250°. Boiling HCl → *o*-toluidine·HCl and NH_4Cl.
3-*N*-Acetyl m.p. 168°. 3-*N*-Benzoyl m.p. 210°. Xanthyl m.p. 228°.

192°d   Biuret $NH_2CONHCONH_2$
Sol. in warm H_2O. Heating above m.p. → NH_3. NaOH soln. +
trace $CuSO_4$ → red colour; with excess $CuSO_4$ → violet colour.
Boiling $Ba(OH)_2$ soln. → NH_3, CO_2 (as $BaCO_3$) and urea. Conc.
HNO_3 + conc. H_2SO_4 at 0° → nitro deriv. (m.p. 165°d) when
poured onto ice. Hot aniline → diphenylbiuret m.p. 210° and
NH_3. Xanthyl m.p. 260°.

M.P.°C

202°d Isatin $C_8H_5O_2N$, Indole-2,3-dione.

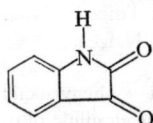

Yellowish-red prisms. Sol. in cold dil. NaOH. Sol. in MeOH, EtOH, Me_2CO, C_6H_6. Spar. sol. in cold H_2O, ether. 3-Semi-carbazone m.p. 266°d. 3-Oxime m.p. 225°d. N-Acetyl m.p. 141°.

219°d Phthalamide $C_6H_4(CONH_2)_2$ (1,2)

Heating above m.p. → phthalimide m.p. 233°, and NH_3. Boiling HCl → phthalic acid m.p. 195° and NH_4Cl. Hot aniline→ phthalanil m.p. 205° and NH_3.

233° Phthalimide $C_6H_4(CO)_2NH$

Almost insol. in cold H_2O. Boiling dil. NaOH → phthalic acid (on acidification) and NH_3. Warm $Ba(OH)_2$ soln. → phthal-amidic acid m.p. 148°. Alc. soln. + alc. KOH → cryst. ppt. of K deriv, which with hot benzyl chloride → benzylphthalimide m.p. 115°. Hot aniline → phthalanil m.p. 205° and NH_3. Xanthyl m.p. 177°.

— Alloxan $(CO)_3(NH)_2CO·H_2O$

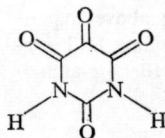

M.p. varies with conditions; often yellow at 180°, melting with decomposition at about 240–254°. Heat → NH_3, CO_2 and CO. V. sol. in cold H_2O, soln. develops red colour on skin. Boiling NaOH soln. → NH_3. Ferrous salts → blue colour. Warm $NH_2OH·HCl$ soln. → violuric acid (purple soln. with aq. Na_2CO_3). $SnCl_2$ in HCl → insol. alloxantin (blue ppt. with $Ba(OH)_2$ soln.).

245°d Barbituric acid $C_4H_4O_3N_2$

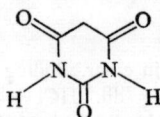

Decomposes on heating → NH_3. Spar. sol. cold H_2O, sol. in hot H_2O. Sol. in alkali. Boiling NaOH soln. → Na malonate + NH_3. NH_4 salt spar. sol. in cold H_2O. $AgNO_3$ soln. → yellow ppt. $NaNO_2$ soln. → Na violurate (purple soln.). Br water → dibromo deriv. m.p. 235°d. Acid diazobenzene soln. → spar. sol. alloxan phenylhydrazone m.p. 284° (this is sol. in K_2CO_3 soln. → reddish yellow colour and on boiling → NH_3 + yellow soln. which on acidification → product m.p. 164°).

260°d Succinamide $H_2NCOCH_2CH_2CONH_2$
(273°d)

Sl. sol. in cold H_2O, more so on warming. Insol. abs. EtOH and ether. Heating above m.p. → succinimide m.p. 125° and NH_3. Boiling dil. NaOH → succinic acid (on acidification) m.p. 185° and NH_3. Hot aniline → succinanil m.p. 156° and NH_3. Xanthyl m.p. 275°.

315°d Creatine $H_2NC(:NH)N(CH_3)CH_2COOH$

Sol. in H_2O; insol. in EtOH and ether. Boiling alkalis → NH_3. Warm HCl → creatinine (q.v.). Addition of $AgNO_3$ then KOH → white ppt. (sol. in excess of KOH); this soln. gelatinises on standing and darkens (rapidly on warming). $HgCl_2$ + KOH → white ppt. insol. in excess of KOH but similarly darkening on standing.

M.P.°C

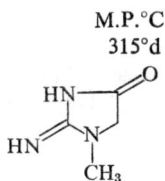

315°d   Creatinine $C_4N_3H_7O$

Sol. in H_2O, spar. sol. in EtOH. $KMnO_4$ oxdtn. $\rightarrow$ oxalic acid and methylguanidine. Soln. saturated with NO_2 then adding NH_4OH $\rightarrow$ insol. nitroso deriv. m.p. 210°. Reduces Fehling's soln. $\rightarrow$ white flocculent ppt. of creatinine-cuprous oxide. Nitroprusside+ $NaOH \rightarrow$ red colour changing to yellow; addtn. acetic acid $\rightarrow$ blue colour on warming. Picric acid $+$ $NaOH \rightarrow$ yellow colour. Picrate m.p. 220° (insol. in H_2O).

418°   Oxamide $H_2NCOCONH_2$

M.p. in sealed tube, *subl.* in open tube. Almost insol. in H_2O, EtOH and ether. $CuSO_4 + NaOH$ dil. $\rightarrow$ red colour. Boiling NaOH dil. $\rightarrow$ oxalic acid (*q.v.*) $+$ NH_3. Hot aniline $\rightarrow$ oxanilide m.p. 254° $+$ NH_3.

N-SUBSTITUTED AMIDES AND IMIDES

LIQUID

B.P.°C

153°   *N,N*-Dimethylformamide $HCON(CH_3)_2$
d_4^{22} 0.948. n_D^{22} 1.429.

165°   *N,N*-Dimethylacetamide $CH_3CON(CH_3)_2$
d_4^{20} 0.943. n_D^{22} 1.437.

221°   *N*-Formylpiperidine $C_6H_{11}ON$
d_4^{23} 1.019. Miscible with H_2O. Weak base. B·HBr m.p. 103°. Boiling $HCl \rightarrow HCOOH$ and piperidine·HCl (*q.v.*). $HgCl_2 \rightarrow$ cmpd. m.p. 148°.

226°   *N*-Acetylpiperidine $C_7H_{13}ON$
d_4^9 1.011. Miscible with H_2O. Weak base. B·HCl m.p. 95°. B·HBr m.p. 131°. Boiling $HCl \rightarrow CH_3COOH$ and piperidine HCl (*q.v.*).

SOLID

M.P.°C

47°   Formanilide C_6H_5NHCHO

B.p. 271°. Sl. sol. in cold H_2O. Boiling $HCl \rightarrow HCOOH$ and aniline HCl (*q.v.*). Hot Zn dust $\rightarrow$ CO, CO_2, aniline and benzonitrile (*q.v.*). Alcoholic $NaOH \rightarrow$ ppt. of Na deriv. $\rightarrow$ benzyl formanilide m.p. 48° on warming with benzyl chloride. Warm alcoholic KOH $+$ $CHCl_3 \rightarrow$ odour of phenyl isocyanide.

48°   *N*-Benzoylpiperidine $C_{12}H_{15}ON$

B.p. 320°. Insol. in H_2O. Boiling $HCl \rightarrow$ benzoic acid m.p. 121° and piperidine HCl (*q.v.*). Slow distillation with PCl_5 (1 mole) $\rightarrow$ 1,5-dichloropentane, b.p. 177° (insol. in cold conc. H_2SO_4), benzonitrile b.p. 190° (sol. in cold conc. H_2SO_4) and $POCl_3$ (decomp. by cold H_2O).

52°   Ethyl carbanilate $C_6H_5NHCOOC_2H_5$

B.p. 237°d. Almost insol. in cold H_2O. Boiling $HCl \rightarrow$ EtOH and aniline HCl (*q.v.*). Boiling aniline $\rightarrow$ EtOH $+$ carbanilide m.p. 241°. Htg. with $P_2O_5 \rightarrow$ phenyl isocyanate b.p. 166°.

M.P.°C

54°   *N*-Ethylacetanilide $C_6H_5N(C_2H_5)COCH_3$
B.p. 249°. Boiling HCl → acetic acid and ethylaniline HCl (*q.v.*).
Conc. $HNO_3 + H_2SO_4$ at 40° → *p*-nitro deriv. m.p. 118° → *p*-nitroethylaniline m.p. 95° on hydrolysis.

66°   Ethyl oxanilate $C_6H_5NHCOCOOC_2H_5$
Sol. in hot H_2O. Boiling dil. HCl → oxalic acid (*q.v.*), aniline HCl (*q.v.*) and EtOH. Htg. with aniline → EtOH and oxanilide m.p. 254°. Boiling alcoholic KOH → K oxanilate → K oxalate + aniline, on continued boiling.

73°   *N,N'*-Diethylcarbanilide $C_6H_5N(C_2H_5)CON(C_2H_5)C_6H_5$
Boiling HCl → ethylaniline HCl (*q.v.*). Explosion regulator.

73°   *N*-Formyldiphenylamine $(C_6H_5)_2NCHO$, (*N,N*-diphenylformamide).
Boiling HCl → formic acid and diphenylamine m.p. 54°. Soln. in conc. H_2SO_4 → blue colour when treated with trace of HNO_3.

85°   Acetoacetanilide $CH_3COCH_2CONHC_6H_5$
Sol. in hot H_2O. $FeCl_3$ → violet colour. Boiling HCl → acetic acid and aniline HCl (*q.v.*). Htg. with aniline → acetone and carbanilide m.p. 241°. Warm conc. H_2SO_4 → lepidone m.p. 224°.

95°   n-Butyranilide $CH_3CH_2CH_2CONHC_6H_5$
Boiling HCl → n-butyric acid and aniline HCl (*q.v.*).

95°   Stearanilide $CH_3(CH_2)_{16}CONHC_6H_5$
Sol. in EtOH, MeOH, ether, benzene. Spar. sol. in pet. ether. Insol. in cold H_2O. Boiling HCl → stearic acid and aniline HCl (*q.v.*).

101°   *N*-Acetyldiphenylamine $(C_6H_5)_2NCOCH_3$, (*N,N*-diphenylacetamide).
Boiling HCl → acetic acid and diphenylamine m.p. 54°. Soln. in conc. H_2SO_4 + trace of HNO_3 → blue colour.

102°   *N*-Methylacetanilide $C_6H_5N(CH_3)COCH_3$
B.p. 237°. Sol. in ether. Boiling HCl → acetic acid and methylaniline·HCl (*q.v.*). Conc. $HNO_3 + H_2SO_4$ at 40° → *p*-nitro deriv. m.p. 153° which on hydrolysis → *p*-nitromethylaniline m.p. 152°. Hot HNO_3 → methyl-2,4-dinitroaniline, m.p. 178°.

105°   Propionanilide $CH_3CH_2CONHC_6H_5$
Mod. sol. in hot H_2O, EtOH, ether. Spar. sol. in cold H_2O. Boiling HCl → propionic acid, b.p. 140° and aniline HCl (*q.v.*). Conc. $HNO_3 + H_2SO_4$ at 0° → *p*-nitro deriv. m.p. 182° which on hydrolysis → *p*-nitroaniline m.p. 147°.

112°   Acet-*o*-toluide $CH_3C_6H_4NHCOCH_3$ (1,2)
B.p. 303°. Sol. in hot H_2O. Boiling HCl → acetic acid and *o*-toluidine·HCl (*q.v.*). Hot $KMnO_4$ soln. → acetylanthranilic acid m.p. 185°.

113°   Antipyrine. See Hydrazine derivatives.

114°   Acetanilide $CH_3CONHC_6H_5$
B.p. 304°. Sol. in hot H_2O. Sol. in EtOH, ether. Boiling HCl → acetic acid + aniline·HCl (*q.v.*). Bromine (1 mole) in acetic

M.P.°C

        acid → *p*-bromo deriv. m.p. 167° → *p*-bromoaniline m.p. 63° on hydrolysis. NaOCl soln. → *N*-chloro deriv. m.p. 91° which with acetic acid + trace of HCl → *p*-chloro deriv. m.p. 179°. Conc. $HNO_3 + H_2SO_4$ → *p*-nitro deriv. m.p. 210° which on hydrolysis → *p*-nitroaniline m.p. 147°.

117°   α-Phenylacetanilide $C_6H_5CH_2CONHC_6H_5$
        Sol. in EtOH, ether. Boiling HCl → phenylacetic acid m.p. 76° and aniline·HCl (*q.v.*). Conc. HNO_3 in cold → 2,4-dinitrophenylacetic acid m.p. 189°.

122°   *N,N'*-Dimethylcarbanilide $C_6H_5N(CH_3)CON(CH_3)C_6H_5$
        B.p. 350°. Boiling HCl → methylaniline·HCl (*q.v.*). Explosion regulator.

127°   Phenylmethylpyrazolone. See Hydrazine derivatives.

128°   Acetylphenylhydrazine. See Hydrazine derivatives.

129°   Piperine $(CH_2O_2):C_6H_3[CH:CH]_2CONC_5H_{10}$ (1,2,4)
        Sol. in EtOH, benzene. Insol. in cold H_2O, pet. ether, dil. mineral acids. Spar. sol. ether. Conc. H_2SO_4 → red colour. Boiling HCl (or alcoholic KOH) → piperidine and piperic acid m.p. 216°. Acid $KMnO_4$ → piperonal m.p. 37°.

133°   4-Acet-*m*-xylidide $(CH_3)_2C_6H_3NHCOCH_3$ (1,3,4)
        Spar. sol. in H_2O. Sol. in EtOH. Boiling HCl → acetic acid and *m*-xylidine·HCl (*q.v.*). Fuming HNO_3 → 5-nitro deriv. m.p. 172°. Slow addtn. KNO_3 to soln. in conc. H_2SO_4 → 6-nitro deriv. m.p. 159°.

134°   Acet-2-naphthylamide $C_{10}H_7NHCOCH_3$, (*N*-β-naphthylacetamide).
        Mod. sol. in H_2O, EtOH. Boiling HCl → acetic acid and 2-naphthylamine·HCl (*q.v.*). Bromine (1 mole) in acetic acid → 1-bromo deriv. m.p. 140°. Fuming HNO_3 → dinitro derivs. (i) m.p. 185°, sol. in EtOH (ii) m.p. 235°, insol. in EtOH.

135°   Acet-*p*-phenetidide ('Phenacetin') $C_2H_5OC_6H_4NHCOCH_3$ (1,4)
        Spar. sol. in cold H_2O. Sol. in hot H_2O. Sol. in ether, benzene, EtOH. Boiling HCl → acetic acid and *p*-phenetidine·HCl, (*q.v.*). Boiling 48% HBr → acetic acid and *p*-aminophenol·HCl (*q.v.*). Warm 10% HNO_3 (2 mole) → 3-nitro deriv. m.p. 103° which on hydrolysis → nitrophenetidine m.p. 113°.

145°   Benz-*o*-toluidide $CH_3C_6H_4NHCOC_6H_5$ (1,2)
        Sol. in acetone, hot acetic acid. Boiling HCl → benzoic acid m.p. 121° and *o*-toluidine·HCl (*q.v.*). Hot $KMnO_4$ soln. → benzoylanthranilic acid m.p. 177°.

145°   α-Triphenylguanidine $C_6H_5N:C(NHC_6H_5)_2$
        Insol. in H_2O. Weak base. B·HCl m.p. 241°. Picrate m.p. 180°. Nitrate and oxalate are spar. sol. in cold H_2O. Hot conc. KOH → aniline. Distln. in CO_2 → carbanilide m.p. 241°. Htg. → aniline and carbodiphenylimide b.p. 330°.

147°   *N*-Phenylurea. See Simple Amides.

147°   Diphenylguanidine. See Simple Amides.

M.P.°C

149° *N*-Benzylurea. See Simple Amides.

149° Oxanilic acid. See Amino Acids.

153° Acet-*p*-toluidide $CH_3C_6H_4NHCOCH_3$ (1,4)
Spar. sol. in H_2O. Sol. in EtOH. Boiling HCl → acetic acid and
p-toluidine·HCl (*q.v.*). Adding to conc. $HNO_3 + H_2SO_4$ at
30°–40° → 3-nitro deriv. m.p. 144° which on hydrolysis →
nitrotoluidine m.p. 116°. Bromine (1 mole) in acetic acid →
3-bromo deriv. m.p. 117°. Hot $KMnO_4$ soln. → *p*-acetamido-
benzoic acid m.p. 256°.

153° Cinnamanilide $C_6H_5CH:CHCONHC_6H_5$
Boiling HCl → cinnamic acid m.p. 133° and aniline·HCl (*q.v.*).
Acid $KMnO_4$ soln. → benzaldehyde and benzoic acid m.p.
121°.

156° Succinanil CH_2—CO
  | >N—C_6H_5
  CH_2—CO

Distils at 400° without decomp. Insol. in H_2O. Sol. in ether, hot
EtOH. Boiling HCl → succinic acid m.p. 185° and aniline·HCl
(*q.v.*). Warm $Ba(OH)_2$ soln. → succinanilic acid m.p. 148°.

158° Benz-*p*-toluidide $C_6H_5CONHC_6H_4CH_3$ (1,4)
Sol. in EtOH. Boiling HCl → benzoic acid m.p. 121° and *p*-
toluidine HCl (*q.v.*). Fuming HNO_3 at 0° → dinitro deriv. m.p.
186°. CrO_3 in acetic acid → *p*-benzoylaminobenzoic acid m.p.
278°.

160° Acet-1-naphthylamide $C_{10}H_7NHCOCH_3$, (*N*-α-naphthylaceta-
mide).
Mod. sol. in boiling H_2O. V. sol. in EtOH. Boiling HCl →
acetic acid and 1-naphthylamine HCl (*q.v.*). Fuming HNO_3 in
acetic acid → dinitro deriv. m.p. 250°. Bromine (1 mole) in
acetic acid → 4-bromo deriv. m.p. 193°.

163° Benzanilide $C_6H_5CONHC_6H_5$
Distils without decomp. Spar. sol. in ether. Sol. in warm EtOH.
Boiling HCl → benzoic acid m.p. 121° and aniline·HCl (*q.v.*).
Adding to 77% HNO_3 at 0° pouring at once into H_2O → *o*-nitro
deriv. m.p. 94° (sol. in cold EtOH) and *p*-nitro deriv. mp. 199°
(insol. in cold EtOH). Bromine (1 mole in acetic acid → *p*-bromo
deriv. m.p. 204°.

165° *p*-Aminoacetanilide $CH_3CONHC_6H_5NH_2$ (1,4)
Spar. sol. in cold H_2O. Sol. in EtOH, ether. Boiling HCl conc. →
acetic acid and *p*-phenylenediamine·HCl (*q.v.*). Boiling acetic
acid or cold acetic anhydride → diacetyl-*p*-phenylenediamine
m.p. 304°. Benzal deriv. m.p. 165°. Azo-2-naphthol deriv. m.p
261°.

167° *N,N'*-Dibenzylurea $CO(NHCH_2C_6H_5)_2$
Insol. in H_2O. Sol. in EtOH, acetic acid. Boiling HCl → benzyl-
amine·HCl (*q.v.*).

168° Benzoylphenylhydrazine. See Hydrazine derivs.

M.P.°C

186°   *N,N'*-Diacetyl-*o*-phenylenediamine $C_6H_4(NHCOCH_3)_2$ (1,2)
V. sol. in hot H_2O, insol. in benzene. Boiling HCl → acetic acid
and *o*-phenylenediamine·HCl (*q.v.*). Htg. above m.p. → acetic
acid and methylbenziminazole m.p. 175°.

187°   Hippuric Acid. See Amino Acids.

191°   *N,N'*-Diacetyl-*m*-phenylenediamine $C_6H_4(NHCOCH_3)_2$ (1,3)
Sol. in hot H_2O. Sol. in EtOH. Boiling HCl → acetic acid and
m-phenylenediamine·HCl (*q.v.*). Conc. HNO_3 at 0° in presence
of urea → dinitro deriv. m.p. 228°.

199°   Carbostyril. See Aminophenols.

210°   Phthalanil

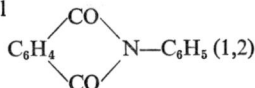

$$C_6H_4 \underset{\diagdown CO}{\overset{\diagup CO}{\diagup\diagdown}} N\!\!-\!\!C_6H_5 \ (1,2)$$

Insol. in H_2O. Sol. in hot H_2O, $CHCl_3$. Boiling HCl → phthalic
acid m.p. 195° and aniline·HCl (*q.v.*). Boiling $Ba(OH)_2$ soln. →
phthalanilic acid m.p. 169° on acidification.

225°   *N,N'*-Di-*m*-tolylurea $CH_3C_6H_4NHCONHC_6H_4CH_3$ (1,3::1',3')
Insol. in H_2O. B·HCl, m.p. 162°. Boiling HCl → *m*-toluidine
HCl (*q.v.*). Boiling acetic anhydride + Na acetate → acet-*m*-
toluidide m.p. 65°.

230°   Succinanilide $C_6H_5NHCOCH_2CH_2CONHC_6H_5$
Insol. in H_2O. Sol. in hot EtOH, ether. Boiling HCl → succinic
acid m.p. 185° and aniline·HCl (*q.v.*). Htg. above m.p. → aniline
and succinanil m.p. 156°.

241°   Carbanilide $CO(NHC_6H_5)_2$
B.p. 270°d. Almost insol. in H_2O. Boiling HCl → aniline·HCl
(*q.v.*). Boiling with acetic anhydride + Na acetate → acetanilide
m.p. 114°. Conc. HNO_3 in cold → di-*m*-nitro deriv. m.p. 248°
which on hydrolysis → *m*-nitroaniline m.p. 114°.

254°   Oxanilide $C_6H_5NHCOCONHC_6H_5$
Insol. in hot H_2O. Sl. sol. in cold EtOH and ether. Boiling
HCl → oxalic acid (*q.v.*) and aniline·HCl (*q.v.*). Boiling with
acetic anhydride + Na acetate → vinylideneoxanilide m.p. 208°.
Fuming HNO_3 in acetic acid → di-*p*-nitro deriv. m.p. 260° which
on hydrolysis → *p*-nitroaniline m.p. 147°.

255°d   *N,N'*-Di-*o*-tolylurea $CH_3C_6H_4NHCONHC_6H_4CH_3$ (1,2::1',2')
Almost insol. in H_2O. Boiling Cl → *o*-toluidine·HCl (*q.v.*).
Boiling with acetic anhydride + Na acetate → aceto-*o*-toluidide
m.p. 112°.

268°   *N,N'*-Di-*p*-tolylurea $CH_3C_6H_4NHCONHC_6H_4CH_3$ (1,4::1',4')
Almost insol. in H_2O. Boiling HCl → *p*-toluidine·HCl (*q.v.*).
Boiling with acetic anhydride + Na acetate → acet-*p*-toluidide
m.p. 147°.

312°   *N,N'*-Diacetyl-*p*-phenylenediamine $C_6H_4(NHCOCH_3)_2$ (1,4)
Sol. in acetic acid; spar. sol. in other organic solvents. Boiling
HCl → acetic acid and *p*-phenylenediamine·HCl (*q.v.*). Fuming
HNO_3 at 0° → dinitro deriv. m.p. 258°.

AMINO ACIDS, ESTERS AND AMIDES

LIQUID

B.P.°C

260°d Methyl anthranilate m.p. 25°. See under Solids.

267° Ethyl anthranilate $H_2NC_6H_4COOC_2H_5$ (1,2)
M.p. 13°. Boiling dil. acid or alkali → EtOH + anthranilic acid
m.p. 144°. Hydrochloride m.p. 170°. Acetyl m.p. 61°. Benzoyl,
m.p. 98°. p-Tosyl m.p. 112°. Cmpd. with T.N.B. m.p. 71°.

294° Ethyl m-aminobenzoate $H_2NC_6H_4COOC_2H_5$ (1,3)
Boiling dil. acid or alkali → EtOH + m-aminobenzoic acid m.p.
174°. Hydrochloride m.p. 185°. Benzoyl m.p. 114° (148°).
Acetyl m.p. 110°. Cmpd. with T.N.B. m.p. 84°.

SOLID

M.P.°C

25° Methyl anthranilate $H_2NC_6H_4COOCH_3$ (1,2)
Almost insol. in cold H_2O. Soln. in EtOH has blue fluorescence.
Volatile in steam. Boiling dil. acid or alkali → MeOH + anth-
ranilic acid m.p. 144°. Hydrochloride m.p. 178°. Acetyl m.p. 101°.
Benzoyl m.p. 100°. Benzenesulphonyl m.p. 107°. Picrate m.p.
Picrate m.p. 106°. Cmpd. with T.N.B. m.p. 106°.

92° Ethyl p-aminobenzoate $H_2NC_6H_4COOC_2H_5$ (1,4)
Insol. in H_2O. Sol. in EtOH, ether. Boiling dil. acid or alkali →
EtOH + p-aminobenzoic acid m.p. 186°. Acetyl m.p. 110°.
Benzoyl m.p. 148°. Picrate m.p. 131°. Cmpd. with T.N.B. m.p.
85°.

127° N-Phenylglycine $C_6H_5NHCH_2COOH$
Sol. in hot H_2O; spar. sol. in ether. Sol. in cold alkali and
mineral acid. NaOH fusion → blue colour of indigo (in air).
Acetyl m.p. 194°. Benzoyl m.p. 63°. Phenylhydantoic acid m.p.
195°. Nitrosamine m.p. 105°d.

147° Anthranilic acid $H_2NC_6H_4COOH$ (1,2)
Sol. in H_2O, EtOH (blue fluorescence). Htg. → aniline. Gentle
htg with $CaCl_2$ and dissolving product in EtOH gives red soln.
showing violet fluorescence on standing. Bromine/H_2O →
dibromo deriv. m.p. 227°. Hydrochloride m.p. 193°. Acetyl m.p.
185°. Benzoyl m.p. 182°. p-Tosyl m.p. 217°. Benzal deriv. m.p.
128°. p-Nitrobenzyl ester m.p. 205°d. Phenacyl ester m.p. 181°.
Amide m.p. 108°. Anilide m.p. 126°. Phenythydantoic acid m.p.
181°. 1-Naphthylhydantoic acid m.p. 193°. Cmpd. with T.N.B.
m.p. 192°.

149° Oxanilic acid $C_6H_5NHCOCOOH$
Sol. in hot H_2O. V. sol. in EtOH, ether. Soln. in conc. H_2SO_4 +
trace $K_2Cr_2O_7$ → blue-violet colour. Boiling acid or alkali →
aniline + oxalic acid (q.v.). Boiling aniline → oxanilide m.p.
254°. Na, NH_4 salts spar. sol. in cold H_2O. Pyridine salt m.p.
132°. Quinoline salt m.p. 122°.

M.P.°C

174°   *m*-Aminobenzoic acid $H_2NC_6H_4COOH$ (1,4)
Sol. in EtOH, hot H_2O. Gentle htg. with $CaCl_2$ and dissolving
product in EtOH gives red soln. (no fluorescence, contrast
anthranilic acid). Acetyl m.p. 250°. Benzoyl m.p. 248°. Benzal
deriv. m.p. 119°. *p*-Nitrobenzyl ester m.p. 201°. Phenacyl ester
m.p. 202°. Amide m.p. 75°. Anilide m.p. 129°. Phenylhydantoic
acid m.p. 270°. Cmpd. with T.N.B. m.p. 118°.

186°   *p*-Aminobenzoic acid $H_2NC_6H_4COOH$ (1,4)
Sol. in hot H_2O, EtOH. Htg. with $CaCl_2$ → as for *m*-amino-
benzoic acid. Acetyl m.p. 252°. Benzoyl m.p. 278°. Benzal deriv.
m.p. 193°. *p*-Nitrobenzyl ester m.p. 248°. Phenacyl ester m.p.
211°. Amide m.p. 183°. Phenylhydantoic acid m.p. 300°. 1-
Naphthylhydantoic acid m.p. 151°.

187°   Hippuric acid $C_6H_5CONHCH_2COOH$
Sol. in hot H_2O, EtOH. Insol. in benzene, ligroin. Boiling HCl →
benzoic acid m.p. 121° + glycine·HCl. *p*-Nitrobenzyl ester m.p.
136°. *p*-Bromophenacyl ester m.p. 151°. Conc. HNO_3 + H_2SO→
m-nitro deriv. m.p. 162°.

198°d   *β*-Alanine $H_2NCH_2CH_2COOH$
Sol. in H_2O. Spar. sol. in EtOH. Insol. in ether, acetone. Heat →
acrylic acid. Benzoyl m.p. 120°. Phenylhydantoic acid m.p. 174°.

198°   L-(+)-Glutamic acid $HOOCCH_2CH_2CH(NH_2)COOH$
M.p. 224° if heated slowly. Spar. sol. in H_2O. Insol. in EtOH.
$[\alpha]_D^{25}$ +11° in H_2O. $[\alpha]_D^{20}$ +34.9° in 10% aq. HCl. Hydrochloride m.p. 203° (insol. cold conc. HCl). Htg. at 150°–160° →
laevorotatory pyrrolidone carboxylic acid m.p. 160°. Hydantoic
acid m.p. 150° (spar. sol. in H_2O). Acetyl m.p. 199°. Benzoyl,
m.p. 138°. *p*-Tosyl m.p. 131°. 1-Naphthylhydantoic acid m.p.
236°.

200d°   *p*-Aminophenylacetic acid $H_2NC_6H_4CH_2COOH$ (1,4)
Mod. sol. in hot H_2O. Hydrochloride m.p. 215°–240°d. *N*-
Acetyl m.p. 168°–170°. *N*-Benzoyl m.p. 205°.

207°   L-(+)-Arginine $H_2NC(:NH)NH(CH_2)_3CH(NH_2)COOH$
Sol. in hot H_2O. Spar. sol. in EtOH. $[\alpha]_D^{20}$ +11.8° in 2% aq.
NaOH. Hydrochloride sinters at 218°, melts 235°d. Benzoyl
(mono) m.p. 298°, (di) m.p. 235°d. Picrate (mono) m.p. 217°,
(di) m.p. 190°.

212°d   Sarcosine CH_3NHCH_2COOH
Sol. in H_2O. Insol. in EtOH. Htg. → anhydride m.p. 149° and
dimethylamine. Hydrochloride m.p. 169°. Hydrobromide m.p.
186°. Acetyl m.p. 135°. Benzoyl m.p. 104°. *N-p*-Tosyl m.p. 102°.
Phenylhydantoic acid m.p. 102°.

222°   L-(−)-Proline H_2C——CH_2

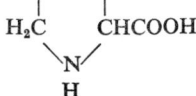

M.P.°C

Sol. in EtOH. $[\alpha]_D^{20}$ −80.9° in H_2O, −52.6° in M/2 HCl. Benzoyl m.p. 156°. p-Tosyl m.p. 132°. Phenylhydantoic acid m.p. 144°. Picrate m.p. 154°.

224°   L-(+)-Lysine $H_2N(CH_2)_4CH(NH_2)COOH$
V. sol. in H_2O. Almost insol. in EtOH. $[\alpha]_D^{20}$ +14.6° in H_2O. With HCl at 165°–170° → (±)-Lysine. Heat → pentamethylene diamine. KOH fusion → propionic acid + acetic acid. Benzoyl (mono) m.p. 235°, (di) m.p. 149°. Phenylhydantoic acid m.p. 184°. Picrate m.p. 266°.

226°d L-(−)-Asparagine $HOOCCH(NH_2)CHCONH_2.H_2O$
Loses water of crystn. above 100°. Rapid htg. gives m.p. of 234°–5°. Mod. sol. in cold H_2O. Insol. in EtOH. $[\alpha]_D^{20}$ −5.42° in H_2O. Triboluminescent. Acid reaction. HNO_2 → L-malic acid. Boiling dil. alkali → NH_3 + L-aspartic acid (q.v.). Dil. NaOH + trace $CuSO_4$ → blue colour. Benzoyl m.p. 189°. p-Tosyl m.p. 175°. Phenylhydantoic acid m.p. 164°. 1-Naphthylhydantoic acid m.p. 199°. Picrate m.p. 180°d.

232°   Glycine H_2NCH_2COOH
Sol. in H_2O. Insol. in EtOH. $CuSO_4$ (trace) → blue colour. $FeCl_3$ soln. → red colour. Acetyl m.p. 206°. Benzoyl m.p. 187°. p-Tosyl m.p. 147°. Phenylhydantoic acid m.p. 197°. Hydantoic acid m.p. 163° (sol. in H_2O; spar. sol. in EtOH). 1-Naphthylhydantoic acid m.p. 191°.

246°d (±)-Serine $HOCH_2CH(NH_2)COOH$
Sol. in H_2O. Insol. in EtOH, ether. $FeSO_4$ + H_2O_2 → $HOCH_2$-CHO. $FeCl_3$ soln. → red colour. Benzoyl m.p. 149°d. p-Tosyl m.p. 213°. Phenylhydantoic acid m.p. 169°. Picrate m.p. 169°.

248°d N-(p-Hydroxyphenyl)glycine $HOC_6H_4NHCH_2COOH$
(241°)   Spar. sol. in H_2O, EtOH. Insol. in ether. Sol. in dil. alkali and mineral acid. $FeCl_3$ soln. → blue colour. $AgNO_3$ soln. → black ppt. in cold, turning purple on heating. Acetyl (mono) m.p. 203°, (di) m.p. 175°. Benzoyl m.p. 117°.

271°   L-Aspartic acid $HOOCCH_2CH(NH_2)COOH$
Sol. in warm H_2O. Insol. in EtOH. $[\alpha]_D^{20}$ +4.36° in H_2O, −1.86° at 90°. Optical rotation is (+) in acid and (−) in alkaline soln. HNO_2 → L-malic acid. Benzoyl m.p. 185°. p-Tosyl m.p. 140°. Phenylhydantoic acid m.p. 162°.

273°   (±)-Phenylalanine $C_6H_5CH_2CH(NH_2)COOH$
Spar. sol. in cold H_2O, EtOH. Insol. in ether. Sublimes part decomp. Heat alone → phenyl lactimide m.p. 290° and β-phenylethylamine b.p. 197°. Oxdtn. → benzoic acid m.p. 121°. Acetyl m.p. 152°. Benzoyl m.p. 187°. p-Tosyl m.p. 134°. Phenylhydantoic acid m.p. 182°d.

277°   L-(−)-Histidine
$$\begin{array}{c} N{=}CH \\ | \qquad\quad NH \\ CH{=}C \\ | \\ CH_2CH(NH_2)COOH \end{array}$$

M.P.°C

Sol. in H_2O. Mod. sol. in EtOH. $[\alpha]_D^{20}$ $-39.74°$ in H_2O. Dextro-rotatory in HCl soln. Bromine/H_2O → red colour. Gives biuret test. Hydrochloride m.p. 252°. Benzoyl, (mono) m.p. 230°. *p*-Tosyl m.p. 203°. Picrate m.p. 86°.

280° *subl.* α-Aminoisobutyric acid $(CH_3)_2C(NH_2)COOH$
Readily sol. in cold H_2O. Almost insol. in EtOH. NaOCl (1 mole) → acetone. Hydantoic acid m.p. 162°. 1-Naphthyl-hydantoic acid m.p. 198°. Benzoyl m.p. 198°.

280° (±)-Aspartic acid $HOOCCH_2CH(NH_2)COOH$
(338°) Spar. sol. in cold H_2O. Insol. in EtOH, ether. Benzoyl m.p. 119° (*hyd.*), 176° (*anhyd.*).

283° 5-Aminosalicylic acid $HOOCC_6H_4(OH)NH_2$ (1,2,5)
Spar. sol. in cold H_2O. Insol. in EtOH. FeCl$_3$ soln. → red colour changing to brown ppt. Acetyl (mono) m.p. 218°, (di) m.p. 184°. Benzoyl m.p. 252°. *p*-Nitrobenzyl ester m.p. 245°d. Azo-2-naphthol deriv. m.p. 201°.

289° L-(−)-Tryptophan $C_{11}H_{12}O_2N_2$

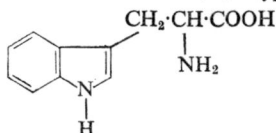

$$CH_2 \cdot CH \cdot COOH$$
$$\quad\quad\quad | $$
$$\quad\quad\quad NH_2$$

Spar. sol. in cold H_2O, EtOH. Sol. in hot H_2O, EtOH. Acid reaction. $[\alpha]_D^{20}$ $-33.4°$ in EtOH. $[\alpha]_D^{20}$ $+6.1°$ in N NaOH. 25% HCl at 170° → (±) cmpd. Bromine/H_2O → red/violet colour. Hydrochloride m.p. 257°d. Acetyl m.p. 189°. Benzoyl m.p. 104°. *p*-Tosyl m.p. 176°. 1-Naphthylhydantoic acid m.p. 158°. Picrate m.p. 195°.

292° (±)-Isoleucine $CH_3CH_2CH(CH_3)CH(NH_2)COOH$
Mod. sol. in H_2O. Sol. in EtOH. Formyl m.p. 121°. Benzoyl m.p. 118°. *p*-Tosyl m.p. 140°. Phenylhydantoic acid m.p. 120°.

293°d (±)-Tryptophan $C_{11}H_{12}O_2N_2$

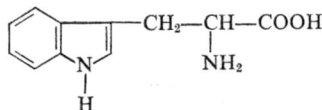

$$CH_2 — CH — COOH$$
$$\quad\quad\quad\quad | $$
$$\quad\quad\quad\quad NH_2$$

Sol. in H_2O-EtOH mixtures. Acetyl m.p. 206°. Benzoyl m.p. 189°. *p*-Tosyl m.p. 176°. Picrate m.p. 186°.

295°d (±)-Alanine $CH_3CH(NH_2)COOH$
Sol. in H_2O. Insol. in EtOH. Above 180° conc. H_2SO_4 → CO. Warm PbO_2 in H_2O → acetaldehyde + NH_3. Benzoyl m.p. 165°. *p*-Tosyl m.p. 139°. Phenylhydantoic acid m.p. 190°d. 1-Naphthyl-hydantoic acid m.p. 198°' Hydantoic acid m.p. 157° (Sol. in 46 parts H_2O and in 100 parts EtOH at 20°).

298°d (±)-Valine $(CH_3)_2CHCH(NH_2)COOH$
Sol. in cold H_2O. Insol. in cold EtOH, ether. Heat alone →

M.P.°C

anhydride m.p. > 300°. Hydrochloride m.p. 189°. Acetyl m.p. 148°. Benzoyl m.p. 132°. *p*-Tosyl m.p. 110°. Hydantoic acid m.p. 176° (Readily sol. in EtOH; spar. sol. in H_2O). 1-Naphthyl-hydantoic acid m.p. 204°.

307°d   (±)-α-Amino-n-butyric acid $CH_3CH_2CH(NH_2)COOH$
Sol. in cold H_2O. Almost insol. in EtOH. Cu salt spar. sol. in H_2O. Benzoyl m.p. 147°. Hydantoic acid m.p. 177°. Phenyl-hydantoic acid. m.p. 170°d. 1-Naphthylhydantoic acid m.p. 194°.

309°d   L-Alanine $CH_3CH(NH_2)COOH$
Sol. in H_2O. Spar. sol. in EtOH. $[\alpha]_D^{22}$ +2.7° in H_2O. HNO_2 → lactic acid. Hydrochloride m.p. 204°. Acetyl m.p. 116°. Benzoyl m.p. 152°. *p*-Tosyl m.p. 133°. 1-Naphthylhydantoic acid m.p. 200°.

315°d   L-(+)-Valine $(CH_3)_2CHCH(NH_2)COOH$
M.p. in sealed tube. Sol. in H_2O. Spar. sol. in EtOH. $[\alpha]_D^{20}$ +6.3° in H_2O, +28.8° in 6M HCl. $Ba(OH)_2$ soln. at 180° → (±) cmpd. Formyl m.p. 156°. Acetyl m.p. 164°. Benzoyl m.p. 127°. *p*-Tosyl m.p. 150°. Phenylhydantoic acid m.p. 147°.

320°   L-(−)-Phenylalanine $C_6H_5CH_2CH(NH_2)COOH$
M.p. (rapid heat) 283°. Sublimes without racemisation. Mod. sol. in cold H_2O. $[\alpha]_D^{20}$ −35.1° in H_2O. Formyl m.p. 167°. Acetyl m.p. 170°. Benzoyl m.p. 146°. *p*-Tosyl m.p. 164°. Phenyl-hydantoic acid m.p. 181°.

332°   (±)-Leucine $(CH_3)_2CHCH_2CH(NH_2)COOH$
M.p. in sealed tube, 293°. Spar. sol. in cold H_2O. V. spar. sol. in EtOH. Acetyl m.p. 157°. Benzoyl m.p. 137°–141°. Phenyl-hydantoic acid m.p. 165°.

337°   L-(−)-Leucine $(CH_3)_2CHCH_2CH(NH_2)COOH$
Mod. sol. in cold H_2O, EtOH. $[\alpha]_D^{20}$ −10.42° in H_2O. $[\alpha]_D^{15}$ +17.3 in 20% HCl. Acetyl m.p. 189°. Benzoyl m.p. 60° (*hyd.*), 107° (*anhyd.*). *p*-Tosyl m.p. 124°. Phenylhydantoic acid m.p. 115°. 1-Naphthylhydantoic acid m.p. 163°.

344°d   L-(−)-Tyrosine $C_9H_{11}O_3N$

$HO-\langle \bigcirc \rangle-CH_2CH(NH_2)COOH$

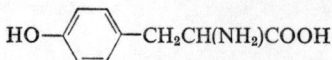

Slow heat, *decomp.* at 290°. Triboluminescent. Spar. sol. in cold H_2O, EtOH, acetic acid. Sol. in mineral acids and alkalis. Insol. in ether. $[\alpha]_D^{20}$ −8.07° in 21% HCl, −9.01° in 11.6% KOH. Cu salt (blue needles) spar. sol. in cold H_2O. $Ba(OH)_2$ soln. at 170° → (±) cmpd. With hot acetic acid + $NaNO_2$ → violet/red colour. Hot HNO_3 → yellow dinitro deriv. Conc. H_2SO_4 + acetaldehyde (little) → red colour. NaOH fusion → *p*-hydroxy-benzoic acid m.p. 213°. *N*-Acetyl m.p. 153°. *O*,*N*-Diacetyl m.p. 172°. *N*-Benzoyl m.p. 165°. *O*,*N*-Dibenzoyl m.p. 211°. *N*-*p*-Tosyl m.p. 187°. Phenylhydantoic acid m.p. 104°. 1-Naphthyl-hydantoic acid m.p. 205°.

M.P.°C

— (±)-Lysine H$_2$N(CH$_2$)$_4$CH(NH$_2$)COOH
Monohydrochloride m.p. 260°d. Benzoyl (mono) m.p. 248°, (di) m.p. 145°. Phenylhydantoic acid m.p. 196°. Picrate m.p. 225° (mono).

PURINES

SOLID

M.P.°C

235° Caffeine (1,3,7-Trimethylxanthine), C$_8$H$_{10}$O$_2$N$_4$

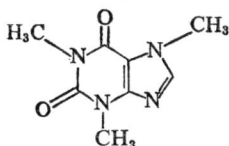

Readily sol. in H$_2$O, CHCl$_3$. Fairly sol. in EtOH, benzene. Spar. sol. in ether, CCl$_4$. May be sublimed unchanged. Alcoholic KOH (boiling) → NH$_3$ + CO$_2$ + CH$_3$NH$_2$ + HCOOH + sarcosine m.p. 210°d. Unchanged by boiling conc. HCl. K$_4$Fe(CN)$_6$ + HNO$_3$ → Prussian Blue, on warming. Boil caffeine (0.1 g) + 2 cm^3 H$_2$O + 1 cm^3 conc. HCl, then add 10 cm^3 saturated bromine water and boil until colourless, dilute with H$_2$O to original volume, add to 2 cm^3 of this soln., 1 drop of 5% aq. FeSO$_4$ + 3 drops NH$_4$OH → blue colour. Iodine in KI soln. → ppt. m.p. 215°. HgCl$_2$ soln. → white ppt. (cryst). Styphnate m.p. 199°.

264° Theophylline (1,3-Dimethylxanthine) C$_7$H$_8$O$_2$N$_4$

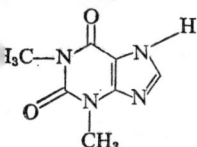

V. sol. in warm H$_2$O, spar. sol. in cold EtOH. Warm in CH$_3$OH + KOH + CH$_3$I → caffeine m.p. 235°. Gives blue colour when treated with HCl + H$_2$O then bromine water etc. (see caffeine). HgNO$_3$ soln. → white ppt. Acetyl m.p. 158°. Benzoyl m.p. 202°.

290° subl. Theobromine (3,7-Dimethylxanthine) C$_7$H$_8$O$_2$N$_4$

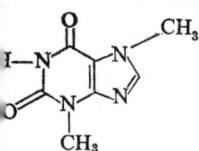

M.p. 329° in sealed tube. Almost insol. in cold H$_2$O, EtOH, ether, CHCl$_3$, ligroin. Sl. sol. in hot H$_2$O. Not decomp. by hot aq. KOH. Warm in CH$_3$OH + KOH + CH$_3$I → caffeine m.p. 235°. Gives blue colour when treated with HCl + H$_2$O then bromine water etc. (see caffeine). Gives cryst. white ppt. with AgNO$_3$ in dil. HNO$_3$.

360°–365° Adenine (6-Aminopurine) C$_5$H$_5$N$_5$

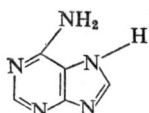

Sublimes at 220°. M.p. that of anhyd. cmpd. (rapid heat). Sol. in hot H$_2$O. Spar. sol. in EtOH. Insol. in ether, CHCl$_3$. HNO$_2$ → hypoxanthine m.p. 150°d. FeCl$_3$ soln. → red colour. Acetyl (di) m.p. 195°d. Benzoyl (mono) m.p. 234°. Picrate m.p. 280°d.

M.P.°C

d   Uric acid (2,6,8-Trihydroxypurine) $C_5H_4O_3N_4$

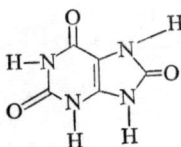

V. spar. sol. in cold H_2O. Insol. in EtOH, ether. Readily sol. in alkali. Spar. sol. in mineral acids. Sol. in conc. H_2SO_4 without decomp. Heating alone → urea, cyanuric acid, HCN + NH_3. Sol in glycerol. Oxdtn. by 7% HNO_3 at 60°–70° → alloxan (*q.v.*) + urea. Oxdtn. by $KMnO_4$ → allantoin. Acts as a weak dibasic acid. Reduces ammon. $AgNO_3$. Reduces Fehling's soln. on long boiling. With $CuSO_4$ + $NaHSO_3$ → white ppt. (cuprous urate). Gives murexide reaction.

NITRILES

LIQUID

B.P.°C

78°   Acrylonitrile CH_2:CHCN, Vinyl cyanide.

Sol. in H_2O with which it gives constant boiling mixture containing 12.5% H_2O. d_4^{20} 0.806. n_D^{20} 1.393. Readily polymerises; gives solid polymer on addtn. of conc. $NaOCH_3$ soln.

81°   Acetonitrile CH_3CN

Sol. in H_2O, separates on addtn. of $CaCl_2$. Hot dil. acid or alkali → acetic acid. d_4^{20} 0.783. n_D^{20} 1.344. Zn + dil. HCl → ethylamine·HCl (*q.v.*). EtOH (1 mole) + HCl gas → acetiminoethyl ether hydrochloride m.p. 98°d.

97°   Propionitrile CH_3CH_2CN

Sol. in H_2O, separates on addtn. of $CaCl_2$. Hot dil. acid or alkali → propionic acid. d_4^{25} 0.783. n_D^{25} 1.366. EtOH (1 mole) + HCl gas → propioniminoethyl ether hydrochloride m.p. 92°d.

108°   Isobutyronitrile $(CH_3)_2CHCN$

Sol. in H_2O. Hot. dil. acid or alkali → isobutyric acid.

118°   n-Butyronitrile $CH_3CH_2CH_2CN$

Sol. in H_2O. Hot dil. acid or alkali → n-butyric acid. d_4^{15} 0.791. n_D^{24} 1.384. Hot EtOH + H_2SO_4 → ethyl n-butyrate b.p. 120°.

119°   β-Butenonitrile CH_2:CHCH$_2$CN, Allyl cyanide.

Sol. in H_2O. Alcoholic KOH → crotonic acid. d_4^{20} 0.838. n_D^{20} 1.406. Oxdtn. → acetic acid. Conc. HCl at 60° → 2-chlorobutyric acid.

120°d   α-Hydroxyisobutyronitrile $(CH_3)_2C(OH)CN$, Acetone cyanohydrin.

Sol. in H_2O, EtOH, ether. Insol. light petroleum. d_4^{20} 0.93. n_D^{20} 1.399. Conc. HCl → α-Hydroxyisobutyric acid m.p. 78°. KOH → acetone + KCN. Evaporation of aq. soln. → HCN + diacetone cyanohydrin m.p. 162°. CH_3COCl → acetyl deriv. b.p. 181°. Warm alcoholic NH_3 soln. → α-aminoisobutyronitrile (b.p. 49°/12 mmHg) which gives hydrochloride, m.p. 145°.

130°   Isovaleronitrile $(CH_3)_2CHCH_2CN$

d_4^{20} 0.788. n_D^{20} 1.392. Hot dil. acid or alkali → isovaleric acid b.p. 176°. Hot EtOH + H_2SO_4 → ethyl isovalerate b.p. 135°.

B.P.°C

141° n-Valeronitrile $CH_3(CH_2)_3CN$
Sl. sol. in H_2O. d_4^{20} 0.799. n_D^{20} 1.397. Hot dil. acid or alkali →
n-valeric acid b.p. 186°. Hot EtOH + H_2SO_4 → ethyl n-valerate
b.p. 144°.

155° Isocapronitrile $(CH_3)_2CH(CH_2)_2CN$, 4-Methylpentanonitrile.
d_4^{20} 0.803. n_D^{20} 1.406. Hot dil. acid or alkali → isocaproic acid
b.p. 199°.

165° n-Capronitrile $CH_3(CH_2)_4CN$
Sol. in EtOH and ether. Insol. in H_2O. d_4^{20} 0.805. n_D^{20} 1.411.
Hot dil. acid or alkali → n-caproic acid b.p. 205°.

182°d (±)-Lactonitrile $CH_3C(OH)CN$, Acetaldehyde cyanohydrin.
Sol. in H_2O, EtOH, ether. Insol. in light petroleum, CS_2. d_4^{20}
0.988. n_D^{18} 1.406. Conc. HCl → lactic acid. KOH → acetaldehyde
resin. CH_3COCl → acetyl deriv. b.p. 172°. NH_3 in EtOH →
unstable aminopropionitrile (hydrochloride m.p. 115°, picrate
m.p. 141°d) which with boiling $Ba(OH)_2$ or HCl → (±)-alanine.

183°d Glycolonitrile $HOCH_2CN$, Formaldehyde cyanohydrin.
B.p. 98°/10 mmHg. Benzoyl, m.p. 195°.

190° Benzonitrile C_6H_5CN
Insol. in H_2O. Sol. in EtOH, ether. d_4^{20} 1.006. n_D^{20} 1.528. Odour of
bitter almonds. Conc. H_2SO_4 at 100° or dil. NaOH + H_2O_2 in
cold (with vigorous stirring) → benzamide m.p. 128°. Boiling
75% H_2SO_4 → benzoic acid m.p. 121°. Conc. HNO_3 + H_2SO_4 →
m-nitro deriv., m.p. 117°. Boiling CH_3OH +H_2SO_4 (conc.) →
methyl benzoate b.p. 198°.

204° o-Tolunitrile $CH_3C_6H_4CN$ (1,2)
Insol. in H_2O. d_4^{20} 0.996. n_D^{20} 1.529. Conc. H_2SO_4 at 100° or
on shaking with dil. NaOH + H_2O_2 in cold → o-toluamide m.p.
142°. Boiling 75% H_2SO_4 → o-toluic acid m.p. 102°. Conc.
HNO_3 + H_2SO_4 → nitro deriv. m.p. 105°. Boiling CH_3OH +
H_2SO_4 → methyl o-toluate b.p. 207°.

207° Ethyl cyanoacetate $NCCH_2COOC_2H_5$
Insol. in H_2O. Sol. in dil. NaOH. Soln. in NH_4OH, on evaporation → cyanoacetamide m.p. 118°. d_4^{20} 1.063. n_D^{20} 1.418. Boiling
EtOH + H_2SO_4 → ethyl malonate b.p. 198°. Warm benzaldehyde + drop of morpholine → ethyl α-cyanocinnamate m.p. 51°.

212° m-Tolunitrile $CH_3C_6H_4CN$ (1,3)
d_4^{20} 1.032. n_D^{20} 1.525. With conc. H_2SO_4 at 100° or on shaking with
H_2O_2 + dil. NaOH in cold → m-toluamide m.p. 97°. Boiling
75% H_2SO_4 → m-toluic acid m.p. 111°.

220° β-Hydroxypropionitrile $HO(CH_2)_2CN$, Hydracrylonitrile.
Sol. in H_2O, EtOH. HCl → β-hydroxypropionic acid + acrylic
acid. CH_3COCl → acetyl deriv. b.p. 206°. P_2O_5 → acrylonitrile
b.p. 77°.

234° Phenylacetonitrile $C_6H_5CH_2CN$, Benzyl cyanide.
Insol. in H_2O. d_4^{20} 1.016. n_D^{20} 1.523. Conc. H_2SO_4 at 100° →
phenylacetamide m.p. 154°. Boiling 75% H_2SO_4 → phenylacetic

B.P.°C

acid m.p. 76°. Conc. HNO_3 + H_2SO_4 → nitro deriv. m.p. 116°.
Boiling CH_3OH + H_2SO_4 → methyl phenylacetate b.p. 220°.

286° Glutaronitrile $NC(CH_2)_3CN$
d_4^{20} 0.985. n_D^{20} 1.429. Boiling mineral acid → glutaric acid m.p. 97°

295° Adiponitrile $NC(CH_2)_4CN$
Insol. in H_2O, CS_2, ether. Sol. in EtOH, $CHCl_3$. d_{19}^{19} 0.951. n_D^{20}
1.459. Boiling mineral acid → adipic acid m.p. 153°.

SOLID

M.P.°C

20° Cinnamonitrile $C_6H_5CH:CHCN$, Styryl cyanide.
B.p. 254°. Sol. in EtOH. Dil. NaOH + H_2O_2 → cinnamamide
m.p. 147°. Hot 70% H_2SO_4 → cinnamic acid m.p. 133°.

22° (±)-Mandelonitrile $C_6H_5CH(OH)CN$, Benzaldehyde cyanohydrin.
B.p. 170°d. Insol. in H_2O. Sol. in EtOH and ether. Red soln. in
conc. H_2SO_4. Heat → benzaldehyde + HCN. Warm HCl → (±)-
mandelic acid m.p. 118°. Acetyl b.p. 138°/11 mmHg. Benzoyl
m.p. 63°. Carbanilate m.p. 105°. 3-Nitrobenzoyl m.p. 83°.

29° p-Tolunitrile $H_3CC_6H_4CN$ (1,4)
B.p. 218°. Sol. in EtOH. d_{30}^{30} 0.9805. Conc. H_2SO_4 at 100°, or
shaking with dil. NaOH + H_2O_2 in cold → p-toluamide m.p.
158°. Boiling 75% H_2SO_4 → p-toluic acid m.p. 180°. Conc.
HNO_3 + H_2SO_4 → nitro deriv., m.p. 107°. Boiling CH_3OH +
H_2SO_4 → methyl p-toluate m.p. 32° (b.p. 217°).

36° 1-Naphthonitrile $C_{10}H_7CN$
B.p. 299°. Sol. in EtOH. Hot 75% H_2SO_4 (rapidly) → 1-naphtho-
amide m.p. 202°. Prolonged boiling with 75% H_2SO_4 → 1-
naphthoic acid m.p. 160°.

54° Succinonitrile $NC(CH_2)_2CN$
B.p. 265–7°. Sol. in H_2O, EtOH, $CHCl_3$. Spar. sol. in ether, CS_2.
Boiling mineral acid or alkali → succinic acid, m.p. 186°.

66° 2-Naphthonitrile $C_{10}H_7CN$
B.p. 306°. Sol. in EtOH, ether, hot ligroin. Spar. sol. in H_2O.
Hot 75% H_2SO_4 (rapidly) → 2-Naphthoamide m.p. 192°.
Prolonged boiling with 75% H_2SO_4 → 2-naphthoic acid m.p.
184°.

69° Cyanoacetic acid $NCCH_2COOH$
V. sol. in H_2O. Htg. to 165° → acetonitrile b.p. 81°. Boiling
NaOH soln. → Na malonate + NH_3. Warm conc. HCl →
malonic acid m.p. 133°d (decomp. to CH_3COOH + CO_2). Warm
benzaldehyde → α-cyanocinnamic acid m.p. 180°.

141° Phthalonitrile NCC_6H_4CN (1,2)
Sol. in EtOH, $CHCl_3$, ether. Spar. sol. in H_2O. Volatile in steam.
Boiling HCl → phthalic acid (slowly) m.p. 206°d.

ISOCYANATES

LIQUID

B.P.°C

166° Phenyl isocyanate C_6H_5NCO
Sharp odour. Insol. in H_2O. With H_2O or $C_6H_5NH_2$, on standing, gives carbanilide, m.p. 241°. With EtOH → ethyl carbanilate, m.p. 52°. With NH_3 → phenylurea, m.p. 147°. With C_6H_5OH → phenyl carbanilate, m.p. 126°. Boiling HCl → $C_6H_5NH_3Cl$.

269° 1-Naphthyl isocyanate $C_{10}H_7NCO$
M.p. 5°. Insol. in H_2O. With H_2O, on standing, gives di-1-naphthylurea, m.p. 314°. With EtOH → ethyl-1-naphthyl carbamate, m.p. 79°. With NH_3 → 1-naphthylurea, m.p. 213°. With $C_6H_5NH_2$ → phenyl-1-naphthylurea, m.p. 222°. With C_6H_5OH → 1-naphthyl carbamate, m.p. 136°. Boiling HCl → $C_{10}H_7NH_3Cl$.

ALIPHATIC NITROHYDROCARBONS

LIQUID

B.P.°C

101° Nitromethane CH_3NO_2
Sol. in 9 parts cold H_2O, soln. is acid to litmus. Sol. in common organic solvents. d_4^{20} 1.137. n_D^{20} 1.381. Reduction by $SnCl_2$ → CH_3NH_2 (q.v.). With benzaldehyde (sl. excess) and NaOH soln. on shaking → benzylidene deriv., m.p. 58°. With chloral hydrate + dil. K_2CO_3 → trichloronitroisopropyl alcohol m.p. 42°.

114° Nitroethane $CH_3CH_2NO_2$
Sol. in 24 parts cold H_2O, readily in dil. NaOH. Addtn. of excess of HCl to alk. soln. → acetaldehyde b.p. 21°. d_4^{20} 1.0497. n_D^{20} 1.392. Reduction by Sn + HCl → $C_2H_5NH_2$ (q.v.). With benzaldehyde (sl. excess) and NaOH soln., on shaking → benzylidene deriv., m.p. 64°. Soln. in NaOH with diazobenzene→ benzeneazonitroethane m.p. 141°.

120° 2-Nitropropane $CH_3CH(NO_2)CH_3$
Insol. in H_2O. d_4^{20} 0.988. n_D^{20} 1.394. Reduction by Sn + HCl → isopropylamine (q.v.).

126° Tetranitromethane $C(NO_2)_4$
M.p. 13°. Insol. in H_2O. Sol. in EtOH, ether. $d_4^{21.4}$ 1.638. $n_D^{21.2}$ 1.434. With pyridine → orange colour. With unsaturated compds → yellow complexes. With C_2H_5ONa in EtOH → trinitromethane m.p. 15° (gives yellow soln. in H_2O, decolourised by strong acids) and ethyl nitrate b.p. 87°.

132° 1-Nitropropane $CH_3CH_2CH_2NO_2$
Insol. in H_2O. d_4^{20} 1.01. n_D^{20} 1.401. Reduction by Sn + HCl → propylamine (hydrochloride m.p. 157°).

226°d Phenylnitromethane $C_6H_5CH_2NO_2$
B.p. 118°/16 mmHg. Insol. in H_2O, sol. in dil. NaOH. d_0^{20} 1.160

B.P.°C

n_D^{20} 1.532. Yellow. Reduction by Sn + HCl → benzylamine b.p. 185°. With benzaldehyde + trace methylamine in EtOH → benzylidene deriv. (α-nitrostilbene) m.p. 75°. With NaOH in EtOH → Na deriv. which with excess HCl → phenylisonitromethane, m.p. 84°.

AROMATIC NITROHYDROCARBONS

LIQUID

B.P.°C

210° Nitrobenzene $C_6H_5NO_2$
M.p. 5°. Almost insol. in H_2O. Sol. in organic liquids. V. pale yellow. Odour of bitter almonds. d_4^{20} 1.203. n_D^{20} 1.553. Reduction by Sn + HCl → aniline b.p. 183° (hydrochloride m.p. 198°). Warm fuming HNO_3 + H_2SO_4 (conc.) → m-dinitrobenzene m.p. 90°. Warming with bromine + $FeCl_3$ (or $FeBr_3$) → m-bromo deriv. m.p. 54°. Boiling Na arsenite soln. → azoxybenzene m.p. 36°.

222° o-Nitrotoluene $H_3CC_6H_4NO_2$ (1,2)
V. pale yellow. Odour like nitrobenzene. Almost insol. in H_2O. Sol. in organic solvents. d_4^{20} 1.168. n_D^{20} 1.547. Reduction by Sn + HCl → o-toluidine b.p. 197°. Unattacked by boiling $K_2Cr_2O_7$ + dil. H_2SO_4. Boiling dil. $KMnO_4$ → o-nitrobenzoic acid m.p. 147°. Conc. HNO_3 + H_2SO_4 at 100° for 3 min. → 2,4-dinitrotoluene m.p. 70°.

233° m-Nitrotoluene $H_3CC_6H_4NO_2$ (1,3)
M.p. 16°. V. pale yellow. Odour like nitrobenzene. Almost insol. in H_2O. Sol. in organic solvents. d_4^{20} 1.157. n_D^{20} 1.547. Reduction by Sn + HCl → m-toluidine b.p. 199°. Readily oxidised by $K_2Cr_2O_7$ + dil. H_2SO_4 or by $KMnO_4$ soln. to m-nitrobenzoic acid m.p. 140°.

241° 1,4-Dimethyl-2-nitrobenzene $(CH_3)_2C_6H_3NO_2$ (1,4,2)
B.p. 101–102°/10 mmHg. Pale yellow. d^{15} 1.132. With conc. HNO_3 + H_2SO_4 at 100° for 5 min. → 1,4-dimethyl-2,3,5-trinitrobenzene m.p. 139°. Reduction by Sn + HCl → 2,5-dimethyl aniline b.p. 214° (m.p. 15°), mod. sol. in H_2O.

243° p-Nitroethylbenzene $H_5C_2C_6H_4NO_2$ (1,4)
d^{25} 1.124. n_D^{19} 1.546. Reduction by Sn + HCl → p-ethylaniline b.p. 214°, of which the acetyl deriv. has m.p. 94°.

244° 1,3-Dimethyl-4-nitrobenzene $(CH_3)_2C_6H_3NO_2$ (1,3,4)
M.p. 2°. $d^{1.75}$ 1.126. Reduction by Sn + HCl → 2,4-dimethylaniline b.p. 212°. Fuming HNO_3 in cold → 4,6-dinitro-m-xylene m.p. 93°. With conc. HNO_3 + H_2SO_4 at 100° for 5 min. → 2,4,6-trinitro-m-xylene m.p. 182°. Oxidation by $KMnO_4$ soln. → 4-nitroisophthalic acid m.p. 258°.

250° 1,2-Dimethyl-3-nitrobenzene $(CH_3)_2C_6H_3NO_2$ (1,2,3)
M.p. 15°. Pale yellow. Fuming HNO_3 + H_2SO_4 (conc.) at 100°

B.P.°C

for 5 min. → 1,2-dimethyl-3,4-dinitrobenzene m.p. 82°. Oxidation by $KMnO_4$ soln. → 3-nitrophthalic acid m.p. 218°.

264° 2-Nitro-p-cymene $(CH_3)_2CHC_6H_3(CH_3)NO_2$ (4,1,2)
d_4^{20} 1.074. n_D^{20} 1.531. Reduction by Sn + HCl → 2-aminocymene b.p. 241° of which the acetyl deriv. has m.p. 115°, and the benzoyl deriv. m.p. 102°. Warm conc. HNO_3 + H_2SO_4 → 2,6-dinitrocymene m.p. 54°. Oxidation by $KMnO_4$ soln. → 2-nitro-4-hydroxyisopropylbenzoic acid m.p. 168°.

SOLID

M.P.°C

15° 1,2-Dimethyl-3-nitrobenzene $(CH_3)_2C_6H_3NO_2$ (1,2,3)
B.p. 250°. See Aromatic Nitrohydrocarbons (liquid).

15° 1,3-Dimethyl-2-nitrobenzene $(CH_3)_2C_6H_3NO_2$ (1,3,2)
B.p. 226°. Fuming HNO_3 + H_2SO_4 (conc.) at 100° for 5 min. → 1,3-dimethyl-2,4,6-trinitrobenzene. Reduction by Sn + HCl → 1,3-dimethylaniline b.p. 216° of which the acetyl deriv. has m.p. 177°.

30° 1,2-dimethyl-4-nitrobenzene $(CH_3)_2C_6H_3NO_2$ (1,2,4)
B.p. 258°. Fuming HNO_3 + H_2SO_4 (conc.) at 100° for 5 min. → 1,2-dimethyl-3,4-dinitrobenzene, m.p. 82°. Reduction by Sn + HCl → 1,2-dimethylaniline m.p. 50° of which the acetyl deriv. has m.p. 99°.

37° o-Nitrobiphenyl $C_6H_5C_6H_4NO_2$ (1,2)
Fuming HNO_3 + H_2SO_4 (conc.) at 100° for 5 min. → 2,4-dinitrobiphenyl m.p. 93°. Reduction by Sn + HCl → o-aminobiphenyl, m.p. 50° of which the acetyl deriv. has m.p. 121°.

44° Nitromesitylene $(CH_3)_3C_6H_2NO_2$ (1,3,5,2)
B.p. 255°. Equal parts fuming HNO_3 + glacial AcOH at b.p. for 30 sec. → dinitromesitylene m.p. 86°. Fuming HNO_3 + H_2SO_4 (conc.) at 100° for 5 min. → 2,4,6-trinitromesitylene m.p. 235°. Reduction by Sn + HCl → mesidine b.p. 232° of which the acetyl deriv. has m.p. 216°, benzoyl deriv. m.p. 204°. CrO_3 in AcOH (glacial) at 65° → p-nitromesitylenic acid m.p. 220–223°.

52° p-Nitrotoluene $CH_3C_6H_4NO_2$ (1,4)
B.p. 234°. V. pale yellow. Odour like nitrobenzene. Fuming HNO_3 + conc. H_2SO_4 at 100° for 3 min. → 2,4-dinitrotoluene m.p. 70°. Readily oxidised by $K_2Cr_2O_7$ + dil. H_2SO_4 or by $KMnO_4$ soln. to p-nitrobenzoic acid m.p. 241°. Reduction by Sn + HCl → p-toluidine, m.p. 45°.

60° 1-Nitronaphthalene $C_{10}H_7NO_2$
B.p. 304°. Readily sol. in organic liquids. Forms dark red soln. in conc. H_2SO_4. Warm conc. HNO_3 + conc. H_2SO_4 → 1,3,8-trinitronaphthalene m.p. 218° (spar. sol. in organic liquids). Oxidised by CrO_3 in glacial AcOH to 3-nitrophthalic acid m.p.

M.P.°C

218°. Reduction by Sn + HCl → 1-naphthylamine, m.p. 50°. Picrate deriv. m.p. 71°.

66° 2,6-Dinitrotoluene $CH_3C_6H_3(NO_2)_2$ (1,2,6)
Gives purple colour with aqueous NaOH. Oxidised by boiling dil. HNO_3 to 2,6-dinitrobenzoic acid, m.p. 202°. Reduction by excess Sn + HCl → 2,6-diaminotoluene m.p. 105°.

70° 2,4-Dinitrotoluene $CH_3C_6H_3(NO_2)_2$ (1,2,4)
Spar. sol. in cold EtOH, readily sol. in benzene. Warm fuming $HNO_3 + H_2SO_4$ conc. → 2,4,6-trinitrotoluene m.p. 82° (explosive). With acetone + KOH → blue colour changed to violet by AcOH. Oxidised by CrO_3 in conc. H_2SO_4 or by $KMnO_4$ soln. to 2,4-dinitrobenzoic acid m.p. 183°. Reduction by excess Sn + HCl → 2,4-diaminotoluene, m.p. 99°. Benzaldehyde + trace of diethylamine → dinitrostilbene, m.p. 139°. Naphthalene adduct in benzene, m.p. 60°.

82° 2,4,6-Trinitrotoluene $CH_3C_6H_2(NO_2)_3$ (1,2,4,6)
V. pale yellow. Explosive. Readily sol. in organic liquids, sparingly in CS_2. Oxidised by CrO_3 in conc. H_2SO_4 to trinitrobenzoic acid m.p. 220°. Reduction by Sn + HCl → unstable triaminotoluene. Benzaldehyde + a little pyridine → trinitrostilbene, m.p. 158°. Naphthalene adduct (in EtOH) m.p. 97°.

90° m-Dinitrobenzene $C_6H_4(NO_2)_2$ (1,3)
B.p. 302°. V. pale yellow. Volatile in steam. Sl. sol. in hot water; sol. in common organic solvents. Soln. in boiling v. dil. NaOH gives purple colour on addition of trace of dextrose (or $SnCl_2$). Boiling alk. $K_3Fe(CN)_6$ → 2,4-dinitrophenol, m.p. 114°. Reduction by hot $NH_4SH/EtOH$ → m-nitroaniline m.p. 114°; reduction by Sn + HCl → m-phenylenediamine, m.p. 63°. Naphthalene adduct (in benzene), m.p. 52°.

93° 1,3-Dimethyl-4,6-dinitrobenzene $(CH_3)_2C_6H_2(NO_2)_2$ (1,3,4,6)
Spar. sol. in cold EtOH. Warm fuming HNO_3 + conc. H_2SO_4 → 1,3-dimethyl-2,4,6-trinitrobenzene, m.p. 125°. Reduction by hot $NH_4SH/EtOH$ → 2,4-dimethyl-5-nitroaniline m.p. 123°; reduction by Sn + HCl → 1,2-dimethyl-4,6-diaminobenzene, m.p. 105°. With aqueous NaOH gives violet colour.

114° p-Nitrobiphenyl $C_6H_5C_6H_4NO_2$ (1,4)
Fuming HNO_3 + conc. H_2SO_4 → 4,4'-dinitrobiphenyl m.p. 233°. Reduction by Sn + HCl → p-aminobiphenyl, m.p. 53° of which the acetyl deriv. has m.p. 171°.

118° o-Dinitrobenzene $C_6H_4(NO_2)_2$ (1,2)
B.p. 319°. Volatile in steam. Sl. sol. in hot H_2O; mod. sol. in common organic solvents. Reduction by hot $NH_4SH/EtOH$ → o-nitroaniline, m.p. 71°; reduction by Sn + HCl → o-phenylenediamine, m.p. 102°. Hot dil. NaOH soln. → o-nitrophenol, m.p. 45°.

122° 1,3,5-Trinitrobenzene $C_6H_3(NO_2)_3$ (1,3,5)
V. pale yellow. Sl. sol. in hot H_2O; mod. sol. in common organic

M.P.°C

solvents. Dil. NaOH soln. → red colour discharged by acid. $K_3Fe(CN)_6 + Na_2CO_3$ soln. → picric acid m.p. 122°. Reduction by Sn + HCl → unstable triaminobenzene (triacetyl deriv. m.p. 208°). Naphthalene adduct (in EtOH) m.p. 152°.

170°   1,8-Dinitronaphthalene $C_{10}H_6(NO_2)_2$ (1,8)

Sol. in pyridine; spar. sol. in benzene, $CHCl_3$. Fuming HNO_3 + conc. H_2SO_4 at b.p. for 5 min. → 1,3-8-trinitronaphthalene, m.p. 218°. Reduction by Sn + HCl (in EtOH) → 1,8-diamino-naphthalene, m.p. 66°.

172°   p-Dinitrobenzene $C_6H_4(NO_2)_2$ (1,4)

B.p. 299°. Volatile in steam. Sl. sol. in hot H_2O; spar. sol. in cold EtOH, mod. sol. in MeOH, $CHCl_3$. Boiling 5% NaOH soln. → p-nitrophenol, m.p. 114°. Reduction by Sn + HCl → p-phenylenediamine, m.p. 140°; reduction by NH_4SH/EtOH → p-nitroaniline, m.p. 147°. Naphthalene adduct (in EtOH) m.p 118°.

214°   1,5-Dinitronaphthalene $C_{10}H_6(NO_2)_2$ (1,5)

Spar. sol. in most organic liquids. Mod. sol. in cold pyridine, more so on heating. Equal vols. of fuming HNO_3 + conc. H_2SO_4 at b.p. for 5 min. → 1,4,5-trinitronaphthalene, m.p. 154°. Reduction by Sn + HCl → 1,5-diaminonaphthalene, m.p. 190°; reduction by cold soln. NH_3 (3 moles) in EtOH → 1-amino-5-nitroanphthalene, m.p. 118°.

NITROETHERS

LIQUID

B.P.°C

265°   o-Nitroanisole $CH_3OC_6H_4NO_2$ (1,2)

M.p. 9°. Insol. in H_2O, sol. in organic liquids. d_4^{20} 1.254. n_D^{20} 1.562. Reduction by Sn + HCl → o-anisidine b.p. 218°. Conc. $HNO_3 + H_2SO_4$ at 0° → 2,4-dinitroanisole m.p. 88°; on warming reaction mixture, trinitroanisole m.p. 68° is formed. Chlorine in acetic acid → 4-chloro deriv. m.p. 97°. Boiling 48% HBr → o-nitrophenol m.p. 44° and MeBr b.p. 4°. Prolonged boiling with conc. NaOH → o-nitrophenol + MeOH.

268°   o-Nitrophenetole $C_2H_5OC_6H_4NO_2$ (1,2)

Insol. in H_2O, sol. in organic liquids. d^{15} 1.19. n_D^{20} 1.542. Reduction by Sn + HCl → o-phenetidine b.p. 229°. Cold conc. $HNO_3 + H_2SO_4$ → 2,4-dinitrophenetole m.p. 86°; on warming the reaction mixture, trinitrophenetole m.p. 78°. Boiling 48% HBr → o-nitrophenol m.p. 44° and EtBr b.p. 38°. Prolonged boiling with conc. NaOH → o-nitrophenol + EtOH.

SOLID

M.P.°C

38°   m-Nitroanisole $CH_3OC_6H_4NO_2$ (1,3)

B.p. 258°. Almost insol. in H_2O, readily sol. in organic liquids.

M.P.°C

Sn + HCl → m-anisidine b.p. 251° (acetyl deriv. m.p. 80°). Boiling 48% HBr → m-nitrophenol m.p. 97° and MeBr b.p. 4°. Unchanged on boiling with 10% NaOH.

54° p-Nitroanisole $CH_3OC_6H_4NO_2$ (1,4)
B.p. 274°. Volatile in steam. Almost insol. in H_2O, readily sol. in EtOH and ether, spar. sol. in cold light petroleum. Sn + HCl → p-anisidine m.p. 58°. Fuming HNO_3 at 0° → 2,4-dinitroanisole m.p. 88°. On warming with conc. HNO_3 + H_2SO_4 → trinitroanisole m.p. 68°. Boiling 48% HBr → p-nitrophenol m.p. 114° and MeBr b.p. 4°. Prolonged boiling with conc. NaOH → p-nitrophenol + MeOH.

59° p-Nitrophenetole $C_2H_5OC_6H_4NO_2$ (1,4)
B.p. 283°. Insol. in H_2O, spar. sol. in cold EtOH, readily sol. in ether. Sn + HCl → p-phenetidine, b.p. 254°. Conc. HNO_3 + H_2SO_4 in cold → 2,4-dinitrophenetole m.p. 86°, in warm → trinitrophenetole m.p. 78°. $KClO_3$ + HCl → 2-chloro deriv. m.p. 78°. Boiling 40% HBr → p-nitrophenol m.p. 114° and EtBr b.p. 38°. Prolonged boiling with conc. NaOH → p-nitrophenol + EtOH.

68°   2,4,6-Trinitroanisole $CH_3OC_6H_2(NO_2)_3$ (1,2,4,6)
Insol. in H_2O, spar. sol. in cold EtOH, sol. in benzene and acetic acid. Boiling dil. NaOH → picric acid m.p. 122° and MeOH. C_2H_5ONa in EtOH (absolute) → addition cmpd. (red needles) which with acid yields a mixture of trinitroanisole and trinitrophenetole. NH_3 in EtOH → picramide m.p. 188°. Warm aniline → trinitrodiphenylamine m.p. 177°. Naphthalene adduct m.p. 188°.

78°   2,4,6-Trinitrophenetole $C_2H_5OC_6H_2(NO_2)_3$ (1,2,4,6)
Insol. in H_2O, readily sol. in ether and benzene. Boiling dil. NaOH → picric acid m.p. 122° and EtOH. NH_3 in EtOH → picramide m.p. 188°. Warm aniline → trinitrodiphenylamine m.p. 177°.

95°   2,4-Dinitroanisole $CH_3OC_6H_3(NO_2)_2$ (1,2,4)
Insol. in H_2O, readily sol. in EtOH and ether. Boiling conc. NaOH → 2,4-dinitrophenol m.p. 114° and MeOH. Warm conc. HNO_3 + H_2SO_4 → trinitroanisole m.p. 68°. Hot aniline → dinitrodiphenylamine m.p. 156°.

NITROALCOHOLS

SOLID

M.P.°C

27° m-Nitrobenzyl alcohol $O_2NC_6H_4CH_2OH$ (1,3)
On heating to 250° → di-m-nitrobenzyl ether m.p. 114°. Zn dust + dil. $CaCl_2$ soln. → m-aminobenzyl alcohol m.p. 97°. Acid $KMnO_4$ → m-nitrobenzaldehyde m.p. 58° and m-nitrobenzoic acid m.p. 140°. Benzoate m.p. 72°.

M.P.°C

74° *o*-Nitrobenzyl alcohol $O_2NC_6H_4CH_2OH$ (1,2)

B.p. 270°d. On heating at 230° slowly decomposes giving hydroxy-indazylbenzoic lactone m.p. 295°. Spar. sol. in H_2O, readily sol. in EtOH and ether. Zn dust + dil. $CaCl_2$ soln. → *o*-aminobenzyl alcohol m.p. 82°. Oxidised by conc. HNO_3 to *o*-nitrobenzalde-hyde m.p. 46°, and by $K_2Cr_2O_7$ + dil. H_2SO_4 to *o*-nitrobenzoic acid m.p. 144°. Acetate m.p. 35°. Benzoate m.p. 102°.

93° *p*-Nitrobenzyl alcohol $O_2NC_6H_4CH_2OH$ (1,4)

B.p. 185°/12 mmHg. Decomposes on heating to 250°. Readily sol. in hot H_2O, spar. in cold. Zn dust + dil. $CaCl_2$ soln. → *p*-aminobenzyl alcohol m.p. 65°. Oxidation → *p*-nitrobenzoic acid m.p. 241°. Acetate m.p. 78°. Benzoate m.p. 94°.

NITROPHENOLS

SOLID

M.P.°C

45° *o*-Nitrophenol $O_2NC_6H_4OH$ (1,2)

B.p. 216°. Bright yellow. Volatile with steam. Characteristic odour. Readily sol. in hot H_2O and most organic liquids. Sol. in dil. NaOH with orange colour: red Na salt is readily sol. in cold 5% NaOH. Zn dust + $CaCl_2$ (dil.) → *o*-aminophenol m.p. 174°. Gentle warming with 60% HNO_3 → 2,4-dinitrophenol m.p. 114°; with conc. HNO_3 + H_2SO_4 → picric acid m.p. 122°. Bromine (2 moles) added to soln. in dil. NaOH → 4,6-dibromo deriv. m.p. 117°. *p*-Toluenesulphonate m.p. 83°. 3,5-Dinitro-benzoate m.p. 155°.

97° *m*-Nitrophenol $O_2NC_6H_4OH$ (1,3)

B.p. 186°/12 mmHg. Pale yellow. Odourless. Not volatile with steam. Readily sol. in hot H_2O and most organic liquids. Sol. in dil. NaOH with orange-yellow colour. Zn dust + $CaCl_2$ (dil.) → *m*-aminophenol m.p. 122°. With warm bromine (2 moles) → dibromo deriv. m.p. 91°. Benzoate m.p. 95°. *p*-Toluenesulphonate m.p. 113°.

114° *p*-Nitrophenol $O_2NC_6H_4OH$ (1,4)

Colourless. Odourless and not volatile with steam. Readily sol. in hot H_2O (melts), EtOH, ether. Sol. in dil NaOH with bright yellow colour; yellow Na salt is spar. sol. in cold 5% NaOH. Zn dust + $CaCl_2$ (dil.) → *p*-aminophenol m.p. 184°d. Gentle warming with 60% HNO_3 → 2,4-dinitrophenol m.p. 114°; with conc. HNO_3 + H_2SO_4 → picric acid m.p. 122°. Bromine (2 moles) in acetic acid at 100° → 2,6-dibromo deriv. m.p. 142°. Benzoate m.p. 142°. *p*-Toluenesulphonate m.p. 97°. 3,5-Dinitro benzoate m.p. 186°.

114° 2,4-Dinitrophenol $(O_2N)_2C_6H_3OH$ (2,4,1)

Colourless. Sol. in hot H_2O, EtOH, ether. Aq. soln. is yellow, discharged by HCl. Bright yellow soln. in dil. NaOH; yellow Na

M.P.°C

salt is spar. sol. in cold 5% NaOH. Sn + HCl → 2,4-diamino-phenol m.p. 79°d. NH₄SH → 4-nitro-2-aminophenol m.p. 142° (violet soln. in dil. NaOH). Warm conc. HNO₃ + H₂SO₄ → picric acid m.p. 122°. Bromine water → 6-bromo deriv. m.p. 118°. Benzoate m.p. 132°. *p*-Toluenesulphonate m.p. 121°.

122°   2,4,6-Trinitrophenol (Picric acid) (O₂N)₃C₆H₂OH (2,4,6,1)
Light yellow. Sol. in warm water; spar. sol. in cold. Sol. in EtOH, benzene. Spar. sol. in ether. Reduction by NH₄SH (or by Zn dust + NH₄OH) → 4,6-dinitro-2-aminophenol (picramic acid), m.p. 168°. Cold alk. NaOCl → chloropicrin b.p. 112°. Benzoate m.p. 163°. Naphthalene adduct m.p. 150°.

138°   2,4-Dinitro-1-naphthol (O₂N)₂C₁₀H₅OH (2,4,1)
Yellow. Insol. in H₂O. Spar. sol. in ether, benzene, EtOH. Sn + HCl → 2,4-diamino-1-naphthol of which the triacetyl deriv. has m.p. 280°d. Benzoate m.p. 174°.

NITROCARBONYL COMPOUNDS

SOLID

M.P.°C

44°   *o*-Nitrobenzaldehyde O₂NC₆H₄CHO (1,2)
B.p. 153°/23 mmHg. Yellow. Volatile with steam. Sl. sol. in H₂O, readily in most organic liquids. Sol. (with gradual decomp.) in dil. NaOH; pptd. by acids. With conc. NaHSO₃ → cryst. addtn. cmpd. sol. in H₂O. KMnO₄ soln. → *o*-nitrobenzoic acid m.p. 144°. Sn + HCl → *o*-anthranil (b.p. 210–215°d). Boiling Ac₂O + NaOAc → *o*-nitrocinnamic acid m.p. 240°. In dil. acetone with trace of NaOH → blue ppt. of indigotin. Anil m.p. 69°. Oxime m.p. 103°. *p*-Nitrophenylhydrazone m.p. 263°. Semicarbazone m.p. 256°. Phenylhydrazone m.p. 156°.

58°   *m*-Nitrobenzaldehyde O₂NC₆H₄CHO (1,3)
B.p. 164°/23 mmHg. Pale yellow. Sl. sol. in H₂O, readily in organic liquids. Sol. in dil. NaOH (with gradual decomp.); pptd. by acids. With NaHSO₃ soln. → addition cmpd. sol. in H₂O. KMnO₄ soln. → *m*-nitrobenzoic acid m.p. 140°. Sn + AcOH → yellow condensation product of *m*-aminobenzaldehyde which with Ac₂O → *m*-acetaminobenzaldehyde m.p. 84°. Hot NaOAc + Ac₂O → *m*-nitrocinnamic acid m.p. 199°. Boiling Ac₂O alone → diacetate m.p. 72°. Anil m.p. 61°. Oxime m.p. 120°. *p*-Nitro-phenylhydrazone m.p. 247°. Semicarbazone m.p. 246°. Phenyl-hydrazone m.p. 120°.

80°   *m*-Nitroacetophenone O₂NC₆H₄COCH₃ (1,3)
Sol. in EtOH. Volatile in steam. Oxime m.p. 132°. Semicarbazone m.p. 261°. 2,4-Dinitrophenylhydrazone m.p. 233°.

106°   *p*-Nitrobenzaldehyde O₂NC₆H₄CHO (1,4)
Slightly sol. in H₂O, light petroleum; spar. sol. in ether, readily sol. in EtOH, benzene. Sol. in dil. NaOH with gradual decomp.;

M.P.°C

pptd. by acids. Not oxidised by dil. HNO_3; $K_2Cr_2O_7$ + dil. $H_2SO_4 \rightarrow p$-nitrobenzoic acid m.p. 241°. Hot $NaHSO_3$ soln. $\rightarrow$ p-aminobenzaldehyde m.p. 70°. Sn + HCl $\rightarrow$ a mixture. Boiling Ac_2O + NaOAc $\rightarrow p$-nitrocinnamic acid m.p. 285°. Boiling Ac_2O alone $\rightarrow$ diacetate m.p. 126°. Anil m.p. 93°. Oxime m.p. 130°. Semicarbazone m.p. 220°. p-Nitrophenylhydrazone m.p. 249°.

NITROCARBOXYLIC ACIDS

SOLID

M.P.°C

125° 3-Nitrosalicylic (Hydrate) $O_2NC_6H_3(OH)COOH$ (3,2,1)
Anhydrous acid has m.p. 144°. Yellow. Spar. sol. in cold H_2O, readily in organic liquids. With $FeCl_3 \rightarrow$ red colour. Heating with lime $\rightarrow o$-nitrophenol m.p. 44°. Sn + AcOH $\rightarrow$ 3-amino-salicylic acid m.p. 235°d. (Hydrochloride m.p. 250°). $PCl_5 \rightarrow$ chloride, m.p. 60°. Amide m.p. 145°. Ethyl ester m.p. 118°.

140° m-Nitrobenzoic $O_2NC_6H_4COOH$ (1,3)
Colourless. Sol. in boiling H_2O (melts); spar. sol. in cold. Hot soda-lime $\rightarrow$ nitrobenzene. Sn + HCl $\rightarrow m$-aminobenzoic acid m.p. 174°. Chloride m.p. 30°. Amide m.p. 143°. Anilide m.p. 154°. p-Toluidide m.p. 162°. p-Nitrobenzyl ester m.p. 141°.

144° 3-Nitrosalicylic (Anhydrous). See hydrated acid, m.p. 125°.

146° o-Nitrobenzoic $O_2NC_6H_4COOH$ (1,2)
Colourless. Sol. in boiling H_2O. Hot soda-lime $\rightarrow$ nitrobenzene. Sn + HCl $\rightarrow$ anthranilic acid m.p. 144°. Chloride may explode under reduced pressure distillation. Amide m.p. 176°. Anilide m.p. 155°. p-Toluidide m.p. 203°. p-Nitrobenzyl ester m.p. 112°.

154° p-Nitrophenylacetic $O_2NC_6H_4CH_2COOH$ (1,4)
Yellow. Spar. sol. in cold H_2O. Sol. in EtOH, ether, benzene. Sn + HCl $\rightarrow p$-aminophenylacetic acid. Oxdtn. $\rightarrow p$-nitrobenzoic acid. Amide m.p. 198°. Anilide m.p. 212°. p-Bromophenacyl ester m.p. 207°.

165° 4-Nitrophthalic acid $O_2NC_6H_3(COOH)_2$ (4,2,1)
Pale yellow. Sol. in hot H_2O, EtOH. Amide m.p. 200°d. Anilide m.p. 192°. p-Toluidide m.p. 172°. Imide m.p. 202°. p-Phenyl-phenacyl ester m.p. 120°.

174° 3,5-Dinitrosalicylic $(O_2N)_2C_6H_2(OH)COOH$ (3,5,2,1)
Colourless. Readily sol. in hot H_2O (yellow soln.), EtOH and ether. With $FeCl_3 \rightarrow$ red colour. Soln. in dil. NaOH + dextrose $\rightarrow$ orange-red colour. Methyl ester m.p. 127°. $PCl_5 \rightarrow$ chloride, m.p. 69°. Amide m.p. 181°. Anilide m.p. 194° (closed tube).

183° 2,4-Dinitrobenzoic $(O_2N)_2C_6H_3COOH$ (2,4,1)
Colourless. Sol. in hot H_2O. Sn + HCl $\rightarrow m$-phenylenediamine m.p. 63°. Chloride m.p. 41°. Amide m.p. 203°. p-Nitrobenzyl ester m.p. 142°. p-Bromophenacyl ester m.p. 158°.

M.P.°C
- 199° *m*-Nitrocinnamic (*trans*) $O_2NC_6H_4CH:CHCOOH$ (1,3)
Colourless. Unsaturated. $KMnO_4 \rightarrow$ *m*-nitrobenzoic acid m.p. 140°. $Sn + HCl \rightarrow$ *m*-aminocinnamic acid m.p. 180°. Amide m.p. 196°. Methyl ester m.p. 123°. *p*-Nitrobenzyl ester m.p. 174°.

- 205° 3,5-Dinitrobenzoic $(O_2N)_2C_6H_3COOH$ (3,5,1)
Colourless. Sol. in hot H_2O. $Sn + HCl \rightarrow$ 3,5-diaminobenzoic acid m.p. 236°. Chloride m.p. 70°. Amide m.p. 183°. Anilide m.p. 234°. *p*-Toluidide m.p. 280°. Methyl ester m.p. 107°. *p*-Nitrobenzyl ester m.p. 157°.

- 218° 3-Nitrophthalic $O_2NC_6H_3(COOH)_2$ (3,1,2)
Colourless. Sol. in hot H_2O. Readily sol. in EtOH, ether. Insol. in $CHCl_3$. Heating above m.p. $\rightarrow$ anhydride m.p. 162°. Amide m.p. 201°d. Anilide m.p. 234°. *p*-Toluidide m.p. 224°. Anil m.p. 138°.

- 220°d 2,4,6-Trinitrobenzoic $(O_2N)_3C_6H_2COOH$ (2,4,6,1)
Very pale yellow. Sol. in hot H_2O and in most organic liquids. When boiled with H_2O, or melted, gives CO_2 and trinitrobenzene m.p. 122°. NaOH soln. $\rightarrow$ red colour, discharged by acids. Chloride m.p. 158°. Amide m.p. 264°d. Methyl ester m.p. 157°. Ethyl ester m.p. 155°.

- 230° 5-Nitrosalicylic $O_2NC_6H_3(OH)COOH$ (5,2,1)
Colourless. Almost insol. in cold H_2O. Sol. in hot H_2O; readily sol. in EtOH. With $FeCl_3 \rightarrow$ red colour. Heating with soda-lime $\rightarrow$ *p*-nitrophenol m.p. 114°. Boiling conc. $HNO_3 \rightarrow$ picric acid m.p. 122°. $Sn + HCl \rightarrow$ 5-aminosalicylic acid m.p. 283°. Methyl ester m.p. 117°. Methyl ether m.p. 148°. Amide m.p. 225°. Anilide m.p. 224°.

- 240° *o*-Nitrocinnamic (*trans*) $O_2NC_6H_4CH:CHCOOH$ (1,2)
Colourless. Almost insol. in cold H_2O. Unsaturated; with bromine $\rightarrow$ dibromide (slowly), m.p. 180°d. $KMnO_4 \rightarrow$ *o*-nitrobenzaldehyde + *o*-nitrobenzoic acid. $Sn + HCl$ (in EtOH) $\rightarrow$ *o*-aminocinnamic acid m.p. 158°d. Chloride m.p. 64°. Amide m.p. 185°. Methyl ester m.p. 73°. *p*-Nitrobenzyl ester m.p. 132°. *p*-Bromophenacyl ester m.p. 141°.

- 241° *p*-Nitrobenzoic $O_2NC_6H_4COOH$ (1,4)
Colourless. Spar. sol. in cold H_2O and EtOH. Sol. in ether; almost insol. in benzene. Heating with soda-lime $\rightarrow$ nitrobenzene. $Sn + HCl \rightarrow$ *p*-aminobenzoic acid m.p. 186°. Chloride m.p. 75°. Amide m.p. 201°. Anilide m.p. 211°. *p*-Toluidide m.p. 204°. Methyl ester m.p. 96°. *p*-Nitrobenzyl ester m.p. 168°. *p*-Bromophenacyl ester m.p. 137°.

- 285° *p*-Nitrocinnamic (*trans*) $O_2NC_6H_4CH:CHCOOH$ (1,4)
Colourless. Almost insol. in cold H_2O, EtOH and ether. Unsaturated; with bromine $\rightarrow$ dibromide (slowly), m.p. 217°. Conc. $HNO_3 + H_2SO_4 \rightarrow$ *p*-nitrobenzaldehyde, m.p. 106°. Alk. $KMnO_4 \rightarrow$ *p*-nitrobenzoic acid, m.p. 241°. $Sn + HCl$ (in EtOH) $\rightarrow$ *p*-aminocinnamic acid, m.p. 175°d. (acetyl deriv. m.p. 259°).

M.P.°C

Chloride m.p. 124°. Amide m.p. 204°. Methyl ester m.p. 161°.
p-Nitrobenzyl ester m.p. 186°. *p*-Bromophenacyl ester m.p. 191°.

NITROSO COMPOUNDS

SOLID

M.P.°C

67° *N*-Nitrosodiphenylamine $(C_6H_5)_2NNO$
Pale yellow. Insol. in H_2O and dil. acids; sol. in organic liquids.
A trace with conc. $H_2SO_4 \rightarrow$ blue colour. Boiling Sn + HCl in
EtOH → diphenylamine m.p. 54°. Warm Zn dust + AcOH →
diphenylhydrazine m.p. 34°.

68° Nitrosobenzene C_6H_5NO
Colourless but turns green on melting. Odourless. Insol. in H_2O;
sol. in organic liquids with green colour. Sn + HCl → aniline.
NaOH in EtOH → azoxybenzene. Alk. H_2O_2 → nitrobenzene.
Aniline in AcOH → azobenzene m.p. 68°.

84° *p*-Nitroso-*N*,*N*-diethylaniline $(C_2H_5)_2NC_6H_4NO$ (1,4)
Dark green. Insol. in H_2O; sol. in dil. acids and organic liquids.
Forms yellow cryst. hydrochloride. Sn + HCl → *p*-amino-
diethylaniline. $KMnO_4$ → *p*-nitro-*N*,*N*-diethylaniline m.p. 77°.

86° *p*-Nitroso-*N*,*N*-dimethylaniline $(CH_3)_2NC_6H_4NO$ (1,4)
Dark green. Insol. in H_2O; sol. in dil. acids and organic liquids.
Forms yellow cryst. hydrochloride. Sn + HCl → *p*-amino-
dimethylaniline. $KMnO_4$ → *p*-nitro-*N*,*N*-dimethylaniline m.p.
163°.

109° 1-Nitroso-2-naphthol $ONC_{10}H_6OH$ (1,2)
Dull yellow. Almost insol. in H_2O; sol. in dil. NaOH and most
organic liquids. Volatile with steam. Bromine in $CHCl_3$ →
addition product m.p. 130°. Cold dil. HNO_3 → 1-nitro-2-
naphthol m.p. 103°. Fuming HNO_3 at 40°–50° → 1,6-dinitro-2-
naphthol m.p. 195°d.

116° *p*-Nitroso-*N*-methylaniline $CH_3NHC_6H_4NO$ (1,4)
Green with blue reflex. Insol. in H_2O; sol. in dil. acids and most
organic liquids. Forms yellow cryst. hydrochloride. Sn + HCl →
p-amino-*N*-methylaniline (dibenzoyl deriv. m.p. 164°). Boiling
conc. NaOH → *p*-nitrosophenol m.p. 125°d.

125°d *p*-Nitrosophenol HOC_6H_4NO (1,4)
Almost colourless. Sol. in cold H_2O, EtOH, ether, giving green
solutions. Spar. sol. in benzene (yellow soln.). Gives red-brown
soln. in dil. NaOH. Conc. HNO_3 → *p*-nitrophenol m.p. 114°.
Ac_2O at 100° → yellow acetyl deriv. m.p. 107°. 2,4-Dinitro-
chlorobenzene + Na acetate in EtOH → 2,4-dinitrophenyl deriv.
m.p. 165°. Me_2SO_4 + NaOH → quinone methoxime m.p. 83°.

145° *p*-Nitrosodiphenylamine $C_6H_5NHC_6H_4NO$ (1,4)
Green. Insol. in H_2O; sol. in dil. NaOH and in organic liquids.

M.P.°C

Zn dust + AcOH → p-aminodiphenylamine m.p. 66°. Boiling conc. NaOH → aniline + p-nitrosophenol m.p. 125°d.

AZOXY, AZO AND HYDRAZINE COMPOUNDS

LIQUID

B.P.°C
227° N-Methyl-N-phenylhydrazine $C_6H_5N(CH_3)NH_2$
Almost insol. in H_2O. Reduces Fehling's soln. on warming. Boiling with Zn + HCl → methylaniline. Acetyl m.p. 92°. Benzoyl m.p. 153°. Benzal deriv. m.p. 106°.

SOLID

M.P.°C
19°   Phenylhydrazine $C_6H_5NHNH_2$
B.p. 243°. Slightly sol. in H_2O. Hydrochloride m.p. 240°. Reduces cold Fehling's soln. Warm soln. in HCl, with $CuSO_4$ → chlorobenzene b.p. 132°. Excess of iodine → iodobenzene, b.p. 188°. Acetyl m.p. 128°. Benzoyl m.p. 168°. Benzal deriv. m.p. 158°

34° N,N-Diphenylhydrazine $(C_6H_5)_2NNH_2$
Insol. in H_2O. Readily sol. in EtOH. Forms blue soln. in conc. H_2SO_4. Hydrochloride m.p. 167°. Reduces ammoniacal $AgNO_3$. Sn + HCl → diphenylamine m.p. 54°. Acetyl m.p. 184°. Benzoyl m.p. 192°. Benzal deriv. m.p. 122°.

36°   Azoxybenzene $C_6H_5N:N(O)C_6H_5$
Light yellow. $SnCl_2$ + HCl → aniline. Zn dust + ethanolic NaOH → azobenzene m.p. 68° → hydrazobenzene m.p. 131°. Warm conc. H_2SO_4 → p-hydroxyazobenzene m.p. 152°.

55° o-Azotoluene $CH_3C_6H_4N:NC_6H_4CH_3$ (2,2′)
Red. Zn dust + ethanolic NaOH → o-hydrazotoluene m.p. 165° which with conc. HCl → o-tolidine m.p. 129°.

59° o-Azoxytoluene $CH_3C_6H_4N:N(O)C_6H_4CH_3$ (2,2′)
Light yellow. Zn dust + ethanolic NaOH → o-azotoluene m.p. 55° → o-hydrazotoluene m.p. 165°.

65° p-Tolylhydrazine $CH_3C_6H_4NHNH_2$ (1,4)
B.p. 240°–244°d. Reduces cold Fehling's soln. Acetyl m.p. 121°. Benzoyl m.p. 146°. Benzal deriv. m.p. 125°.

68°   Azobenzene $C_6H_5N:NC_6H_5$
B.p. 293°. Orange-red. Zn dust + ethanolic NaOH → hydrazobenzene m.p. 131°. SO_2 in EtOH → benzidine m.p. 127°. Bromine (2 moles) in AcOH → dibromo deriv. m.p. 187°. Benzaldehyde at 200° → benzanilide m.p. 164°.

75° p-Azoxytoluene $CH_3C_6H_4N:N(O)C_6H_4CH_3$ (4,4′)
Light yellow. Zn dust + ethanolic NaOH → p-azotoluene m.p. 144° → p-hydrazotoluene m.p. 133°. $SnCl_2$ + HCl → p-toluidine m.p. 45°.

M.P.°C

96° Semicarbazide $H_2NCONHNH_2$
Very sol. in H_2O and EtOH. Insol. in ether, benzene, $CHCl_3$.
Neutral to litmus. Hydrochloride m.p. 173°d. Boiling HCl →
hydrazine hydrochloride + NH_4Cl. Reduces Fehling's soln. and
ammon. $AgNO_3$. Acetone → isopropenyl deriv. m.p. 188°.

100° Aminoazo-o-toluene $CH_3C_6H_4N:NC_6H_3(CH_3)NH_2$ (2,3′,4′)
Purple. Sol. in organic liquids (yellow soln.) and in dil. acids
(red soln.). Sn + HCl → o-toluidine + p-toluylenediamine m.p.
64° (diacetyl m.p. 220°). Acetyl m.p. 185°. Azo-2-naphthol
deriv. m.p. 186°.

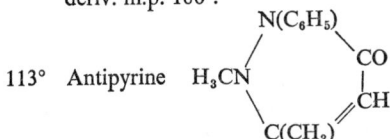

113° Antipyrine H_3CN

$$N(C_6H_5)$$
$$\diagdown\quad CO$$
$$\diagup\quad\vert$$
$$C(CH_3)\quad CH$$

Readily sol. in cold H_2O and EtOH. Spar. sol. in ether. Fairly
strong base. Bromine in $CHCl_3$ → dibromide (pptd by ether),
m.p. ca. 150° which with cold H_2O → bromoantipyrine m.p. 117°.
$NaNO_2$ added to soln. in dil. HCl → nitroso deriv. (green) m.p.
200°d. Orange colour with $FeCl_3$. Nitration → p-nitro deriv.
m.p. 273°. Picrate m.p. 188°.

117° p-Dimethylaminoazobenzene $C_6H_5N:NC_6H_5N(CH_3)_2$ (4)
Orange. Sol. in organic liquids (yellow soln.) and in dil. acids
(red soln.). Sn + HCl → aniline + p-aminodimethylaniline m.p.
41° (b.p. 262°). MeI → methiodide m.p. 174°.

126° p-Aminoazobenzene $C_6H_5N:NC_6H_4NH_2$(4)
Yellow with blue reflex. Sol. in organic liquids (yellow soln.) and
in dil. acids (red soln.). Sn + HCl → aniline + p-phenylenediamine m.p. 140°. Acetyl m.p. 146°. Benzoyl m.p. 211°. Benzal
deriv. m.p. 127°.

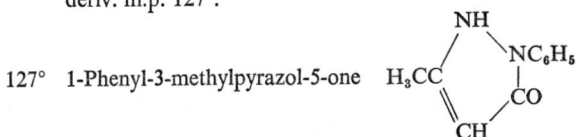

127° 1-Phenyl-3-methylpyrazol-5-one H_3CC

$$NH$$
$$\diagup\quad\diagdown\, NC_6H_5$$
$$\vert\qquad\vert$$
$$CH\qquad CO$$

Insol. in cold H_2O, ether, petroleum ether. Sol. in hot H_2O, cold
dil. acids or alkalis, benzene. Boiling acetone → red colour.
Boiling aq. $FeCl_3$ → purple ppt. of pyrazole blue. $FeCl_3$ in
EtOH → red colour. Benzoyl m.p. 75°. Benzal deriv. m.p. 107°.
Bromine (2 moles) in hot AcOH → dibromo deriv. m.p. 80°.
N_2O_3 → isonitroso deriv. m.p. 157°.

128° Acetylphenylhydrazine $C_6H_5NHNHCOCH_3$
Sol. in hot H_2O. Boiling conc. HCl → phenylhydrazine hydro-
chloride. Reduces Fehling's soln. Bromine (2 moles) in conc.
HCl → dibromo deriv. m.p. 146°.

128° Benzeneazo-o-cresol $C_6H_5N:NC_6H_3(CH_3)OH$ (3,4)
Yellow. Sol. in organic liquids (yellow soln.) and in dil. NaOH

M.P.°C

(orange soln.). Sn + HCl → aniline + 5-amino-*o*-cresol m.p. 176°d. Acetyl m.p. 87°. Benzoyl m.p. 110°.

131° Hydrazobenzene $C_6H_5NHNHC_6H_5$
Insol. in H_2O. Sol. in EtOH, ether. Oxidised rapidly by air (in alkaline soln.) to azobenzene m.p. 68°. Weak base. Hot conc. HCl → benzidine hydrochloride. Diacetyl m.p. 105°. Dibenzoyl m.p. 162°.

132° Benzeneazo-2-naphthol $C_6H_5N:NC_{10}H_6OH$ (1,2)
Orange. Sol. in organic liquids (yellow soln.). Insol. in dil. NaOH. Sol. in conc. H_2SO_4 (magenta soln.). Sn + HCl → aniline + 1-amino-2-naphthol (picrate m.p. 109°). Acetyl m.p. 117°. Benzoyl m.p. 125°.

144° *p*-Azotoluene $CH_3C_6H_4N:NC_6H_4CH_3$ (4,4′)
Orange. Zn dust + dil. ethanolic NaOH → *p*-hydrazotoluene m.p. 133°. Sn + HCl → *p*-toluidine m.p. 45°.

152° *p*-Hydroxyazobenzene $C_6H_5N:NC_6H_4OH$ (4)
Yellow. Sol. in organic liquids (yellow soln.) and in dil. NaOH (orange soln.). Sn + HCl → aniline + *p*-aminophenol m.p. 184°d. Acetyl m.p. 84°. Benzoyl m.p. 138°.

168° Benzoylphenylhydrazine $C_6H_5NHNHCOC_6H_5$
Slightly sol. in hot H_2O, ether. Readily sol. in warm dil. KOH. Oxidation by HgO → benzoylazobenzene (red) m.p. 80°. Boiling conc. HCl → phenylhydrazine hydrochloride + benzoic acid m.p. 122°.

NITROAMINO COMPOUNDS

SOLID

M.P.°C

71° *o*-Nitroaniline $H_2NC_6H_4NO_2$ (1,2)
Yellow. Volatile with steam. Sol. in hot H_2O. Zn dust + NaOH in dil. EtOH → *o*-phenylenediamine m.p. 102°. Acetyl m.p. 93°. Benzoyl m.p. 98°. *p*-Tosyl m.p. 142°.

76° 2,4-Dimethyl-6-nitroaniline $H_2NC_6H_2(CH_3)_2NO_2$ (1,2,4,6)
Orange. Acetyl m.p. 176°. Benzoyl m.p. 185°.

78° 4-Methyl-3-nitroaniline $H_2NC_6H_3(CH_3)NO_2$ (1,4,3)
Yellow. Sol. in hot H_2O. Sn + HCl → *m*-toluylenediamine m.p. 99°. Acetyl m.p. 148°. Benzoyl m.p. 172°. *p*-Tosyl m.p. 164°.

92° 2-Methyl-3-nitroaniline $H_2NC_6H_3(CH_3)NO_2$ (1,2,3)
Light yellow. Sl. sol. in hot H_2O. Sn + HCl → toluylene-2,6-diamine m.p. 104°. Acetyl m.p. 158°. Benzoyl m.p. 168°.

97° 2-Methyl-6-nitroaniline $H_2NC_6H_3(CH_3)NO_2$ (1,2,6)
Orange. Almost insol. in hot H_2O. Sn + HCl → toluylene-2,3-diamine m.p. 61°. Acetyl m.p. 158°. Benzoyl m.p. 167°. *p*-Tosyl m.p. 122°.

M.P.°C

107° 2-Methyl-5-nitroaniline $H_2NC_6H_3(CH_3)NO_2$ (1,2,5)
Yellow. Sn + HCl → m-toluylenediamine m.p. 90°. Acetyl m.p.
151°. Benzoyl m.p. 183°. Benzenesulphonyl m.p. 172°.

114° m-Nitroaniline $H_2NC_6H_4NO_2$ (1,3)
B.p. 285°. Yellow. Volatile with steam. Sol. in hot H_2O. Stable
towards hot NaOH soln. Sn + HCl → m-phenylenediamine m.p.
63°. Acetyl m.p. 155°. Benzoyl m.p. 155°. p-Tosyl m.p. 138°.

117° 4-Methyl-2-nitroaniline $H_2NC_6H_3(CH_3)NO_2$ (1,4,2)
Red. Volatile with steam. Almost insol. in hot H_2O. Zn dust +
NaOH in dil. EtOH → toluylene-3,4-diamine m.p. 88° (diacetyl
deriv. m.p. 210°). Acetyl m.p. 96°. Benzoyl m.p. 148°. p-Tosyl
m.p. 166°.

123° 2,4-Dimethyl-5-nitroaniline $H_2NC_6H_2(CH_3)_2NO_2$ (1,2,4,5)
Acetyl m.p. 159°. Benzoyl m.p. 200°. p-Tosyl m.p. 192°.

130° 2-Methyl-4-nitroaniline $H_2NC_6H_3(CH_3)NO_2$ (1,2,4)
Light yellow. Almost insol. in hot H_2O. Readily sol. in EtOH.
Sn + HCl → p-toluylenediamine m.p. 64° (diacetyl deriv. m.p.
220°). Acetyl m.p. 202°. Benzoyl m.p. 178°. p-Tosyl m.p. 174°.

147° p-Nitroaniline $H_2NC_6H_4NO_2$ (1,4)
Yellow. Not volatile with steam. Sol. in hot H_2O. Sn + HCl →
p-phenylenediamine m.p. 140°. Prolonged boiling with conc.
NaOH → p-nitrophenol m.p. 114° and NH_3. Acetyl m.p. 215°.
Benzoyl m.p. 199°. p-Tosyl m.p. 191°. Picrate m.p. 100°.

157° p-Nitrophenylhydrazine $O_2NC_6H_4NHNH_2$ (1,4)
Orange-red. Sn + HCl → p-phenylenediamine m.p. 140°. Acetyl
m.p. 205°. Benzoyl m.p. 193°. Picrate m.p. 119°. p-Nitrophenylhydrazone of acetone m.p. 148°.

169° Picramic acid $(O_2N)_2C_6H_2(OH)NH_2$ (4,6,1,2)
Red. Sl. sol. in H_2O. Sol. in dil. NaOH with orange colour.
Sol. in EtOH, benzene. Sn + HCl → triaminophenol (readily
oxidisable by air). N-Acetyl m.p. 201°. O-Acetyl m.p. 193°.
N-Benzoyl 230°. O-Benzoyl m.p. 220°. p-Tosyl m.p. 191°.

180° 2,4-Dinitroaniline $H_2NC_6H_3(NO_2)_2$ (1,2,4)
Yellow, with blue reflex. Sl. sol. in hot H_2O. Does not form salts
with acids. Gives red colour with acetone + dil. NaOH. Boiling
conc. NaOH → 2,4-dinitrophenol m.p. 114° + NH_3. Acetyl m.p.
120°. Benzoyl m.p. 202°. p-Tosyl m.p. 219°.

190° Picramide $H_2NC_6H_2(NO_2)_3$ (1,2,4,6)
Yellow with blue reflex. Boiling dil. NaOH → picric acid m.p.
122°. Sn + HCl → triaminophenol (readily oxidised by air).
Diazotisable. Acetyl m.p. 230°. Benzoyl m.p. 196°. Benzenesulphonyl m.p. 211°.

198°d 2,4-Dinitrophenylhydrazine $(O_2N)_2C_6H_3NHNH_2$ (2,4,1)
Red. Spar. sol. in cold EtOH, ether, benzene. Readily sol. in hot
ethyl acetate. Almost insol. in cold dil. acids, salts being hydrolysed by H_2O. Boiling conc. NaOH → 2,4-dinitrophenol m.p.
90° + NH_3. Acetyl m.p. 198°. Benzoyl m.p. 206°. 2,4-Dinitro-
phenylhydrazone of acetone m.p. 128°.

NITROCARBOXYLIC AMIDES

SOLID

M.P.°C

92°   o-Nitroacetanilide O₂NC₆H₄NHCOCH₃ (1,2)
Yellow. Sol. in hot H₂O. Sol. in 10% soln. of KOH in H₂O/
EtOH (1:12); soln. on standing deposits o-nitroaniline m.p. 71°,
of which the benzoyl deriv. has m.p. 98°.

96°   4-Methyl-2-nitroacetanilide CH₃C₆H₃(NO₂)NHCOCH₃ (4,2,1)
Yellow. Sol. in hot H₂O. Boiling dil. NaOH → 4-methyl-2-
nitroaniline m.p. 117°. Boiling dil. KMnO₄ → 3-nitro-4-acetamino-
benzoic acid m.p. 220°.

143°   m-Nitrobenzamide O₂NC₆H₄CONH₂ (1,3)
Colourless. Sol. in hot H₂O. Boiling dil. NaOH → m-nitrobenzoic
acid m.p. 140°. Warm PCl₅ → m-nitrobenzonitrile m.p. 117°.

154°   m-Nitrobenzanilide O₂NC₆H₄CONHC₆H₅ (1,3)
Colourless. Slightly sol. in hot H₂O. Boiling conc. HCl or 50%
H₂SO₄ → m-nitrobenzoic acid m.p. 140° and aniline.

155°   o-Nitrobenzanilide O₂NC₆H₄CONHC₆H₅ (1,2)
Sol. in hot H₂O. Boiling conc. HCl or 50% H₂SO₄ → o-nitrobenzoic acid m.p. 146° and aniline.

155°   m-Nitroacetanilide O₂NC₆H₄COCH₃ (1,3)
Colourless. Insol. in H₂O and in 30% KOH. Boiling dil. NaOH→
m-nitroaniline m.p. 114°. Boiling HCl + Sn → m-phenylenediamine m.p. 64°.

176°   o-Nitrobenzamide O₂NC₆H₄CONH₂ (1,2)
Colourless. Sol. in hot H₂O. Boiling dil. NaOH → o-nitrobenzoic
acid m.p. 146°. Sn + HCl → anthranilic acid m.p. 147°. Warm
PCl₅ → o-nitrobenzonitrile m.p. 110°.

201°   p-Nitrobenzamide O₂NC₆H₄CONH₂ (1,4)
Colourless. Sol. in hot H₂O. Boiling dil. NaOH → NH₃ + p-
nitrobenzoic acid m.p. 241°. Boiling HCl + Sn → p-aminobenzoic acid m.p. 187°. Warm PCl₅ → p-nitrobenzonitrile m.p.
148°. Xanthyl deriv. m.p. 232°.

202°   2-Methyl-4-nitroacetanilide CH₃C₆H₃(NO₂)NHCOCH₃ (2,4,1)
Colourless. Spar. sol. in hot H₂O. Boiling dil. NaOH → 2-methyl-
4-nitroaniline m.p. 130°. Prolonged boiling with conc. NaOH →
5-nitro-o-cresol m.p. 94°. Boiling dil. KMnO₄ → 5-nitro-2-
acetaminobenzoic acid m.p. 215°.

216°   p-Nitroacetanilide O₂NC₆H₄NHCOCH₃ (1,4)
Colourless. Insol. in 10% soln. of KOH in H₂O/EtOH (1:12).
Sol. with slow decomp. in 30% KOH. Boiling dil. NaOH → p-
nitroaniline m.p. 148°. Prolonged boiling with conc. NaOH →
p-nitrophenol m.p. 114°. Boiling HCl + Sn → p-phenylenediamine m.p. 141°. Fe filings + dil. AcOH → p-aminoacetanilide
m.p. 165°.

ALKYL NITRITES AND NITRATES

B.P.°C		d_4^{20}	n_D^{20}
17°	Ethyl nitrite	0.900 (15°)	1.331 (10°)
45°	Isopropyl nitrite	0.856	—
48°	n-Propyl nitrite	0.886	1.360
65°	Methyl nitrate	1.208	1.375
67°	Isobutyl nitrite	0.871	1.373
76°	n-Butyl nitrite	0.882	1.377
88°	Ethyl nitrate	1.108	1.385
99°	Isoamyl nitrite	0.871	1.387
102°	Isopropyl nitrate	1.035	1.391
104°	n-Amyl nitrite	0.882	1.389
110°	n-Propyl nitrate	1.054	1.397
124°	Isobutyl nitrate	1.015	1.403
124°	sec-Butyl nitrate	1.026	1.402
136°	n-Butyl nitrate	1.023	1.407
148°	Isoamyl nitrate	0.998	1.413
157°	n-Amyl nitrate	0.996	—

HALOGEN SUBSTITUTED AMINES

LIQUID

B.P.°C

209° o-Chloroaniline $ClC_6H_4NH_2$ (1,2)
Hydrochloride m.p. 235°. Solution in conc. H_2SO_4 when treated with 1 mole $KNO_3 \rightarrow$ 5-nitro deriv., m.p. 117°. Sandmeyer's reaction $\rightarrow$ o-dichlorobenzene b.p. 99°. Acetyl m.p. 88°. Benzoyl m.p. 99°. p-Tosyl m.p. 105°. Picrate m.p. 134°.

223° 2-Chloro-4-methylaniline $Cl(CH_3)C_6H_3NH_2$ (2,4,1)
M.p. 7°. Nitrate m.p. 189°d. Acetyl m.p. 118°. Benzoyl m.p. 137°. p-Tosyl m.p. 103°.

230° m-Chloroaniline $ClC_6H_4NH_2$ (1,3)
Sulphate spar. sol. in cold H_2O. Sandmeyer's reaction $\rightarrow$ m-dichlorobenzene b.p. 172°. Acetyl m.p. 73°. Benzoyl m.p. 122°. p-Tosyl m.p. 138°. Picrate m.p. 177°.

245° 3-Chloro-2-methylaniline $Cl(CH_3)C_6H_3NH_2$ (3,2,1)
M.p. 2°. Volatile in steam. Hydrochloride m.p. 251°d. Acetyl m.p. 159°. Benzoyl m.p. 173°.

SOLID

M.P.°C

18° m-Bromoaniline $BrC_6H_4NH_2$ (1,3)
B.p. 251°. Sandmeyer's reaction $\rightarrow$ m-chlorobromobenzene, b.p. 196°. Acetyl m.p. 88°. Benzoyl m.p. 135°. Picrate m.p. 180°.

26° 2-Bromo-4-methylaniline $Br(CH_3)C_6H_3NH_2$ (2,4,1)
B.p. 240°. Volatile in steam. Hydrochloride m.p. 221°. Acetyl m.p. 117°. Benzoyl m.p. 149°.

M.P.°C

26° 3-Chloro-4-methylaniline   Cl(CH₃)C₆H₃NH₂ (3,4,1)

$Cl(CH_3)C_6H_3NH_2$ (3,4,1)

B.p. 237°. Acetyl m.p. 105°. Benzoyl m.p. 122°.

29° 4-Chloro-2-methylaniline $Cl(CH_3)C_6H_3NH_2$ (4,2,1)

Sol. hot EtOH. Acetyl m.p. 140°. Benzoyl m.p. 172° (142°).

32° o-Bromoaniline $BrC_6H_4NH_2$ (1,2)

B.p. 250°. Sandmeyer's reaction → o-chlorobromobenzene, b.p. 195°. Acetyl m.p. 99°. Benzoyl m.p. 116°. p-Tosyl m.p. 90°. Picrate m.p. 129°.

50° 2,5-Dichloroaniline $Cl_2C_6H_3NH_2$ (2,5,1)

Hydrochloride m.p. 191°. Acetyl m.p. 133°. Benzoyl m.p. 120°. Picrate m.p. 86°.

59° 4-Bromo-2-methylaniline $Br(CH_3)C_6H_3NH_2$ (4,2,1)

Sol. in hot EtOH. Volatile in steam. Nitrate m.p. 183°. Acetyl m.p. 157°. Benzoyl m.p. 115°.

60° o-Iodoaniline $IC_6H_4NH_2$ (1,2)

Sol. in common organic solvents. Spar. sol. in H_2O. Volatile in steam. Hydrochloride m.p. 153°. Acetyl m.p. 109°. Benzoyl m.p. 139°. Picrate m.p. 112°.

63° 2,4-Dichloroaniline $Cl_2C_6H_3NH_2$ (2,4,1)

B.p. 245°. Sandmeyer's reaction → 1,2,4-trichlorobenzene m.p. 16°, b.p. 213°. Acetyl m.p. 145°. Benzoyl m.p. 117°. p-Tosyl m.p. 126°. Picrate m.p. 106°.

64° 2-Chloro-p-phenylenediamine $ClC_6H_3(NH_2)_2$ (2,1,4)

Sol. in H_2O. Monoacetyl m.p. 134°. Diacetyl m.p. 196°. Dibenzoyl m.p. 228°. Di-p-Tosyl m.p. 167°.

66° p-Bromoaniline $BrC_6H_4NH_2$ (1,4)

B.p. 245°d. Sandmeyer's reaction → p-chlorobromobenzene m.p. 67°. Acetyl m.p. 167°. Benzoyl m.p. 204°. p-Tosyl m.p. 101°. Picrate m.p. 180°.

67° p-Iodoaniline $IC_6H_4NH_2$ (1,4)

Sol. in EtOH. Mod. sol. in ether; spar. sol. in petroleum ether. Volatile in steam. Acetyl m.p. 184°. Benzoyl m.p. 222°. p-Nitrobenzoyl m.p. 269°.

71° p-Chloroaniline $ClC_6H_4NH_2$ (1,4)

B.p. 230°. Sulphate is spar. sol. in cold H_2O. Sandmeyer's reaction → p-dichlorobenzene m.p. 53°. Acetyl m.p. 179°. Benzoyl m.p. 193°. p-Tosyl m.p. 96°. Picrate m.p. 178°.

76° 4-Chloro-o-phenylenediamine $ClC_6H_3(NH_2)_2$ (4,1,2)

Sol. in EtOH, ether. Spar. sol. in cold H_2O. Mod. volatile in steam. Diacetyl m.p. 208°. Dibenzoyl m.p. 230°.

78° 2,4,6-Trichloroaniline $Cl_3C_6H_2NH_2$ (2,4,6,1)

B.p. 263°. Insol. in HCl; salts hydrolysed by H_2O. Sandmeyer's reaction → 1,2,3,5-tetrachlorobenzene m.p. 50° b.p., 246°. Diazotisation in EtOH + H_2SO_4 and then warming → 1,3,5-trichlorobenzene m.p. 63°, b.p. 208°. Acetyl m.p. 206°. Benzoyl m.p. 174°. Picrate m.p. 83°.

80° 2,4-Dibromoaniline $Br_2C_6H_3NH_2$ (2,4,1)

Sandmeyer's reaction → 2,4-dibromochlorobenzene m.p. 27°

M.P.°C

b.p. 258°. Acetyl m.p. 146°. Benzoyl m.p. 134°. *p*-Tosyl m.p. 134°. Picrate m.p. 124°.

91° 4-Chloro-*m*-phenylenediamine (2,4-Diaminochlorobenzene) $ClC_6H_3(NH_2)_2$ (4,1,3)
Sol. in EtOH. Spar. sol. in H_2O. Diacetyl m.p. 242° (monoacetyl m.p. 170°). Dibenzoyl m.p. 178°. Di-*p*-Tosyl m.p. 215°.

106° *p*-Bromophenylhydrazine $BrC_6H_4NHNH_2$ (1,4)
$SnCl_2 + HCl \rightarrow p$-bromoaniline, m.p. 66° and aniline. Acetyl m.p. 167°. Benzoyl m.p. 156°d. Benzaldehyde → deriv. m.p. 127°. Acetone → deriv. m.p. 98°.

120° 2,4,6-Tribromoaniline $Br_3C_6H_2NH_2$ (2,4,6,1)
Insol. in HCl; salts are hydrolysed by H_2O. Sandmeyer's reaction → 2,4,6-tribromochlorobenzene m.p. 90°. Diazotisation in EtOH + H_2SO_4 and then warming → 1,3,5-tribromobenzene m.p. 119°. Boiling Ac_2O → diacetyl deriv. m.p. 127° which yields monoacetyl deriv. m.p. 232° on gentle hydrolysis with dilute NaOH. Benzoyl m.p. 198°. Formyl m.p. 222°.

HALOGEN SUBSTITUTED AMIDES

SOLID

M.P.°C

88° *o*-Chloroacetanilide $CH_3CONHC_6H_4Cl$ (1,4)
Boiling dil. acids or alkalis → *o*-chloroaniline b.p. 208°, of which the benzoyl deriv. has m.p. 99°.

99° *o*-Bromoacetanilide $CH_3CONHC_6H_4Br$ (1,4)
Boiling dil. acids or alkalis → *o*-bromoaniline, m.p. 32° of which the benzoyl deriv. has m.p. 116°.

119° Chloroacetamide $ClCH_2CONH_2$
B.p. 224°d. Readily sol. in H_2O and EtOH. Sl. sol. in ether. Boiling HCl → chloroacetic acid, m.p. 63°. With hexamethylene-tetramine in acetone → cmpd, m.p. 160°d. Xanthyl deriv., m.p. 209°.

134° *m*-Chlorobenzamide $ClC_6H_4CONH_2$ (1,3)
Boiling alkali → *m*-chlorobenzoic acid, m.p. 158° of which the *p*-nitrobenzyl ester has m.p. 107°.

141° *o*-Chlorobenzamide $ClC_6H_4CONH_2$ (1,2)
Boiling alkali → *o*-chlorobenzoic acid, m.p. 142° of which the *p*-nitrobenzyl ester has m.p. 106°.

155° *o*-Bromobenzamide $BrC_6H_4CONH_2$ (1,2)
Boiling alkali → *o*-bromobenzoic acid, m.p. 150° of which the *p*-nitrobenzyl ester has m.p. 110°.

155° *m*-Bromobenzamide $BrC_6H_4CONH_2$ (1,3)
Boiling alkali → *m*-bromobenzoic acid, m.p. 155°, of which the *p*-nitrobenzyl ester has m.p. 105°.

167° *p*-Bromoacetanilide $CH_3CONHC_6H_4Br$ (1,4)
Boiling conc. HCl or alkali → *p*-bromoaniline, m.p. 66°, of

M.P.°C

which the benzoyl deriv. has m.p. 204°. Fuming HNO_3 in AcOH at 50° → 2-nitro deriv. m.p. 104°.

179°   *p*-Chloroacetanilide $CH_3CONHC_6H_4Cl$ (1,4)
Boiling conc. HCl or alkali → *p*-chloroaniline, m.p. 72°, of which the benzoyl deriv. has m.p. 192°. Fuming HNO_3 in AcOH at 50° → 2-nitro deriv. m.p. 104°.

179°   *p*-Chlorobenzamide $ClC_6H_4CONH_2$ (1,4)
Boiling alkali → *p*-chlorobenzoic acid, m.p. 243° of which the *p*-nitrobenzyl ester has m.p. 130°.

189°   *p*-Bromobenzamide $BrC_6H_4CONH_2$ (1,4)
Boiling alkali → *p*-bromobenzoic acid, m.p. 252° of which the *p*-nitrobenzyl ester has m.p. 141°.

194°   *p*-Chlorobenzanilide $ClC_6H_4CONHC_6H_5$ (1,4)
Boiling conc. HCl or 50% H_2SO_4 → *p*-chlorobenzoic acid m.p. 243°, and aniline b.p. 184°.

HALOGENO NITROHYDROCARBONS

SOLID

M.P.°C

33°   *o*-Chloronitrobenzene $ClC_6H_4NO_2$ (1,2)
Pale yellow. B.p. 245°. Zn dust + dil. AcOH → *o*-chloroaniline, b.p. 209°. Conc. HNO_3 + conc. H_2SO_4 at 100° → 2,4-dinitro-chlorobenzene m.p. 53°. Boiling KOH in MeOH → *o*-nitro-anisole m.p. 9°, b.p. 265°. Slow addition of Na_2S (0.5 mole) to soln. in boiling EtOH → di-*o*-nitrophenyl sulphide, m.p. 122°.

42°   *o*-Bromonitrobenzene $BrC_6H_4NO_2$ (1,2)
Pale yellow. B.p. 264°. Sol. in cold fuming H_2SO_4. HCl + Sn → *o*-bromoaniline, m.p. 32°. Conc. HNO_3 + conc. H_2SO_4 at 100° → 2,4-dinitrobromobenzene m.p. 75°. Reacts with KOH and Na_2S like *o*-chloronitrobenzene (*q.v.*).

46°   *m*-Chloronitrobenzene $ClC_6H_4NO_2$ (1,3)
Pale yellow. B.p. 235°. HCl + Sn → *m*-chloroaniline, b.p. 230°. Fuming HNO_3 + conc. H_2SO_4 → 3,4-dinitrochlorobenzene m.p. 37°. Boiling KOH (or Na_2S) in MeOH → 3,3′-dichloroazoxy-benzene m.p. 97°.

49°   *o*-Nitrobenzylchloride $O_2NC_6H_4CH_2Cl$ (1,2)
Treating with HCl + $SnCl_2$ at 40°–50° and then adding Zn dust → *o*-toluidine b.p. 197°. Hot $KMnO_4$ soln. → *o*-nitro-benzoic acid m.p. 144°. Conc. HNO_3 + conc. H_2SO_4 → 2,4-dinitrobenzylchloride, m.p. 34°. Warm Na_2CO_3 soln. → *o*-nitrobenzyl alcohol m.p. 74°. Phenol + NaOH in EtOH → *o*-nitrobenzylphenyl ether m.p. 63°.

53°   2,4-Dinitrochlorobenzene $ClC_6H_3(NO_2)_2$ (1,2,4)
Pale yellow. B.p. 315°. Spar. sol. in cold EtOH, readily in ether and benzene. HCl + $SnCl_2$ → chloro-*m*-phenylenediamine m.p.

M.P.°C

91° (dibenzoyl m.p. 178°; di-p-tosyl m.p. 215°). Fuming HNO$_3$ + conc. H$_2$SO$_4$ at 100° → picryl chloride m.p. 83°. Boiling dil. NaOH → 2,4-dinitrophenol m.p. 114°. Boiling KOH in MeOH→ 2,4-dinitroanisole m.p. 95°. Hydrazine → 2,4-dinitrophenyl-hydrazine m.p. 199°.

55°   1,4-Dichloro-2-nitrobenzene   O$_2$NC$_6$H$_3$Cl$_2$ (2,1,4)
Pale yellow. B.p. 266°. Spar. sol. in cold EtOH, readily in benzene. HCl + Sn → 2,5-dichloroaniline, m.p. 50° (acetyl m.p. 133°; benzoyl m.p. 120°). Warm fuming HNO$_3$ + conc. H$_2$SO$_4$→ 1,3-dinitro-2,5-dichlorobenzene m.p. 104°. Boiling KOH soln. in MeOH → 2-nitro-4-chloroanisole m.p. 98°.

56° m-Bromonitrobenzene   O$_2$NC$_6$H$_4$Br (1,3)
Pale yellow. B.p. 257°. Readily sol. in EtOH. HCl + Sn → m-bromoaniline m.p. 18° (benzoyl m.p. 135°). Heating with excess of fuming HNO$_3$ + conc. H$_2$SO$_4$ → 3,4-dinitrobromobenzene m.p. 59°. Boiling KOH (or Na$_2$S) in MeOH → 3,3′-dibromo-azoxybenzene m.p. 111°.

71° p-Nitrobenzyl chloride   O$_2$NC$_6$H$_4$CH$_2$Cl (1,4)
Readily sol. in EtOH. HCl + SnCl$_2$ followed by addition of Zn dust → p-toluidine m.p. 45°. Hot KMnO$_4$ soln. → p-nitrobenzoic acid m.p. 241°. Cold conc. HNO$_3$ + conc. H$_2$SO$_4$ → 2,4-dinitrobenzyl chloride m.p. 34°. Boiling Na$_2$CO$_3$ soln. → p-nitrobenzyl alcohol m.p. 93°. Phenol + NaOH in EtOH → p-nitrobenzylphenyl ether m.p. 91°.

75°   2,4-Dinitrobromobenzene   BrC$_6$H$_3$(NO$_2$)$_2$ (1,2,4)
Pale yellow. HCl + Sn → m-phenylenediamine m.p. 63°. Reacts with aq. NaOH, KOH in MeOH and hydrazine like 2,4-dinitrochlorobenzene ($q.v.$).

83° p-chloronitrobenzene   ClC$_6$H$_4$NO$_2$ (1,4)
V. pale yellow. B.p. 242°. HCl + Sn → p-chloroaniline m.p. 71°. Conc. HNO$_3$ + conc. H$_2$SO$_4$ at 100° → 2,4-dinitrochlorobenzene m.p. 53°. Prolonged boiling with conc. NaOH soln. → p-nitrophenol m.p. 114°. Slow addtn. of Na$_2$S (0.5 mole) to boiling soln. in EtOH → di-p-nitrophenyl sulphide m.p. 154°.

83°   Picryl chloride   ClC$_6$H$_2$(NO$_2$)$_3$ (1,2,4,6)
Yellow. Spar sol. in ether; readily sol. in hot EtOH and benzene. Boiling dil. NaOH → picric acid, m.p. 122°. Boiling KOH in MeOH → trinitroanisole, m.p. 68°; boiling KOH in EtOH → trinitrophenetole, m.p. 78°. NH$_3$ in EtOH → picramide, m.p. 188°. Hot aniline → trinitrodiphenylamine, m.p. 177°. Naphthalene adduct, m.p. 150°.

84°   1,4-Dibromo-2-nitrobenzene   O$_2$NC$_6$H$_3$Br$_2$ (2,1,4)
Pale yellow. HCl + Sn → 2,5-dibromoaniline m.p. 51° (Acetyl m.p. 171°). Boiling KOH in MeOH → 2-nitro-4-bromoanisole m.p. 86°.

100° p-Nitrobenzyl bromide   O$_2$NC$_6$H$_4$CH$_2$Br (1,4)
For oxidation and replacement reaction, cf. p-nitrobenzyl chloride m.p. 71°.

M.P.°C
127° p-Bromonitrobenzene $BrC_6H_4NO_2$ (1,4)
Very pale yellow. B.p. 259°. Spar. sol. in EtOH. HCl + SnCl$_2$ →
p-bromoaniline m.p. 66°. Conc. HNO$_3$ + conc. H$_2$SO$_4$ at 100° →
2,4-dinitrobromobenzene m.p. 75°. Reacts with NaOH soln.
and with Na$_2$S like p-chloronitrobenzene ($q.v.$).

MERCAPTANS (THIOLS)

LIQUID

B.P.°C
6° Methyl CH_3SH, Methanethiol.
Soln. in dil. NaOH gives purple colour with nitroprusside. Hg
salt m.p. 175°. 2,4-Dinitrophenyl thioether m.p. 128°, sulphone
m.p. 190°.

36° Ethyl C_2H_5SH, Ethanethiol.
Almost insol. in H$_2$O. Soln. in dil. NaOH gives purple colour
with nitroprusside. Yellow ppt. with AgNO$_3$. Pb acetate →
yellow Pb salt m.p. 150°. Hg salt m.p. 76°. I$_2$/H$_2$O → disulphide
b.p. 153°. 2,4-Dinitrophenyl thioether m.p. 115°, sulphone m.p.
160°.

67° n-Propyl C_3H_7SH, Propanethiol.
Soln. in dil. NaOH gives purple colour with nitroprusside. Hg
salt m.p. 70°. I$_2$/H$_2$O → disulphide b.p. 192°. 2,4-Dinitrophenyl
thioether m.p. 81°, sulphone m.p. 128°.

97° n-Butyl $CH_3(CH_2)_3SH$, Butanethiol.
Soln. in dil. NaOH gives purple colour with nitroprusside. Hg
salt m.p. 86°. I$_2$/H$_2$O → disulphide b.p. 226°. 2,4-Dinitrophenyl
thioether m.p. 66°, sulphone m.p. 92°.

169° Thiophenol C_6H_5SH, Phenyl mercaptan.
Soln. in dil. NaOH gives transient purple colour with nitroprusside followed immediately by a turbidity. Soln. in conc.
H$_2$SO$_4$ becomes red turning blue on warming. Yellow ppt. with
Cu, Pb, Ag salts. I$_2$/dil. NaOH (or with H$_2$O$_2$) → disulphide m.p.
61°. Me$_2$SO$_4$/alkali → methyl phenyl sulphide b.p. 188° which
with KMnO$_4$/AcOH → sulphone m.p. 88°. 2,4-Dinitrophenyl
thioether m.p. 121°, sulphone m.p. 161°. 3,5-Dinitrobenzoyl
m.p. 149°.

194° Benzyl $C_6H_5CH_2SH$
Soln. in dil. NaOH gives purple colour with nitroprusside which
slowly fades. HNO$_3$ oxdtn. → benzaldehyde and benzoic acid.
I$_2$/H$_2$O → disulphide m.p. 73°. 2,4-Dinitrophenyl thioether m.p.
130°, sulphone m.p. 183°.

194° Thio o-cresol $CH_3C_6H_4SH$, o-Tolyl mercaptan.
M.p. 15°. Hg salt m.p. 170°. 2,4-Dinitrophenyl thioether m.p.
101°, sulphone m.p. 155°.

195° Thio m-cresol $CH_3C_6H_4SH$, m-Tolyl mercaptan.
2,4-Dinitrophenyl thioether m.p. 91°, sulphone m.p. 145°.

SOLID

M.P.°C

43°   Thio p-cresol   CH₃C₆H₄SH, p-Tolyl mercaptan.
      B.p. 195°. Blue colour with warm conc. H₂SO₄. I₂/dil. NaOH (or
      with H₂O₂) → disulphide m.p. 48°. Me₂SO₄/alkali → methyl
      p-tolyl sulphide b.p. 209°, which with KMnO₄/AcOH → sul-
      phone m.p. 86°. 2,4-Dinitrophenyl thioether m.p. 103°, sulphone
      m.p. 190°.

SULPHIDES (THIOETHERS) AND DISULPHIDES

LIQUID

B.P.°C

37°   Dimethyl sulphide $(CH_3)_2S$
      Aq. HgCl₂ → adduct m.p. 150°. Sulphone m.p. 109°.

66°   Ethyl methyl sulphide $C_2H_5SCH_3$
      Aq. HgCl₂ → adduct m.p. 128°. Sulphone m.p. 36°.

84°   Thiophen C_4H_4S
      Soln. in H₂SO₄ → red → deep brown colour. Isatin + H₂SO₄ →
      blue colour. H₂SO₄ + nitrite → blue colour. HgHal₂ → 2-
      mercurihalide m.p. 182° (Cl), 169° (Br), 116° (I).

92°   Diethyl sulphide $(C_2H_5)_2S$
      Aq. HgCl₂ → adduct m.p. 128°. Sulphone m.p. 73°.

110° Dimethyl disulphide $(CH_3)_2S_2$
      Zn/HCl → methyl mercaptan (q.v.).

121° Thiophane C_4H_8S, Thiolane.
      Aq. HgCl₂ → adduct m.p. 128°. Sulphone m.p. 28°.

142° Di n-propyl sulphide $(C_3H_7)_2S$
      Aq. HgCl₂ → adduct m.p. 122°. Sulphone m.p. 29°.

153° Diethyl disulphide $(C_2H_5)_2S_2$
      Zn/HCl → ethyl mercaptan (q.v.).

182° Di n-butyl sulphide $(C_4H_9)_2S$
      Aq. HgCl₂ → adduct m.p. 111°. Sulphone m.p. 44°.

188° Methyl phenyl sulphide $C_6H_5SCH_3$
      Sulphone m.p. 88°.

195° Benzyl methyl sulphide $C_6H_5CH_2SCH_3$
      Sulphone m.p. 127°.

204° Ethyl phenyl sulphide $C_6H_5SC_2H_5$
      Sulphone m.p. 41°.

217° Di-2-chloroethyl sulphide $(ClCH_2CH_2)_2S$, Mustard gas.
      Highly toxic. M.p. 14°. Sulphone m.p. 56°.

295° Diphenyl sulphide $(C_6H_5)_2S$
      KMnO₄/AcOH → sulphone m.p. 128°. Br₂ (2 moles) → 4,4′-
      dibromo m.p. 112°.

SOLID

M.P.°C

48°   Di p-tolyl disulphide $(CH_3C_6H_4)_2S_2$
      Zn/HCl → thio p-cresol (q.v.).

M.P.°C

49°   Dibenzyl sulphide $(C_6H_5CH_2)_2S$
Aq. $HgCl_2 \to$ adduct m.p. 136° (131°). $KMnO_4/AcOH \to$
sulphone m.p. 151°.

57°   Di p-tolyl sulphide $(CH_3C_6H_4)_2S$
$KMnO_4/AcOH \to$ sulphone m.p. 158°.

60°   Diphenyl disulphide $(C_6H_5)_2S_2$
$Zn/HCl \to$ thiophenol ($q.v.$). Conc. $HNO_3 \to$ benzene sulphonic
acid.

64°   Di o-tolyl sulphide $(CH_3C_6H_4)_2S$
Sulphone m.p. 135°.

73°   Dibenzyl disulphide $(C_6H_5CH_2)_2S_2$
$Zn/HCl \to$ benzyl mercaptan ($q.v.$).

SULPHOXIDES AND SULPHONES

LIQUID

B.P.°C

189°   Dimethyl sulphoxide $(CH_3)_2SO$
d_4^{20} 1.1014. n_D^{20} 1.4783. Dipolar aprotic solvent.

SOLID

M.P.°C

27°   Sulpholane $C_4H_8SO_2$, tetrahydrothiophene-1,1-dioxide.
B.p. 285°.

65°   3-Sulpholene $C_4H_6SO_2$, 2,5-dihydrothiophene-1,1-dioxide.
Heat (110°–130°) causes complete decomp. to butadiene $+ SO_2$.

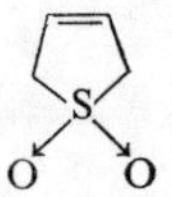

70°   Phenyl sulphoxide $(C_6H_5)_2SO$
Transient blue colour with conc. H_2SO_4. $KMnO_4/AcOH \to$
sulphone m.p. 128°. $Zn/HCl \to$ sulphide b.p. 295°.

95° p-Tolyl sulphoxide $(CH_3C_6H_4)_2SO$
$KMnO_4 \to$ sulphone m.p. 158°. $Zn/HCl \to$ sulphide m.p. 57°.

127°   Sulphonal $(C_2H_5SO_2)_2C(CH_3)_2$
Spar. sol. in H_2O, sl. sol. in alcohol. Unaffected by strong acids,
alkalis, oxidising agents, bromine. Fusion with $KCN \to$ odour
of mercaptan (yields KSCN).

128°   Phenyl sulphone $(C_6H_5)_2SO_2$
Sl. sol. in hot H_2O. Unaffected by oxidising/reducing agents.
Conc. $HNO_3/H_2SO_4 \to m$-nitro (di) m.p. 201°.

134°   Benzyl sulphoxide $(C_6H_5CH_2)_2SO$
Sol. in hot H_2O, and in alcohol. $KMnO_4/AcOH \to$ sulphone
m.p. 151° (some benzoic acid formed). $Zn/HCl \to$ sulphide m.p.
49°.

151°   Benzyl sulphone $(C_6H_5CH_2)_2SO_2$
Insol. in H_2O, spar. sol. in alcohol.

M.P.°C
158° p-Tolyl sulphone $(CH_3C_6H_4)_2SO_2$
       $KMnO_4/AcOH$ (prolonged) → phenyl sulphone pp'-dicarboxylic
       acid m.p. > 300°.

SULPHONIC ACIDS

Most sulphonic acids decompose on heating and the melting
points quoted are primarily intended as a working guide to the
decomposition temperatures. Where no such temperatures are
applicable the compounds are listed in alphabetical order.

SOLID

M.P.°C
20°   Methanesulphonic CH_3SO_3H
      B.p. 167°/10 mmHg. Amide m.p. 90°. Anilide m.p. 99°.
43°   Benzenesulphonic $C_6H_5SO_3H$
      Anhydrous m.p. 66°. Very sol. in H_2O. Reflux with strong
      H_2SO_4 → benzene. KOH fusion → phenol. SBT salt m.p. 148°.
      Amide m.p. 153° (156°). Anilide m.p. 110°.
48° m-Nitrobenzenesulphonic $NO_2C_6H_4SO_3H$
      Readily sol. in cold H_2O, less so in conc. HCl. Sn/HCl → metan-
      ilic acid. SBT salt m.p. 146°. Amide m.p. 162°. Anilide m.p. 126°.
57° o-Toluenesulphonic $CH_3C_6H_4SO_3H$
      Hygroscopic leaflets. Ba salt sl. sol. in cold H_2O. $KMnO_4$ →
      o-sulphobenzoic acid. KOH fusion → o-cresol + salicyclic acid.
      SBT salt m.p. 170°. Amide m.p. 153°. Anilide m.p. 136°.
68° p-Chlorobenzenesulphonic $ClC_6H_4SO_3H$
      SBT salt m.p. 175°. Amide m.p. 144°. Anilide m.p. 104°.
88°   Naphthalene 1-sulphonic $C_{10}H_7SO_3H + 2H_2O$
      Anhydrous m.p. 90°. Warm conc. H_2SO_4 → 2-sulphonic acid
      isomer. KOH fusion → 1-naphthol. Na salt in aq. alk. + Raney
      nickel → naphthalene. SBT salt m.p. 137°. Amide m.p. 50°.
      Anilide m.p. 112°.
92° p-Toluenesulphonic $CH_3C_6H_4SO_3H$
      Anhydrous. Hydrate m.p. 105°. Ba salt readily sol. in cold H_2O.
      Boiling strong H_2SO_4 → toluene. KOH fusion → p-cresol. SBT
      salt m.p. 182°. Amide m.p. 137°. Anilide m.p. 103°.
103° p-Bromobenzenesulphonic $BrC_6H_4SO_3H$
      SBT salt m.p. 170°. Amide m.p. 166°, Anilide m.p. 119°.
120° 5-Sulphosalicylic $HOOC(OH)C_6H_3SO_3H$
      Readily sol. in H_2O. Red colour with $FeCl_3$. Reflux with strong
      H_2SO_4, or KOH fusion → salicylic acid. SBT salt m.p. 204°.
124° Naphthalene 2-sulphonic $C_{10}H_7SO_3H$
      Anhydrous m.p. 91°. Unaffected by warm conc. H_2SO_4. KOH
      fusion → 2-naphthol. Na salt in aq. alk. + Raney nickel →
      naphthalene. SBT salt m.p. 191°. Amide m.p. 212° (217°).
      Anilide m.p. 132°.

M.P.°C

134° o-Sulphobenzoic $HOOCC_6H_4SO_3H$
Anhydrous. Hydrate m.p. 69°. Heat with phenol + trace of
$H_2SO_4 \rightarrow$ red. Red soln. in alkalis, yellow in acids. KOH
fusion → salicylic acid. Imide (saccharin) m.p. 223°. Anil m.p.
190°. Anilide (di) m.p. 194°.

141° m-Sulphobenzoic $HOOCC_6H_4SO_3H$
Anhydrous. Hydrate m.p. 96°. Easily sol. in H_2O. KOH fusion →
m-hydroxybenzoic acid. SBT salt m.p. 163°. Amide (di) m.p. 170°

170° 1-Naphthol 4-sulphonic $HOC_{10}H_6SO_3H$, Neville and Winther's.
Transient blue colour with $FeCl_3$. 90% alc. soln. of Na salt on
heating → 1-naphthol. Warm conc. HNO_3 → nitro (di) m.p.
178°. SBT salt m.p. 103°.

193° (+) Camphor 10-sulphonic $[C_{10}H_{15}O]SO_3H$
Soluble in H_2O. $[\alpha]_D$ + 21° (10% aq. soln.). SBT salt m.p. 210°.
Amide m.p. 132°. Anilide m.p. 120°.

245° Naphthalene 1,5-disulphonic $C_{10}H_6(SO_3H)_2$
Anhydrous. SBT salt m.p. 257°. Amide (di) m.p. 310°. Anilide
(di) m.p. 249°.

259° p-Sulphobenzoic $HOOCC_6H_4SO_3H$
Anhydrous. Hydrate m.p. 94°. Soluble in H_2O. KOH fusion →
p-hydroxybenzoic acid. SBT salt m.p. 213°. Amide (di) m.p.
236°. Anilide (di) m.p. 253°.

>300°d Metanilic $NH_2C_6H_4SO_3H(m\text{-})$
Sl. sol. in cold H_2O. Br_2/H_2O → soluble di- and tribromo
derivatives. Na salt on heating with NaOH → m-aminophenol.
SBT salt m.p. 148°. Amide m.p. 142°.

>300°d Sulphanilic $NH_2C_6H_4SO_3H(p\text{-})$
Spar. sol. in cold, sol. in hot H_2O. Na salt on heating with
NaOH → aniline. Boiling MnO_2/dil. H_2SO_4 → p-benzoquinone.
Br_2/H_2O → tribromoaniline m.p. 119°. Azo 2-naphthol derivative
is an orange dye. SBT salt m.p. 182°. Amide m.p. 165°. Anilide
m.p. 200°. N-Acetyl m.p. 214°.

— Benzene m-disulphonic $C_6H_4(SO_3H)_2$ + $2\frac{1}{2}H_2O$
Very hygroscopic crystals. KOH fusion → resorcinol. SBT salt
m.p. 214°. Amide m.p. 229°. Anilide m.p. 144°.

— Ethanesulphonic $C_2H_5SO_3H$
SBT salt m.p. 115°. Amide m.p. 60°. Anilide m.p. 58°.

— Naphthalene 2,6-disulphonic $C_{10}H_6(SO_3H)_2$
Hygroscopic leaflets. KOH fusion → 2,6-dihydroxynaphthalene
m.p. 218° [acetyl (di) m.p. 175°; methyl ether (di) m.p. 150°].
SBT salt m.p. 256°. Amide m.p. 305°.

— Naphthalene 2,7-disulphonic $C_{10}H_6(SO_3H)_2$
Very hygroscopic needles. Ba salt sl. sol. in cold H_2O. KOH
fusion → 2,7-dihydroxynaphthalene m.p. 186° [acetyl (di) m.p.
136°; benzoyl (di) m.p. 139°; methyl ether (di) m.p. 139°]. SBT
salt m.p. 212°. Amide m.p. 242°.

— 2-Naphthol 3,6-disulphonic $HOC_{10}H_5(SO_3H)_2$, R acid.
Deliquescent needles, v. sol. in H_2O. Na salt (insol. in 90% alc.)

M.P.°C

reduces $AgNO_3$ immediately. Distil with $PCl_5 \rightarrow$ 2,3,6-trichloro-naphthalene m.p. 90°. Benzenediazonium salt $\rightarrow$ orange dye. SBT salt m.p. 233°. Anilide m.p. 202°.

— 2-Naphthol 6,8-disulphonic $HOC_{10}H_5(SO_3H)_2$, G acid. Hygroscopic crystals. Na salt (sol. in 90% alc.) reduces $AgNO_3$ slowly. Distil with $PCl_5 \rightarrow$ 1,3,7-trichloronaphthalene m.p. 113°. Benzenediazonium salt $\rightarrow$ yellow-orange dye. SBT salt m.p. 228°. Anilide m.p. 195°.

— Phenol p-sulphonic $HOC_6H_4SO_3H$ Hygroscopic liquid. Heat with $PCl_5 \rightarrow p$-dichlorobenzene m.p. 53°. Warm conc. $HNO_3 \rightarrow$ 2,4-dinitrophenol. Warm $Br_2/H_2O \rightarrow$ tribromophenol m.p. 95°. SBT salt m.p. 169°. Amide m.p. 177°. Anilide m.p. 141°.

SULPHONIC ESTERS

LIQUID

B.P.°C

150°/15 mmHg Methyl benzenesulphonate $C_6H_5SO_2OCH_3$
Boiling dil. NaOH $\rightarrow$ methanol + Na salt. Boiling conc. KI $\rightarrow$ methyl iodide. Warming with phenol + NaOH $\rightarrow$ anisole; with 2-naphthol $\rightarrow$ 2-naphthyl methyl ether m.p. 72°. Adduct with dimethylaniline m.p. 180°.

156°/15 mmHg Ethyl benzenesulphonate $C_6H_5SO_2OC_2H_5$
Spar. sol. in H_2O. Boiling dil. NaOH $\rightarrow$ ethanol + Na salt. Boiling conc. KI $\rightarrow$ ethyl iodide. Warming with phenol + NaOH $\rightarrow$ phenetole; with 2-naphthol $\rightarrow$ 2-naphthyl ethyl ether m.p. 37°.

SOLID

M.P.°C

28°   Methyl p-toluenesulphonate $CH_3C_6H_4SO_2OCH_3$
B.p. 161°/10 mmHg. Insol. in H_2O. Adduct with dimethylaniline m.p. 160°.

33°   Ethyl p-toluenesulphonate $CH_3C_6H_4SO_2OC_2H_5$
B.p. 173°/15 mmHg. Adduct with dimethylaniline m.p. 48°.

35°   Phenyl benzenesulphonate $C_6H_5SO_2OC_6H_5$
Heating (about 200°C) with very conc. NaOH $\rightarrow$ phenol. Conc. $HNO_3/H_2SO_4 \rightarrow$ nitro (di) m.p. 132°, which on boiling with NaOH $\rightarrow p$-nitrophenol + Na m-nitrobenzenesulphonate.

40°   Ethyl p-bromobenzenesulphonate $BrC_6H_4SO_2OC_2H_5$
58°   Benzyl p-toluenesulphonate $CH_3C_6H_4SO_2OCH_2C_6H_5$
60°   Methyl p-bromobenzenesulphonate $BrC_6H_4SO_2OCH_3$
88°   1-Naphthyl p-toluenesulphonate $CH_3C_6H_4SO_2OC_{10}H_7$
91°   Ethyl p-nitrobenzenesulphonate $NO_2C_6H_4SO_2OC_2H_5$
96°   Phenyl p-toluenesulphonate $CH_3C_6H_4SO_2OC_6H_5$
125° 2-Naphthyl p-toluenesulphonate $CH_3C_6H_4SO_2OC_{10}H_7$

ALKYL SULPHATES AND ACID SULPHATES

LIQUID

B.P.°C

188°   Methyl sulphate $(CH_3)_2SO_4$
d_4^{25} 1.321. Insol. in cold H_2O, hydrolysed on boiling. Odourless, toxic vapour. Boiling dil. NaOH → methanol + Na methyl sulphate. Boiling conc. KI → methyl iodide. Warming with phenol + NaOH → anisole; with 2-naphthol → 2-naphthyl methyl ether m.p. 72°. With aniline → dimethylaniline.

208°   Ethyl sulphate $(C_2H_5)_2SO_4$
d_4^{25} 1.172. Insol. in cold H_2O, slowly hydrolysed on boiling. Boiling dil. NaOH → ethanol + Na ethyl sulphate. Boiling conc. KI → ethyl iodide. Warming with phenol + NaOH → phenetole; with 2-naphthol → 2-naphthyl ethyl ether m.p. 37°. With aniline → mono and diethylaniline.

—   Methyl sulphuric acid CH_3OSO_3H, Methyl hydrogen sulphate. The free acid is a hygroscopic liquid. Na, K salts are sol. in alc. NH_4 salt m.p. 135°. Ca salt extremely sol. in H_2O. No ppt. in cold with $BaCl_2$/dil. HCl, $BaSO_4$ ppt. on boiling.

—   Ethyl sulphuric acid $C_2H_5OSO_3H$, Ethyl hydrogen sulphate. The free acid is a hygroscopic liquid. Na, K salts are sol. in alc. NH_4 salt m.p. 99° (alc.). Ca, Ba salts extremely sol. in H_2O. No ppt. in cold with $BaCl_2$/dil. HCl, $BaSO_4$ slowly ppt. on boiling.

SULPHINIC ACIDS

SOLID

M.P.°C

83°   Benzenesulphinic $C_6H_5SO_2H$
Decomp. at 100°. Spar. sol. in cold H_2O, easily sol. in ether. $FeCl_3$ → orange ppt. Readily oxidised → benzene sulphonic acid. Zn/acid → thiophenol. KOH fusion → benzene. Aniline salt m.p. 132°. Hydrazine salt m.p. 132°. Amide m.p. 121°. Anilide m.p. 113°.

85°   p-Toluenesulphinic $CH_3C_6H_4SO_2H$
Spar. sol. in cold H_2O, easily sol. in ether. Very easily oxidised → p-toluenesulphonic acid. Zn/acid → p-thiocresol. KOH fusion → toluene. NH_4 salt m.p. 175°d. Amide m.p. 120°. Anilide m.p. 138°.

THIOCARBOXYLIC ACIDS

LIQUID

B.P.°C

—   Thioformic HCOSH
Unstable. Amide m.p. 29°. Anilide m.p. 137°d. p-Toluidide m.p. 173°.

B.P.°C

93° Thioacetic CH_3COSH

Yellow. Unpleasant odour. Fairly sol. in cold H_2O, soln. on boiling gives H_2S and acetic acid. $CuSO_4 \rightarrow$ red-brown ppt. of Cu salt, blackens on heating. Aniline $\rightarrow H_2S$ + acetanilide. Amide m.p. 108°. Anilide m.p. 76°. p-Toluidide m.p. 130°. Anhydride b.p. 63°/20 mmHg.

129°/23 mmHg Thioglycollic $HSCH_2COOH$

Oxidises in air. Anilide m.p. 110°. p-Toluidide m.p. 125°.

— Thio-oxalic $HOOCCOSH$

Free acid does not exist. Anilide (di) m.p. 144°.

SOLID

M.P.°C

24° Thiobenzoic C_6H_5COSH

Yellow. Oxidised in ether soln. by air $\rightarrow$ benzoyl disulphide m.p . 130°. Long boiling with H_2O or dil. NaOH $\rightarrow$ benzoic acid + H_2S. Aniline $\rightarrow$ benzanilide + H_2S. NH_4 salt m.p. 118°. Amide m.p. 115°. p-Toluidide m.p. 129°. Anhydride m.p. 48°.

44° Thio p-toluic $CH_3C_6H_4COSH$

129° 1-Thiophenic $(C_4H_3S)COOH$

Irritating odour. Very sol. hot, insol. cold H_2O. Isatin + $H_2SO_4 \rightarrow$ blue colour. Anilide m.p. 140°.

165° Thiosalicylic $HSC_6H_4COOH(o\text{-})$

Acetyl m.p. 125°.

SULPHONYL CHLORIDES

Most of the derivatives given for sulphonic acids (q.v.) are suitable for the characterisation of the corresponding acid chlorides.

LIQUID

B.P.°C

60°/21 mmHg Methanesulphonyl CH_3SO_2Cl
70°/20 mmHg Ethanesulphonyl $C_2H_5SO_2Cl$

SOLID

M.P.°C

14° Benzenesulphonyl $C_6H_5SO_2Cl$

B.p. 116°/10 mmHg (246°d). Insol. in and very slowly hydrolysed by cold H_2O, on boiling gives HCl + benzene sulphonic acid. Zn/dil. HCl $\rightarrow$ thiophenol.

53° p-Chlorobenzenesulphonyl $ClC_6H_4SO_2Cl$

63° Benzene 1,3-disulphonyl $C_6H_4(SO_2Cl)_2$

64° m-Nitrobenzenesulphonyl $NO_2C_6H_4SO_2Cl$

M.P.°C

67° (+) Camphor 10-sulphonyl   [C$_{10}$H$_{15}$O]SO$_2$Cl
68° Naphthalene 1-sulphonyl   C$_{10}$H$_7$SO$_2$Cl
    Boiling H$_2$O → naphthalene 1-sulphonic acid + HCl. Zn/dil.
    HCl → 1-thionaphthol b.p. 145°/10 mmHg.
69° p-Toluenesulphonyl   CH$_3$C$_6$H$_4$SO$_2$Cl, p-Tosyl.
    Boiling H$_2$O → p-toluenesulphonic acid + HCl. Zn/dil. HCl →
    p-thiocresol.
75° p-Bromobenzenesulphonyl   BrC$_6$H$_4$SO$_2$Cl
76° Naphthalene 2-sulphonyl   C$_{10}$H$_7$SO$_2$Cl
80° p-Nitrobenzenesulphonyl   NO$_2$C$_6$H$_4$SO$_2$Cl
141° p-Hydroxybenzenesulphonyl   HOC$_6$H$_4$SO$_2$Cl

SULPHONAMIDES

SOLID

M.P.°C

132° (+) Camphor 10-sulphonamide   [C$_{10}$H$_{15}$O]SO$_2$NH$_2$
137° p-Toluenesulphonamide   CH$_3$C$_6$H$_4$SO$_2$NH$_2$
    Spar. sol. in cold H$_2$O, sol. in dil. NaOH, alc., ether. Boiling
    strong HCl → NH$_4$ p-toluenesulphonate, which with strong
    H$_2$SO$_4$ → toluene. Me$_2$SO$_4$/dil. NaOH → N-methyl (di) m.p.
    80°. Xanthyl m.p. 197°.
150° Naphthalene 1-sulphonamide   C$_{10}$H$_7$SO$_2$NH$_2$
    Sol. in dil. NaOH, alc., ether. Me$_2$SO$_4$/dil. NaOH → N-methyl
    (di) m.p. 137°. N-Acetyl m.p. 185°. N-Benzoyl m.p. 194°.
153° Benzenesulphonamide   C$_6$H$_5$SO$_2$NH$_2$
    Sol. in dil. NaOH, alc., ether. Boiling strong HCl → NH$_4$
    benzenesulphonate, which with strong H$_2$SO$_4$ → benzene.
    Me$_2$SO$_4$/dil. NaOH → N-methyl (di) m.p. 47°. Xanthyl m.p.
    206°. N-Acetyl m.p. 125°. N-Benzoyl m.p. 147°.
153° o-Toluenesulphonamide   CH$_3$C$_6$H$_4$SO$_2$NH$_2$
    Sol. in dil. NaOH, alc., ether. Boiling strong HCl → NH$_4$
    o-toluenesulphonate, which with strong H$_2$SO$_4$ → toluene.
    With K$_3$Fe(CN)$_6$ soln. → o-sulphonamidobenzoic acid m.p. 154°,
    which on heating → imide (saccharin) m.p. 223°. Xanthyl m.p.
    183°.
162° m-Nitrobenzenesulphonamide
    N-Methyl (di) m.p. 122°. N-Acetyl m.p. 189°.
165° Sulphanilamide   NH$_2$C$_6$H$_4$SO$_2$NH$_2$(p-)
    Acetic anhydride/gl. AcOH → acetyl (di) m.p. 217°. Benzoyl (di)
    m.p. 268°. Xanthyl m.p. 208°. (Sulphaguanidine m.p. 189°,
    sulphapyridine m.p. 191°, sulphamethazine m.p. 198°, sulph-
    athiazole m.p. 202°).
212° Naphthalene 2-sulphonamide   C$_{10}$H$_7$SO$_2$NH$_2$
(217°) Sol. in dil. NaOH, alc., ether. Me$_2$SO$_4$/dil. NaOH → N-methyl
    (di) m.p. 94°. N-Acetyl m.p. 146°.

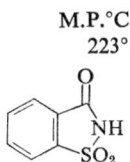

M.P.°C
223°   o-Sulphobenzimide C$_7$H$_5$NSO$_3$, Saccharin.
Almost insol. in H$_2$O, readily sol. in dil. NaOH or Na$_2$CO$_3$.
Na deriv. has intensely sweet taste. Warm with resorcinol +
conc. H$_2$SO$_4$ (trace) → sulphonfluorescein (red soln. in dil.
NaOH with green fluorescence). Me$_2$SO$_4$/dil. NaOH → N-methyl
(di) m.p. 131°. Xanthyl m.p. 198°.

THIOAMIDES

SOLID

M.P.°C
28°   Thioformamide HCSNH$_2$
Very sol. in alc., acetone; sol. in ether; insol. in benzene. p-
Toluidide m.p. 173°.
76°   Thioacetanilide C$_6$H$_5$NHCSCH$_3$
108°   Thioacetamide CH$_3$CSNH$_2$
Sol. in H$_2$O, alc., spar. sol. in ether. Reflux with dil. H$_2$SO$_4$ →
acetic acid.
110°   m-Tolylthiourea CH$_3$C$_6$H$_4$NHCSNH$_2$
116°   Thiobenzamide C$_6$H$_5$CSNH$_2$
N-Benzoyl m.p. 117° (red).
153°   Thiocarbanilide C$_6$H$_5$NHCSNHC$_6$H$_5$, sym-Diphenylthiourea.
Insol. in H$_2$O, sol. in dil. NaOH, alc. Boiling conc. NaOH →
aniline + Na$_2$S. Boiling strong HCl → phenyl isothiocyanate
(sharp odour). Me$_2$SO$_4$/NaOH (15%) soln. → S-methyl m.p.
109°. Boiling alc. soln. + yellow HgO → carbanilide m.p.
238° + HgS (black).
154°   Phenylthiourea C$_6$H$_5$NHCSNH$_2$
Sol. in hot H$_2$O, alc., dil. NaOH. Boiling conc. NaOH →
aniline, NH$_3$, Na$_2$S. Boiling strong HCl → phenyl isothiocyanate
(sharp odour). Aniline (heat) → thiocarbanilide + NH$_3$.
160°   o-Tolylthiourea CH$_3$C$_6$H$_4$NHCSNH$_2$
181°   Thiosemicarbazide NH$_2$CSNHNH$_2$
(ca.)   Very sol. in H$_2$O, alc. Boiling conc. NaOH → hydrazine, NH$_3$,
Na$_2$S. Acetone → isopropylidene m.p. 179°. Benzaldehyde →
benzal m.p. 159°. Hydrochloride m.p. 188°. Acetyl m.p. 165°.
2,4-Dinitrophenyl m.p. 210°d.
182°   Thiourea NH$_2$CSNH$_2$
Fairly sol. in cold H$_2$O, spar. sol. in ether. After melting gives
test for NH$_4$SCN (blood red colour with FeCl$_3$). K$_4$Fe(CN)$_6$ in
dil. AcOH → green colour, soon changing to blue. Boiling conc.
NaOH → NaSCN, NH$_3$, Na$_2$S. Heating with benzyl chloride in
dil. alc. soln. → S-benzyl isothiouronium chloride m.p. 175°.
Me$_2$SO$_4$/H$_2$O (heat) → S-methyl isothiouronium sulphate m.p.
241° (235°d), which with NH$_4$OH gives guanidine sulphate +
CH$_3$SH. On standing with methyl iodide → S-methyl isothiouronium hydriodide m.p. 117°. Hydrochloride m.p. 136°.
Xanthyl m.p. 136°.

M.P.°C
182° p-Tolylthiourea $CH_3C_6H_4NHCSNH_2$
(188°)

THIOCYANATES AND ISOTHIOCYANATES

LIQUID

B.P.°C
130°   Methyl thiocyanate CH_3SCN
Almost insol. in H_2O. Boiling alc. $KOH \rightarrow NH_3$, KCN, K_2CO_3
and dimethyl sulphide. $Zn/HCl \rightarrow CH_3SH + HCN$.
146°   Ethyl thiocyanate C_2H_5SCN
Insol. in H_2O, sol. in alc., ether. Boiling alc. $KOH \rightarrow NH_3$,
KCN, K_2CO_3 and diethyl sulphide.
152°   Allyl isothiocyanate C_3H_5NCS, Allyl mustard oil.
Very sharp odour. Sl. sol. in H_2O, readily in alc. NH_4OH
(warm) $\rightarrow$ allyl thiourea m.p. 74°. Aniline (gentle warming) $\rightarrow$
allyl phenylthiourea m.p. 98°.
221°   Phenyl isothiocyanate C_6H_5NCS
Sharp odour. Insol. in H_2O. NH_4OH (warm) $\rightarrow$ phenylthiourea
m.p. 154°. Aniline (warm) $\rightarrow$ thiocarbanilide m.p. 153°.

SOLID

M.P.°C
41°   Benzyl thiocyanate $C_6H_5CH_2SCN$
B.p. 256°. Insol. in H_2O. Boiling alc. $KOH \rightarrow NH_3$, KCN,
K_2CO_3 and benzyl disulphide m.p. 71°. $Zn/HCl \rightarrow C_6H_5CH_2SH$
$+ HCN$. $HNO_3 \rightarrow$ benzoic acid.
58°   1-Naphthyl isothiocyanate $C_{10}H_7NCS$
Aniline (warm) $\rightarrow$ 1-naphthyl phenylthiourea m.p. 158°.

PHOSPHORUS COMPOUNDS

LIQUID

B.P.°C
41°   Trimethyl phosphine $(CH_3)_3P$
127°   Triethyl phosphine $(C_2H_5)_3P$
197°   Trimethyl phosphate $(CH_3O)_3PO$
215°   Triethyl phosphate $(C_2H_5O)_3PO$
d_4^{20} 1.068. n_D^{20} 1.4055.
222°   Phenyl dichlorophosphine $C_6H_5PCl_2$
235°   Hexamethyl phosphoramide $[(CH_3)_2N]_3PO$
d_4^{20} 1.0253. n_D^{20} 1.4582. Dipolar aprotic solvent.
320°   Diphenyl chlorophosphine $(C_6H_5)_2PCl$
—   Dibenzyl phosphochloridate $(C_6H_5CH_2O)_2P(O)Cl$

SOLID

M.P.°C
  49° Triphenyl phosphate $(C_6H_5O)_3PO$
  80° Triphenyl phosphine $(C_6H_5)_3P$
124°d Pentaphenyl phosphorane $(C_6H_5)_5P$
153° Triphenyl phosphine oxide $(C_6H_5)_3PO$

7 Quantitative Determination of Constituent Elements

Carbon, hydrogen, nitrogen and oxygen may be determined by automatic, semi-automatic or manual techniques. Sulphur and the halogens may be determined manually. In the case of automatic methods, carbon, hydrogen and nitrogen may be determined simultaneously on a single sample, whereas only carbon and hydrogen are determined simultaneously by the manual procedures.

Manual and Semi-Automatic Procedures

Determination of Carbon and Hydrogen

The principal method used in the determination of carbon and hydrogen in an organic compound is the combustion of the substance in the presence of oxygen and the determination of the carbon dioxide and water formed.

C, H, N, O, S, Halogens $\rightarrow CO_2 + H_2O$ + oxides of nitrogen and sulphur

+ halogens

The water and carbon dioxide produced during combustion are absorbed, respectively, in a preweighed tube containing magnesium perchlorate ('anhydrone') and in a preweighed tube containing a zone of sodium hydroxide on asbestos ('ascarite') ($\frac{2}{3}$) followed by a zone of magnesium perchlorate ($\frac{1}{3}$).

$$CO_2 + 2NaOH \rightarrow Na_2CO_3 + H_2O$$

If the original organic compound contains nitrogen, sulphur or a halogen, their combustion products must be removed before they reach the preweighed absorption tubes.

The oxygen used in the combustion process is purified either by heating to 800–900°C, or to 450°C in the presence of platinised asbestos, in order to oxidise any hydrocarbons and hydrogen which may be present; any resultant water and carbon dioxide are removed by passing the gas through tubes containing magnesium perchlorate and sodium hydroxide/asbestos.

The purified gas is passed over the platinum boat containing 3–15 mg of sample and through the combustion furnace, which may contain an oxidant such as silver orthovanadate at 650°C or copper (II) oxide/lead chromate (2/1) at 680°C, or a baffled empty tube (Fig. 4), at 800–900°C. The combustion products are then passed over silver wire or gauze at 400°C, to remove halogens and oxides of sulphur, and over

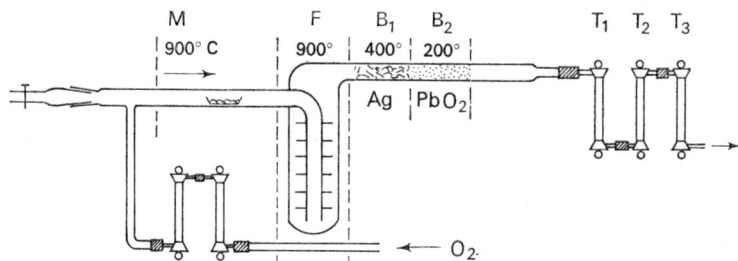

Fig. 4

lead dioxide at 200°C (or magnanese dioxide, placed between the absorption tubes, at room temperature) to remove oxides of nitrogen.

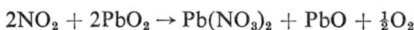

$$2NO + 2PbO_2 \rightarrow Pb(NO_2)_2 \cdot PbO + \tfrac{1}{2}O_2$$

$$2NO_2 + 2PbO_2 \rightarrow Pb(NO_3)_2 + PbO + \tfrac{1}{2}O_2$$

The gas stream is then passed through an absorption tube containing magnesium perchlorate to absorb the water formed, through a tube containing sodium hydroxide–asbestos and magnesium perchlorate to absorb the carbon dioxide formed, and finally through an unweighed guard tube of magnesium perchlorate. This last tube is present to prevent moisture creeping back into the system (T_3).

Procedure.   The absorption tubes T_1 and T_2 (Fig. 4) must first be brought to constant weight by weighing the tubes, connecting to the rest of the apparatus and passing oxygen for about 15 minutes at a flow rate of about 50 cm³ per minute, with the furnace F and heating blocks B_1 and B_2 at their operating temperatures. The tubes are reweighed, reconnected to the apparatus and the operation repeated until successive weighings differ by a maximum of 0.01 mg (6 decimal places) or 0.03 mg (5 decimal places).

Absorption tubes are always handled with chamois leather gloves or finger stalls. It may be necessary to close the ends of the absorption tubes when they are removed from the combustion apparatus. The ends of some tubes have a very narrow bore and can be left open to the atmosphere during the weighing stage, others have ground glass taps which should be closed.

It is important that the tubes are treated in exactly the same manner each time they are weighed and that a rigid time scale is adhered to. When the tubes are removed from the apparatus, it is recommended that they are wiped with a damp chamois leather cloth and then with a dry one. They are then grounded on a water pipe etc. to remove static electricity, placed on a rack next to the balance for 5–15 minutes (but for precisely the same time on each occasion) and weighed in the same order each time. Counterpoises should be used.

3–9 mg of sample is weighed to six decimal places (or 10–15 mg to five decimal places) in a previously ignited platinum boat, placed in the combustion tube, and the apparatus purged with oxygen for about five minutes.

If a phosphorus containing compound is being analysed, the sample is covered with about ten times its weight of tungstic oxide to assist in the complete oxidation of carbon. The movable furnace (M) is then brought to temperature (800–900°C) at a position farthest from the sample, and then slowly moved towards and over the boat. When this furnace has reached F (about 10 minutes later) and remained in contact with the stationary furnace for a few minutes, it is quickly returned to its original position and again moved up to F, but this time more rapidly, to ensure complete combustion. Both furnaces are then switched off and gas allowed to pass for a further five minutes, when the absorption tubes T_1 and T_2 are removed and reweighed. The platinum boat is reweighed to ascertain if any non-combustible residue is left.

Blank determinations must be carried out prior to the sample determination, using an empty boat, or a boat containing tungstic oxide in the case of samples containing phosphorus.

There are many reliable variations of the 'empty-tube' procedure described above, details of which may be found elsewhere.[1] Descriptions of reliable alternative methods involving the use of combustion catalysts and hence lower combustion temperatures, are readily available.[1,2,3,4]

Semi-automatic methods are principally similar to those described above and involve the manual weighing of the sample and of the absorption tubes at the end of the determination. However, the heating and movement of the furnace M, and the various time factors involved, may be automatically controlled to a greater or lesser degree dependent upon the apparatus.

Determination of Nitrogen

The two main methods which are currently used for the determination of nitrogen are based upon the procedures of Dumas and Kjeldahl.

1   Dixon, *Modern Methods in Organic Microanalysis*, Van Nostrand, 1968.
2   Steyermark, *Quantitative Organic Microanalysis*, 2nd edition, Academic Press, 1961.
3   Ingram, *Methods of Organic Elemental Microanalysis*, Chapman & Hall, 1962.
4   Van Nieuwenburg & Van Ligten, *Quantitative Chemical Micro-Analysis*, Elsevier, 1963.

Neither method is applicable to all nitrogen containing compounds. The Dumas method, which involves the conversion of organic nitrogen to nitrogen gas, tends to give low results for compounds containing N-methyl groups, probably due to the formation of some methylamine which partially dissolves in the aqueous potassium hydroxide in the gas burette. This method also frequently gives inaccurate data for pyrimidines and compounds containing long aliphatic chains (due to the production of methane) unless potassium chlorate and/or copper acetate is added to the sample to aid combustion, and for sulphonamides, hydrazines, hydrazones, semicarbazides, nitrates, nitroso and nitro compounds, for which compounds acceptable results may be obtained if the sample is mixed with potassium dichromate or vanadium pentoxide. For many compounds which contain these groups, however, the Kjeldahl method gives superior results.

The Kjeldahl method involves the conversion of organic nitrogen to ammonium hydrogen sulphate by digestion with concentrated sulphuric acid, and determination of the ammonia liberated from the ammonium salt. The normal Kjeldahl procedure is not suitable for use with compounds which contain N—N or N—O linkages, but can be applied to such substances after a reduction step has been carried out.[1,2]*

The Dumas method. Nitrogen containing organic compounds are decomposed in an atmosphere of carbon dioxide by copper (II) oxide at about 700°C to give elemental nitrogen together with some oxides of nitrogen. The oxides of nitrogen are reduced to nitrogen by copper turnings. The volume of total gaseous nitrogen produced is measured in a nitrometer, the carbon dioxide carrier gas being previously removed using aqueous potassium hydroxide.

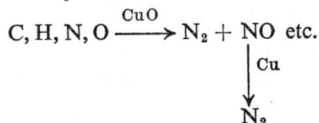

$$\text{C, H, N, O} \xrightarrow{\text{CuO}} \text{N}_2 + \text{NO etc.}$$
$$\Big\downarrow \text{Cu}$$
$$\text{N}_2$$

Procedure. The apparatus is diagrammatically represented in Fig. 5. The carbon dioxide carrier gas is generated from solid material kept in a Dewar flask.

A quantity of sample equivalent to about 0.5–2 mg of nitrogen (depending upon the capacity of the nitrometer used) is weighed to six decimal places in a porcelain boat and placed in the combustion tube at B, between powdered copper (II) oxide. The tube is rotated so that the sample becomes intimately mixed with the oxide, and then connected to the rest of the apparatus. Alternatively, the copper oxide powder at C may be replaced by copper (II) oxide gauze in which case copper oxide powder is added to the sample in the boat.

* See p. 230.

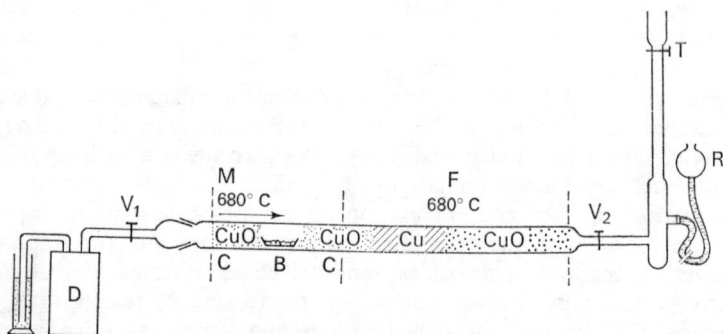

Fig. 5

The system is then purged with carbon dioxide prior to heating being started, in order to prevent oxidation of the metallic copper. The static furnace is then heated to 670–680°C and the tube allowed to attain equilibrium. The valves V_1 and V_2 are now closed, the tube connected to the nitrometer and the nitrometer filled with 50% potassium hydroxide solution to about the tap T, using the reservoir R. The reservoir is lowered and valves V_1 and V_2 opened; carbon dioxide is allowed to pass until micro bubbles only are seen in the nitrometer. The nitrometer is again filled with aqueous potassium hydroxide and the rate of accumulation of gas in the nitrometer noted.

The nitrometer is again filled, the valve V_1 is closed, and the movable furnace M heated. When the furnace reaches 670–680°C, it is moved *slowly* toward the sample (if movement is too rapid, high results will be obtained). Combustion of the sample must be controlled so that bubbles appear in the nitrometer at a rate of not more than 1 in 2 seconds. When the movable furnace has reached the stationary furnace and has remained there some minutes, the valve V_2 is closed, the movable furnace placed in its original position (next to the carbon dioxide generator), and the valve V_2 opened slowly to control gas evolution into the nitrometer (1 bubble/second).

As soon as gas evolution ceases, the valve V_1 is opened and the movable furnace allowed to traverse the appropriate part of the tube in about 10 minutes. It is then again placed next to valve V_1 and allowed to cool. When micro bubbles are again obtained in the nitrometer, the volume of gas in the nitrometer is obtained. The temperature of the nitrometer and the atmospheric pressure are noted.

The volume of the gas so obtained has to be corrected in accordance with the calibration certificate issued with the nitrometer, the volume of gas contained by the nitrometer after a 'blank' run, and the volume of aqueous potassium hydroxide adhering to the wall of the tube. Two per cent of the volume of gas has been suggested as the figure which should be used for the last correction.[2]* The percentage of nitrogen in the compound may then be calculated. The error allowed is $\pm 0.2\%$.

Should it be necessary to use one of the oxidants previously mentioned (p. 231) to assist in the combustion of difficult samples, the powdered oxidant is added to the sample in the boat (20–30 mg copper acetate; 10–20 mg potassium chlorate; 50 mg potassium dichromate) and the boat is not in this case inverted by rotating the tube.

The Kjeldahl method.   The nitrogen containing sample is digested with concentrated sulphuric acid in the presence of a mercuric oxide catalyst. The resultant ammonium hydrogen sulphate is treated with aqueous alkali and the liberated ammonia trapped in aqueous boric acid and then titrated with standard hydrochloric acid.

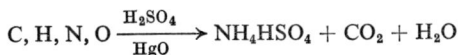

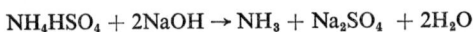

$$C, H, N, O \xrightarrow[\text{HgO}]{\text{H}_2\text{SO}_4} NH_4HSO_4 + CO_2 + H_2O$$

$$NH_4HSO_4 + 2NaOH \rightarrow NH_3 + Na_2SO_4 + 2H_2O$$

$$NH_3 + HCl \rightarrow NH_4Cl$$

Potassium sulphate is added to the digestion mixture in order to raise the temperature of reflux to about 340°C. Sodium thiosulphate is frequently added with the alkali in order to ensure break up of Hg—N compounds formed during digestion e.g. $Hg(NH_3)_2SO_4$.[2]*

Procedure

Reagents

Catalyst mixture—mercuric oxide and potassium sulphate (1:30) intimately ground together.

Boric acid solution—5% w/v aqueous boric acid.

Indicator—a mixture of Bromocresol green and methyl red (4:1), 0.1% in ethanol.

Sodium hydroxide–thiosulphate reagent—48 g sodium hydroxide and 4.8 g sodium thiosulphate dissolved in 100 cm³ water.

The sample size should be equivalent to about 10 cm³ of 0.01M hydrochloric acid (i.e. 1–2 mg nitrogen) but should not exceed 200 mg. The sample is weighed directly into a 30 cm³ Kjeldahl digestion flask or into a capsule which is then transferred to such a flask. (Platinum boats or capsules must not be used for this purpose as all nitrogen is lost in the presence of this metal.) 2 g of the catalyst mixture and 5 cm³ of concentrated sulphuric acid are added and the flask heated in a mantle or on a sand bath in a fume cupboard, care being taken that the reaction mixture does not boil. When initial decomposition is complete (sulphur dioxide evolution ceases after about 30–45 minutes) the flask is heated more vigorously to maintain refluxing at about half way up the neck (Fig. 6) until the mixture becomes pale yellow,

* See p. 230.

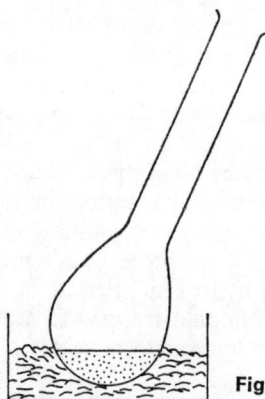

Fig. 6

and then for 30 minutes longer. A total heating time of up to four hours may be necessary.

When cool, the sides of the digestion flask are washed down with about 3 cm³ of water to dissolve the solid matter, heating if necessary. The digest is then transferred to the distillation flask D (Fig. 7) via the funnel F using 4 × 3 cm³ portions of water.

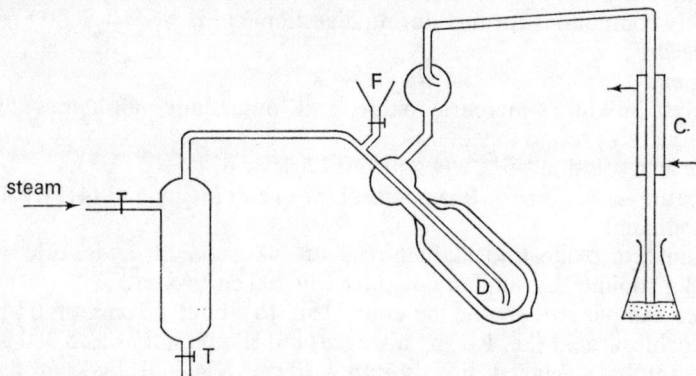

Fig. 7

With water running through the condenser C, the tip of the condenser immersed beneath 10 cm³ of the boric acid solution (containing six drops of indicator), and steam flowing through the tap T, a 20 cm³ portion of the sodium hydroxide–thiosulphate reagent is added to D via the funnel F. As the alkali is introduced at the bottom of the acid layer minimal mixing occurs at this stage. Slowly pass steam, by closing T, to mix the reactants, and then more vigorously to ensure that all the ammonia passes over. When about 25 cm³ of distillate has been collected, the receiver is lowered and condensate allowed to rinse the inside of the condenser tip; the outside of the tip is then rinsed into the flask.

The contents of the flask are titrated with 0.01M hydrochloric acid, the end point being reached when the colour obtained is the same as that given by a mixture of 10 cm³ boric acid solution, water equivalent to that used in the determination, indicator and a slight excess of 0.01M hydrochloric acid. A blank determination should be carried out on the digestion reagents.

Determination of Oxygen

The method used for the direct determination of oxygen is based on the procedure of Unterzaucher and involves the decomposition of the substance to carbon monoxide (at 1120°C in the presence of carbon, or at 900°C using a platinum–carbon catalyst) in a nitrogen or helium atmosphere, conversion of the monoxide to carbon dioxide using copper oxide, and absorption of the dioxide on preweighed sodium hydroxide–asbestos/magnesium perchlorate.

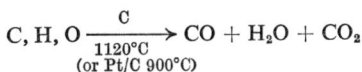

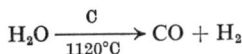

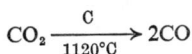

$$C, H, O \xrightarrow[\substack{1120°C \\ (or\ Pt/C\ 900°C)}]{C} CO + H_2O + CO_2$$

$$H_2O \xrightarrow[1120°C]{C} CO + H_2$$

$$CO_2 \xrightarrow[1120°C]{C} 2CO$$

(N.B. All oxygen from the sample is quantitatively converted to carbon monoxide).

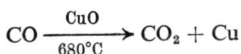

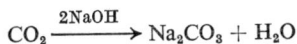

$$CO \xrightarrow[680°C]{CuO} CO_2 + Cu$$

$$CO_2 \xrightarrow{2NaOH} Na_2CO_3 + H_2O$$

The decomposition products from the combustion furnace are passed through a bed of copper turnings at 900°C to absorb any carbon disulphide, carbonyl sulphide or hydrogen sulphide produced from sulphur-containing compounds, and to convert the oxygen in carbonyl sulphide into carbon monoxide.

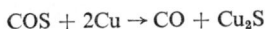

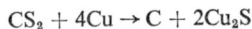

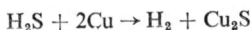

$$COS + 2Cu \rightarrow CO + Cu_2S$$

$$CS_2 + 4Cu \rightarrow C + 2Cu_2S$$

$$H_2S + 2Cu \rightarrow H_2 + Cu_2S$$

The gas stream is then passed through a tube containing potassium hydroxide or 'ascarite' (sodium hydroxide–asbestos) to absorb halogen acid compounds produced from halogen containing organic compounds and finally over copper oxide wire at 680°C to convert the carbon monoxide to the dioxide which is then absorbed on preweighed sodium hydroxide–asbestos/magnesium perchlorate as in the determination of elemental carbon (p. 228). A tube of magnesium perchlorate is placed at the end of the system to prevent moisture creeping back into the apparatus, and a similar tube is placed before the carbon dioxide absorbing tube to ensure that no water reaches this preweighed tube.

The carrier gas is purified by passing over sodium hydroxide–asbestos and magnesium perchlorate and then over copper turnings at 600°C to ensure complete freedom from atmospheric oxygen.

Procedure. Consistent blank determinations should be made prior to carrying out the sample determination, allowing precisely the same quantity of nitrogen carrier gas, say 700 cm³, to pass through the apparatus during the blank and sample determinations.

The fixed furnaces F_1, F_2, F_3 (Fig. 8) should be adjusted to their operating temperatures and, with the movable furnace M at room temperature, an empty platinum boat inserted into the tube through the ground glass joint J and positioned about 8 cm from F_1. Taps T_1, T_2 and T_3 are then turned so that the carrier gas flushes air from the apparatus in the direction T_2, T_3, F_1, T_1. The gas should be allowed to flow for twenty minutes at a rate of about 10 cm³ per minute.

The carbon dioxide absorption tube is weighed (p. 229) and attached to the apparatus. Carrier gas is then diverted along the route T_2, F_1, T_3, F_2 by the closure of tap T_1 and appropriate use of the taps T_2 and T_3. The movable furnace M is brought up to temperature as far from the boat as possible and, when at 1120°C, it is positioned about 4 cm from the boat and very slowly moved over the boat (taking about 15 minutes) until it comes into contact with the fixed furnace F_1. F_2 is then moved slightly forward (towards F_3) for about 10 minutes keeping M in contact with it, in order to ensure complete pyrolysis of any material which may have condensed in the cooler zone between the two furnaces. The furnace F_1 is returned to its original position and M allowed to cool. Gas is allowed to pass until about 700 cm³ has passed through the system. The carbon dioxide absorption tube is weighed as before. Blank determinations can be as little as 0.1 mg carbon dioxide and should not exceed 0.2 mg carbon dioxide.

The process is repeated with an amount of sample equivalent to 1–1.5 mg oxygen.

Details of a variation of the procedure described above, involving reaction of the carbon monoxide formed with iodine pentoxide and titrimetric determination of the iodine liberated, can be found elsewhere.[2]*

* See p. 230.

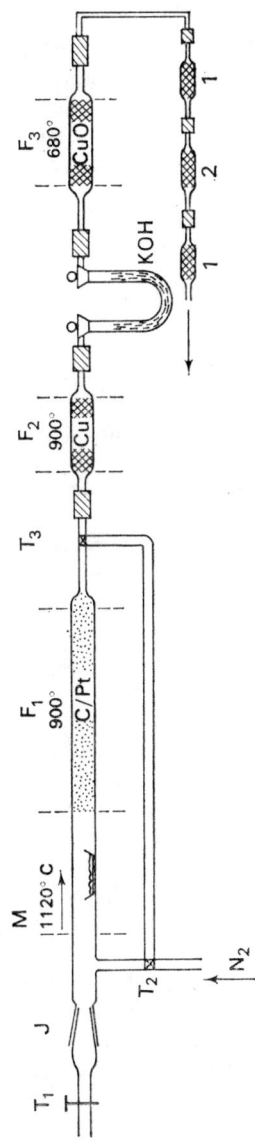

1. magnesium perchlorate
2. CO_2 absorption tube

Fig. 8

Determination of the Halogens and Sulphur

The methods which are available for the determination of halogens and of sulphur fall mainly into three categories; those based upon (a) oxidation by nitric acid, (b) combustion in an oxygen atmosphere, (c) combustion in a stream of hydrogen, followed by determination of the halogen acid, or of the hydrogen sulphide produced. The examples which follow are restricted to methods of types (a) and (b) and are outlines of procedures which have proved reliable in collaborative studies. Details of procedures based on combustion in a hydrogen atmosphere can be found elsewhere.[4]*

Determination of Bromine, Chlorine and Iodine

Carius method. Halogen containing organic compounds are oxidised to carbon dioxide, water and halogen acid by fuming nitric acid. The halogen acid is converted to silver halide, in the presence of silver nitrate, and the halide determined gravimetrically

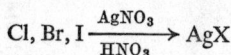

$$Cl, Br, I \xrightarrow[HNO_3]{AgNO_3} AgX$$

5–10 mg of the sample is weighed in a porcelain boat, weighing tube or capillary, and inserted into the bottom of a Carius combustion tube. (These tubes are 20–25 cm in length, about 13 mm in diameter, and may be of the heavy walled or light walled variety.) 20 mg of silver nitrate is added, followed by 0.5–0.7 cm³ fuming nitric acid, s.g. 1.49–1.50, (0.3 cm³ in the case of thin walled tubes), the tube being rotated during the addition of the acid. Cooling may be necessary before and during the addition of the nitric acid if reaction is likely to proceed at room temperature. Alternatively the sample may be inserted into the tube in a weighing bottle after the acid has been introduced and mixed with the latter after the tube has been sealed (CAUTION—high pressures are generated).

The Carius tube is sealed, placed in the cold Carius furnace, slowly heated to 250°C, and allowed to remain at that temperature for 7–8 hours. After cooling, the tube is opened, the contents are diluted with water (care!) to within 3–4 cm of the top of the tube, and the tube is heated on a steam bath, protected from light, for about 20 minutes. In the case of iodine containing compounds this time is increased and the mixture stirred in order to break up the silver nitrate–silver iodide eutectic, by allowing the silver nitrate component to go into the solution. Stirring is continued until the iodide is in the form of a fine powder.

The silver halide precipitate is filtered from the liquid, washed with

* See p. 230.

alternate small portions of nitric acid (1:200) and ethanol via the Carius tube, and dried at 120°C for about 30 minutes and then to constant weight. The permitted error is $\pm 0.3\%$ halogen.

Pregl method. The compound is combusted in a stream of purified oxygen and the products passed over platinum. The halogen produced is converted to sodium halide by an alkaline reducing agent; the sodium salt is converted to silver halide which is then determined gravimetrically.

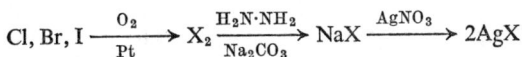

$$\text{Cl, Br, I} \xrightarrow[\text{Pt}]{O_2} X_2 \xrightarrow[\text{Na}_2\text{CO}_3]{\text{H}_2\text{N·NH}_2} \text{NaX} \xrightarrow{\text{AgNO}_3} 2\text{AgX}$$

The apparatus is represented in Fig. 9 and consists essentially of a combustion tube, an absorption tube T_1 containing a glass spiral, a

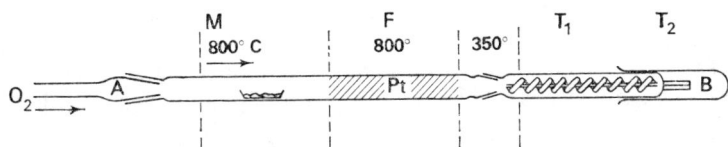

Fig. 9

fixed furnace F at 800°C and a movable furnace M. The spiral is coated with reagent in the following manner. A few drops of a saturated aqueous solution of hydrazine sulphate are added to about 5 cm³ of a 25% aqueous sodium carbonate solution in a hard glass test tube. The tip B of the absorption tube T_1 is immersed in this reagent and the solution sucked up the tube to about 5 mm past the end of the spiral. The reagent solution is then allowed to drain out of the tube under gravity, leaving the spiral wetted with the alkaline reducing agent. The absorption tube is connected to the combustion tube and the wetted test-tube T_2 is fitted over the tip of the absorption tube.

The sample (5–10 mg), in a platinum boat, is inserted into the end A of the combustion tube and oxygen supplied at a rate of 12–15 cm³/minute. The movable furnace is slowly moved towards and across the boat until it touches the static furnace. It is then returned to its original position and again, more rapidly this time, moved up to the static furnace to ensure complete combustion of the sample. In the case of iodine containing samples crystals of iodine may be deposited in the combustion tube beyond the long furnace. This halogen is driven into the absorption tube by slowly moving the combustion tube into the furnace or by the use of a burner. The furnaces are turned off and the system allowed to cool with oxygen still passing through the apparatus.

About 4 cm³ of water is placed in the test-tube T_2 and the spiral, on which sodium halide has been deposited, washed with this water, using suction as before, followed by washing down with several small portions of water from the top of the absorption tube into the same test-tube. Two or three drops of 30% aqueous hydrogen peroxide are added to the washings to oxidise residual reducing agent, the test-tube heated on a steam bath for 5 minutes, cooled, and 2 cm³ of concentrated nitric acid and 2 cm³ of 5% aqueous silver nitrate solution added. The tube is again heated on a steam bath, protected from light, to coagulate the precipitate of silver halide (1–1½ hours). After cooling, the precipitate is filtered, washed (see p. 238), and dried to constant weight at 120°C.

Schöniger flask method for the halogens. The organic compound is combusted by burning in an oxygen-filled flask containing aqueous sodium hydroxide (for the determination of chlorine) or aqueous sodium metabisulphite (for bromine or iodine). The hydrogen halide formed during combustion is converted to alkali metal halide and the halide ion determined titrimetrically.

The 300–500 cm³ conical flask has a ground glass stopper, to the lower end of which is fitted a sample holder of platinum gauze (Fig. 10).

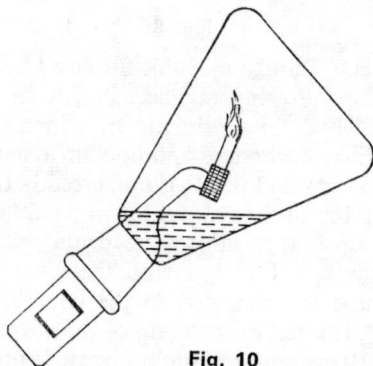

Fig. 10

The sample (4–10 mg) is weighed into a piece of halide-free filter paper, and the folded paper fitted into the sample holder, with the fuse protruding. Liquid samples may be weighed into methylcellulose capsules which are then folded in the filter paper.

10 cm³ of 0.01M sodium hydroxide or 10 cm³ of water and 50 mg of sodium metabisulphite are put into the flask, which is then filled with oxygen, and loosely stoppered with a cork. The fuse is ignited and the stopper inserted into the flask as rapidly as possible and held firmly in place. (Electrical ignition should be used where possible.) The flask is held neck downwards in order to prevent unburned fragments falling into the liquid and being extinguished. The combustion stage should be carried out with the flask held behind a safety screen, the operator wearing goggles and gloves.

When combustion is complete the flask is shaken vigorously for 10–15 minutes until all cloudiness disappears. About 2–3 cm³ of water is then put into the collar surrounding the stopper and the stopper carefully loosened. This operation may be difficult, as the flask is under a partial vacuum, and should be carried out with care and patience. The stopper and wires are washed with water, the contents of the flask transferred to a suitable 100 cm³ vessel, and the halide ion determined titrimetrically.

Classical methods of determining halide ion do not give sufficiently accurate end points to be of use on a micro scale but potentiometric titration has been found to be suitable.[1]* Glass and silver rod electrodes are used and the halide ion is titrated with 0.1M silver nitrate (dispensed from a syringe burette) after acidifying with about five drops of 30% nitric acid and diluting to about 60 cm³ with isopropanol. A blank determination is carried out in an analogous manner, adding 2 cm³ of 0.01M sodium halide to the solution before titrating with the silver nitrate solution.

In the case of bromide or iodide ion, amplification procedures may be employed to allow a conventional titrimetric ending. Bromide ion is oxidised with excess sodium hypochlorite, the excess reagent destroyed with sodium formate, potassium iodide and sulphuric acid added, and the liberated iodine determined with sodium thiosulphate in the normal manner. This procedure represents a sixfold amplification.

$$Br^- + 3NaOCl \rightarrow 3NaCl + BrO_3^-$$

$$BrO_3^- + 6HI \rightarrow Br^- + 3I_2 + 3H_2O$$

A similar procedure can be employed with iodide ion but, in this case, it is preferable to oxidise to iodate with bromine.

Determination of Sulphur

Carius method.   Organic sulphur is oxidised to acid sulphate by fuming nitric acid in the presence of an alkali metal salt such as sodium chloride. The sodium hydrogen sulphate is converted to sodium sulphate and the sulphate determined either by titration with 0.01M barium chloride solution using a tetrahydroxyquinone indicator or gravimetrically as barium sulphate. Alternatively, the sodium chloride is replaced by barium chloride and the barium sulphate produced is determined gravimetrically.

The determination is carried out in a similar manner to that outlined for the Carius method of halogen determination (p. 238) using a sample size equivalent to about 1–1.5 mg sulphur, 15–20 mg pure sodium chloride, and 0.5–0.6 cm³ fuming nitric acid in the Carius tube.

* See p. 230.

After cooling and opening the tube (CARE) the contents are carefully diluted with water and transferred to a 50 cm³ beaker. The nitric acid is evaporated on a steam bath, the residue of sodium hydrogen sulphate is dissolved in water and made alkaline to phenolphthalein (1 drop) with 0.1M sodium hydroxide. The colour is just discharged by back titration with 0.01M hydrochloric acid, the solution is diluted to 15 cm³ with water, 15 cm³ 95% aqueous ethanol and powdered tetrahydroxy-quinone are added, and the sulphate titrated with 0.01M barium chloride.[2] The permitted error is $\pm 0.3\%$.

Pregl method. The organic substance is combusted in a stream of oxygen and the products passed over platinum. The oxides of sulphur produced are converted to sulphuric acid either by hydrogen peroxide or by bromine.

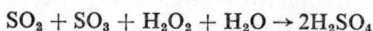

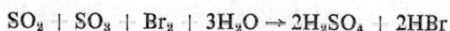

$$SO_2 + SO_3 + H_2O_2 + H_2O \rightarrow 2H_2SO_4$$

$$SO_2 + SO_3 + Br_2 + 3H_2O \rightarrow 2H_2SO_4 + 2HBr$$

The apparatus and procedure is essentially the same as that outlined for the Pregl determination of halogens (p. 239) except that the spiral is coated with either 6% aqueous hydrogen peroxide or saturated aqueous bromine. A sample size equivalent to about 0.5–1.5 mg sulphur is used.

In the absence of phosphorus, halogens and nitrogen, the sulphuric acid produced may be determined with standard alkali using a methyl red indicator. The spiral (previously coated with hydrogen peroxide) and test-tube are washed into a conical flask, two drops of methyl red added and the contents boiled for 30 seconds to remove carbon dioxide. The sulphuric acid is then titrated with 0.01M sodium hydroxide. In the absence of phosphorus, but in the presence of the halogens and nitrogen, the titrimetric procedure using barium chloride and tetra-hydroxyquinone can be used if bromine is used on the spiral. If phosphorus is present the gravimetric procedure must be used. In this case the spiral is washed by filling the tube* with 0.05M hydrochloric acid and washing down from the top with successive small quantities of the acid into a weighed crucible. 1 cm³ of 10% aqueous barium chloride is added and the contents of the crucible reduced on a steam bath. The spiral is again washed several times with hydrochloric acid, the washings being concentrated to about 2 cm³ in the crucible. The liquid is removed from the crucible using a weighed filter stick, the residue is washed three times with 1 cm³ portions of 0.05M hydro-chloric acid and the crucible and stick dried at 120°C, heated to 700°C for five minutes in a muffle furnace, rewashed, dried and ignited to constant weight.

* See p. 239.

Schöniger flask method.   This method is carried out in the manner described for the determination of the halogens (p. 240) using a sample size equivalent to 0.5–1 mg sulphur. In this instance the flask contains 10 cm³ of hydrogen peroxide solution (1 cm³ of 100 volume hydrogen peroxide in 10 cm³ water) which has been neutralised to one drop of methyl red with 0.01M sodium hydroxide. After the combustion has been completed and the stopper and wires rinsed with water, the contents of the flask may be titrated with 0.01M sodium hydroxide using methyl red indicator. Alternatively the solution is evaporated almost to dryness on a steam bath, the residue dissolved in water, and the solution neutralised and titrated with 0.01M barium chloride using a tetrahydroxyquinone indicator, or with 0.01M barium acetate (from a syringe burette) using a conductivity bridge.

The presence of nitrogen, phosphorus, or the halogens in the sulphur containing organic compound interfere in the above procedures. However, sulphur can be determined in the presence of these elements by slight modifications to the conductimetric method.[1]*

Determination of Phosphorus

The determination of phosphorus in organic compounds is commonly carried out by oxidation to orthophosphate ions and determination of these ions by gravimetric or colorimetric means.

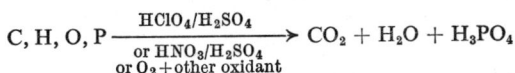

$$C, H, O, P \xrightarrow[\substack{or\ HNO_3/H_2SO_4 \\ or\ O_2 + other\ oxidant}]{HClO_4/H_2SO_4} CO_2 + H_2O + H_3PO_4$$

The following colorimetric procedures have been found to give satisfactory results.[1]* As there is a slight hazard attached to each method, operations should be carried out behind a suitable screen.

Acidic oxidation.   The sample (equivalent to about 1 mg phosphorus) is weighed into a calibrated 100 cm³ hard glass Kjeldahl flask, 3 cm³ of concentrated sulphuric acid and 0.5 cm³ of 70% perchloric acid are added and the flask transferred to a heating device (a mantle or a sand bath capable of maintaining a temperature of 250–300°C) which is already at the operating temperature.

Digestion should continue for two minutes after the evolution of dense white fumes has ceased. The flask is cooled, the contents are diluted with water to about 50 cm³, 5 cm³ of aqueous ammonium vanadate (1% in 2M nitric acid) is added followed by 10 cm³ of ammonium molybdate solution (10% in water), and the mixture diluted with water to the 100 cm³ calibration mark. After mixing, 15 minutes is allowed for the colour to develop, when the absorbance at 430 nm is

* See p. 230.

obtained. A blank determination should be carried out, the absorbance of which should not exceed 0.03.

A calibration graph is required in order to calculate the phosphorus content of the sample. This is obtained by dissolving about 0.44 g of potassium dihydrogen phosphate in one dm^3 of water and carrying out the above procedure on 0, 2.5, 5.0, 7.5, 10, 15, 20 cm^3 portions of this solution. Absorbance is plotted against mg $P/100$ cm^3 solution.

Schöniger flask method. For reliable results to be obtained, the phosphorus in the sample must be completely converted to ortho-phosphate ion. Combustion in oxygen alone may be incomplete and therefore, in this method, an additional oxidant (e.g. bromine water) is present in the flask. Even so, incorrect results may be obtained with some phosphorus containing compounds, for example those containing a $P \rightarrow S$ linkage, due to incomplete combustion.

The combustion is carried out as described for the determination of halogens by this method (p. 240) but with a mixture of 5 cm^3 of 0.5M sodium hydroxide and 4 cm^3 of saturated bromine water in the flask. A sample size equivalent to about 1 mg of phosphorus is used. After the stopper and wires have been washed, 6 cm^3 of 25 % sulphuric acid is added to the flask and the solution boiled until it becomes colourless. It is then transferred to a 100 cm^3 graduated flask, treated with 10 cm^3 of ammonium vanadate solution* followed, after mixing, by 10 cm^3 of aqueous ammonium molybdate (5%). The solution is made up to 100 cm^3, the colour allowed to develop for 30 minutes, and the absorbance at 430 nm obtained. A blank determination should have an absorbance of not more than 0.03.

A calibration graph is required and is obtained using quantities of potassium dihydrogen phosphate similar to those used in the previous procedure.

Automatic Procedures

Semi-automatic methods of elemental analysis were earlier (p. 230) defined as those methods which involve the manual weighing of the sample and of the absorption tubes (or noting the volume of nitrogen in the Dumas method) at the end of the determination, but where the intervening steps are carried out automatically.

For present purposes elemental analysis procedures will be considered to be automatic even if manual weighing of the sample and manual insertion of the sample into the reaction tube are necessary, provided that the apparatus will perform all subsequent steps unattended and display the results as a trace on a chart recorder or in digital form.

---

* 2.5 g in 500 cm^3 of water; cooled, 20 cm^3 concentrated nitric acid added, and diluted to 1 dm^3.

Instruments are available at the present time for the automatic quantitative analysis of carbon, hydrogen and nitrogen (Dumas) on a single sample, carbon and hydrogen on a single sample, oxygen alone, and nitrogen alone (either Dumas or Kjeldahl method).

Carbon, Hydrogen and Nitrogen on a Single Sample

At the present time Carlo Erba, Hewlett Packard Ltd., Perkin–Elmer Ltd., and Technicon Instruments Company Ltd., produce instruments for the automatic determination of carbon, hydrogen and nitrogen (Dumas) based on the chemical principles discussed earlier (p. 228), the analysis time being in the region of 10–15 minutes. These instruments all have a combustion stage in which the sample is combusted in a stream of about 3% oxygen in helium (Carlo Erba make provision for alternative combustion in the absence of oxygen) and the products passed over an oxidant, followed by a reduction stage to reduce the oxides of nitrogen to elemental nitrogen. The Carlo Erba and Hewlett Packard instruments separate the resulting carbon dioxide, water, and nitrogen by gas chromatography, each gas being detected by the same katharometer.

The Perkin–Elmer system has three katharometer detectors (K)—one for each gas (Fig. 11).

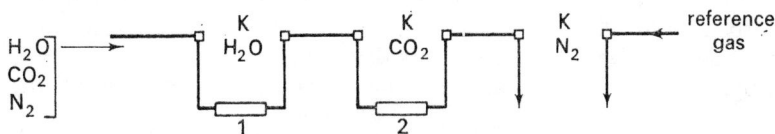

1. H_2O absorption tube
2. CO_2 absorption tube.

Fig. 11

The Technicon system employs one katharometer detector. The water is temporarily retained by silica, the carbon dioxide and nitrogen are measured together, the nitrogen is measured alone after the carbon dioxide has been absorbed (Fig. 12) and the water is then determined after desorption by an automatic heater.

With all these instruments the sample is weighed, placed in the furnace (automatic injection of pre-weighed samples is provided in the Carlo Erba system) and the calculation of percentage carbon, hydrogen, and nitrogen made using data from a digital output or from a trace on a recorder chart.

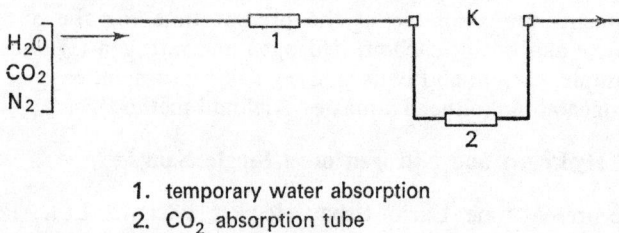

1. temporary water absorption
2. CO_2 absorption tube

Fig. 12

The Hewlett Packard system can be linked to a Cahn Ratio Electrobalance which impresses a signal on the slide wire of the recorder during weighing and so obviates the necessity of manually recording the weight of the sample.

These instruments will accept samples down to about 0.2 mg in weight. If sample sizes of less than 1 mg are used, special care has to be taken to ensure the homogeneity of the bulk sample prior to weighing on a microgram balance.

Carbon and Hydrogen on a Single Sample

The system marketed by the American Instrument Company Inc uses a stream of pure oxygen and an analysis time of $1\frac{1}{2}$ minutes is claimed. The water produced by the combustion is temporarily adsorbed on to silica, the oxides of nitrogen are removed by manganese dioxide, and the carbon dioxide is determined by the katharometer K (Fig. 13). The water is desorbed by a heater, allowed to react with

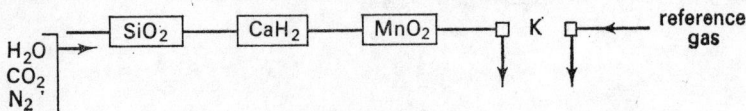

Fig. 13

calcium hydride and the resultant hydrogen determined by the katharometer, the output giving peaks on a recorder chart. Sample weights are limited to the region 0.2–0.8 mg.

Oxygen

By replacing the combustion tubes in the Perkin Elmer and Technicon 'CHN' instruments with tubes containing (a) platinum/carbon or nickel/carbon and (b) copper (II) oxide (see p. 235) elemental oxygen may be determined as carbon dioxide.

The Carlo Erba instrument, referred to above, can also be used for the determination of oxygen. No change of assembly is required as the instrument contains additional reaction tubes, in parallel with those required for C, H, N analysis, making it possible to switch from C, H, N to O operation. Automatic sample introduction can also be used when the instrument is being employed in the oxygen mode.

Sample sizes of about 0.5 mg are employed for automatic oxygen analysis, each determination taking about 10–20 minutes to complete.

Nitrogen

Coleman Instruments Incorporated market two instruments for the determination of nitrogen by the Dumas method (one for 5–50 mg and one for 50–500 mg of sample) which involve inserting a weighed sample into the combustion tube and reading the volume of gas produced, from a digital counter. A single analysis takes about ten minutes.

An automatic instrument based on the Kjeldahl method of nitrogen determination is also available (Technicon Instruments Co. Ltd.). The sample is weighed and placed in the sampler. The whole process is then automatic, the sample being fed to the digester, digestion reagents added, the digestion products diluted and mixed with sodium hydroxide, sodium phenate, and sodium hypochlorite; after a predetermined time the coloured solution is fed to a colorimeter (630 nm) and the output displayed on a chart recorder. The analysis time is about three minutes per sample.

8 Quantitative Determination of Reactive Groups

Determination of Acidity

The neutralisation equivalent of an acid is defined as the number of grams of substance equivalent to one mole of alkali, and is therefore equal to the molecular weight of the substance divided by the number of acid groups present.

The neutralisation equivalent may be determined by weighing a quantity of the acid equivalent to about 5–10 cm³ of 0.01M or 0.1M sodium hydroxide (5–300 mg), dissolving it in neutralised water or 95% aqueous ethanol containing phenolphthalein, and titrating with standard alkali. Water (or ethanol) may be neutralised by adding several drops of phenolphthalein solution (1% in 95% aqueous ethanol), boiling for 30 seconds, and adding alkali dropwise to produce a faint persistent pink colour.

This method may be used to determine certain phenols, such as trinitrophenol; it is applicable to acidic compounds which also contain one aromatic amino group, but not when aliphatic amino groups are present. The presence of amide, imide, and phenolic groups will generally interfere with the determination.

Determination of Basicity

All amines whether aliphatic, aromatic, or heterocyclic, exhibit basic character to a degree which depends upon the environment of the nitrogen atom. These compounds can usually be determined by titration with standard acid, though the procedure chosen will depend largely upon the basicity of the compound under examination.

Water soluble aliphatic amines can be titrated in aqueous solution. 50 cm³ of water is neutralised by adding six drops of bromocresol green–methyl red mixed indicator (5 parts of 1% bromocresol green and 1 part 0.1% methyl red) and adding 0.1M hydrochloric acid until the green colour of the indicator disappears. The sample (equivalent to 10–20 mg nitrogen) is added and titrated with 0.1M hydrochloric acid. If the sample is coloured, or if the scale of the determination is to be reduced, potentiometric titration may be used (glass/calomel electrodes).

Compounds containing basic nitrogen atoms, whether aliphatic or aromatic, can be determined by titration with perchloric acid in acetic acid using 1% crystal violet in acetic acid as indicator. 2–3 drops of indicator are added to 50 cm³ of acetic acid, and 0.1M perchloric acid in acetic acid is added until the first persistent green colour (absence of violet) is obtained. The sample, equivalent to 10–20 mg nitrogen, is added and titrated with 0.1M perchloric acid in acetic acid to the same green end point.

The above procedure can be made much more sensitive by employing potentiometric titration. The sample, equivalent to about 0.5 mg nitrogen (or 0.05 mg nitrogen) is dissolved in 2.5 cm³ chloroform and 2.5 cm³ acetic acid added. The glass and calomel electrodes are wiped with tissue and placed in the solution and the sample titrated with 0.1M (or 0.01M) perchloric acid in acetic acid from a syringe burette (volume of titrant ∼0.3 cm³). The reagent can be standardised against potassium hydrogen phthalate in the chloroform/acetic acid solvent system. An assembly suitable for this determination is diagrammatically shown below (Fig. 14).[1]* The KCl–Agar plug is prepared by immersing the

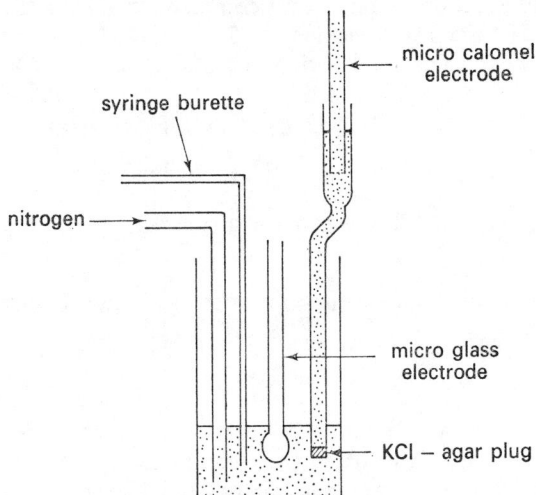

micro calomel electrode

syringe burette

nitrogen

micro glass electrode

KCl – agar plug

Fig. 14

tip of the tube in hot potassium chloride/agar solution (30 g potassium chloride, 3 g agar, 100 cm³ water) removing the tube and allowing the gel to solidify.

Determination of Hydroxyl, Thiol and Primary and Secondary Amino Groups by Acetylation

The sample (equivalent to 0.1–0.3 mM of monofunctional compound) is inserted into the bottom of a weighed tube using either a

* See p. 230.

microsyringe or filler tube. Redistilled acetic anhydride (20–40 mg) is added to the tube and centrifuged down; the tube is reweighed. Dry, redistilled pyridine (4–6 drops) is added followed by further centrifugation. The tube is then sealed and centrifuged in both directions to ensure complete homogeneity of the contents. Duplicate blanks are prepared in like manner. All four tubes are kept at room temperature overnight, or for a shorter time in a heated water bath.

Each tube is broken, using a glass rod, beneath 5 cm³ of water in a conical flask. The acetic acid produced is titrated with 0.05 or 0.1M aqueous sodium hydroxide using phenolphthalein indicator and the result expressed as per cent —OH, —NH₂ etc. in the sample.

$$ROH + (CH_3CO)_2O \rightarrow ROCOCH_3 + CH_3COOH$$

A similar procedure may be carried out on a larger scale (10–30 mM of sample/2–4 g of acetic anhydride, in a suitable flask), refluxing for 30–60 minutes. After the reaction is complete, water is added down the condenser and the flask reheated for 5–10 minutes to ensure complete conversion of the acetic anhydride to acetic acid.

Determination of 1,2-Diols and Epoxides

Periodic acid splits 1,2-diols in the following way:

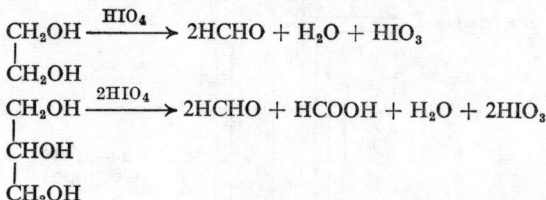

$$\underset{\overset{|}{CH_2OH}}{CH_2OH} \xrightarrow{HIO_4} 2HCHO + H_2O + HIO_3$$

$$\underset{\overset{|}{CH_2OH}}{\overset{|}{CHOH}}{CH_2OH} \xrightarrow{2HIO_4} 2HCHO + HCOOH + H_2O + 2HIO_3$$

If a 1,2-diol is treated with excess periodic acid, then the excess can be determined after the reaction is complete, by adding potassium iodide to the neutralised reaction mixture and determining the liberated iodine with standard sodium thiosulphate.

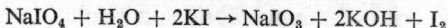

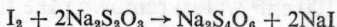

$$NaIO_4 + H_2O + 2KI \rightarrow NaIO_3 + 2KOH + I_2$$

$$I_2 + 2Na_2S_2O_3 \rightarrow Na_2S_4O_6 + 2NaI$$

About 0.7 mM of the sample is weighed into a suitable 150 cm³ flask and 25 cm³ of 1.5% periodic acid in 0.4M aqueous perchloric acid added. The mixture is allowed to stand for an hour with occasional swirling, after which time it is neutralised with excess solid sodium bicarbonate. Either 10 cm³ of 10% aqueous potassium iodide is added and the liberated iodine titrated with 0.2M sodium thiosulphate, or, preferably, 25 cm³ of 0.1M sodium arsenite is added, followed by

10 cm^3 of 10% aqueous potassium iodide, and the excess sodium arsenite back titrated with 0.1M aqueous iodine solution. (This alternative procedure prevents the loss of iodine vapour on addition of the potassium iodide.) Blank determinations are carried out alongside the sample determinations.

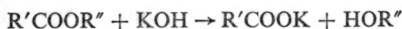

$$I_2 + Na_3AsO_3 + H_2O \rightarrow Na_3AsO_4 + 2HI$$

Epoxides are hydrolysed to 1,2-diols in aqueous acid solution and therefore the above procedure can be used for the determination of epoxides.

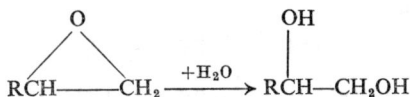

Dixon[1]* gives details of a slight variation of the above procedure for use with sample sizes down to 0.002 mM of 1,2-diol.

Determination of Esters

10 cm^3 of 0.2M 95% aqueous alcoholic potassium hydroxide (prepared by dissolving the hydroxide in water and then adding the appropriate quantity of absolute ethanol) is added to about 0.5 mM of monoester and the mixture refluxed gently for one hour. After cooling, the condenser is washed with water and the excess potassium hydroxide determined with 0.1M aqueous hydrochloric acid after adding phenolphthalein indicator. Blank determinations are carried out alongside the sample determinations.

$$R'COOR'' + KOH \rightarrow R'COOK + HOR''$$

The scale of this determination can be reduced to handle samples down to about 0.05 mM of monoester, by using 5 cm^3 of 0.15M aqueous alcoholic potassium hydroxide.

Determination of Amides

Normal hydrolysis with ethanolic potassium hydroxide seldom results in the complete cleavage of amides. However, if the reaction temperature is raised by using potassium hydroxide in ethane-1,2-diol quantitative hydrolysis can be obtained, but this method is seldom used because of the long reaction times required (about six hours).[5]

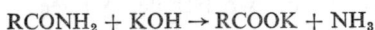

$$RCONH_2 + KOH \rightarrow RCOOK + NH_3$$

Amides are often estimated by formation of their ferric hydroxamate derivatives and spectrophotometric determination of the colour

---

* See p. 230.
5   Hillenbrand and Pentz in *Organic Analysis*, Vol. 3, Interscience, 1956, p. 129.

produced. Certain amides may be determined by potentiometric titration with perchloric acid (see p. 248).

Determination of Methoxyl or Ethoxyl Groups

The following method, first introduced by Zeisel in 1885, is based upon the fact that methyl and ethyl ethers or esters yield iodomethane or iodoethane quantitatively on treatment with boiling hydriodic acid. It is applicable to the analysis of methoxyl or ethoxyl compounds which are not highly volatile.

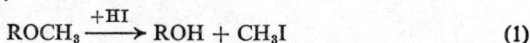

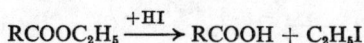

$$ROCH_3 \xrightarrow{+HI} ROH + CH_3I \tag{1}$$

$$RCOOC_2H_5 \xrightarrow{+HI} RCOOH + C_2H_5I$$

The alkyl iodide is converted to iodic acid by an aqueous solution of bromine in the presence of sodium acetate and acetic acid. The iodine produced by treatment of the iodic acid with potassium iodide in acid solution is determined with sodium thiosulphate.

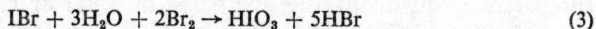

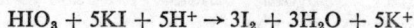

$$CH_3I + Br_2 \rightarrow CH_3Br + IBr \tag{2}$$

$$IBr + 3H_2O + 2Br_2 \rightarrow HIO_3 + 5HBr \tag{3}$$

$$HIO_3 + 5KI + 5H^+ \rightarrow 3I_2 + 3H_2O + 5K^+$$

The apparatus consists of a reaction flask F of about 50 cm³ capacity (Fig. 15) fitted with a side arm A and connected, via a water condenser C, to a scrubber S, which in turn is connected to a receiver R.

The first reaction (1) takes place in F, under reflux, with water in the condenser. The water is then drained out of the condenser and the alkyl iodide produced is allowed to distil through the scrubber S (which removes acid vapours) and into the receiver R, assisted by a flow of inert gas, where reactions (2) and (3) take place. The scrubber is half filled with 25% w/v aqueous sodium acetate solution, and the acetic acid–potassium acetate–bromine reagent (100 cm³:10 g:3 mM) is placed in the reciver.

Procedure. The sample (0.01–0.02 mM of monofunctional compound) is weighed in a platinum or aluminium boat, or in a capillary, and inserted into the reaction flask together with 2–3 g phenol. Hydriodic acid, s.g. 1.7 (5 cm³) is run in through the side arm and a nitrogen or carbon dioxide supply immediately connected to this arm and regulated such that one or two bubbles per second escape from the receiver. The flow rate of inert gas is not altered further throughout the determination.

Steyermark[2]* recommends leaving the reaction mixture at room temperature for thirty minutes followed by heating under reflux for

* See p. 230.

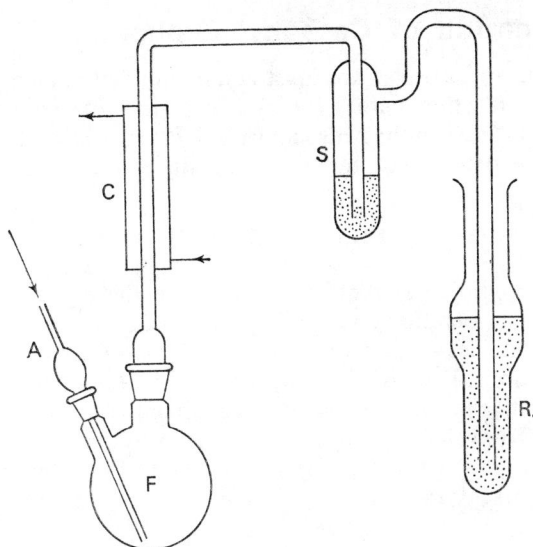

Fig. 15

thirty minutes. Water is then drained from the condenser, and heating is continued for a further thirty minutes. After this time the delivery tube is rinsed into the receiver with water and the contents of the receiver washed into a conical flask containing 5 cm³ of 25% aqueous sodium acetate solution. Formic acid is added, dropwise, to destroy excess bromine and any remaining bromine vapour is removed by blowing gas into the flask. 10% sulphuric acid (5 cm³) and potassium iodide (0.5 g) are added, the flask stoppered, swirled, and the liberated iodine determined with 0.01M sodium thiosulphate.

A blank is carried out in like manner. The maximum error allowed is ±0.3%.

Determination of Nitro Groups

The percentage of nitro groups in a compound may be obtained from the determination of elemental nitrogen.

Alternatively, nitro, nitroso, azo and hydrazo groups may be determined by reduction with titanous chloride in acid solution, and titration of the excess reagent with ferric ammonium sulphate.

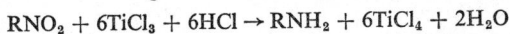

$$RNO_2 + 6TiCl_3 + 6HCl \rightarrow RNH_2 + 6TiCl_4 + 2H_2O$$

The determination has to be carried out under an inert atmosphere as the reagent is readily oxidised by atmospheric oxygen. Details of the method can be found elsewhere.[1]*

* See p. 230.

Determination of Carbonyl Groups

Aldehydes and ketones are most commonly determined by making use of their reaction with hydroxylamine or hydroxylamine hydrochloride. In the case of the hydrochloride, hydrogen chloride is liberated and may be determined by titration with sodium hydroxide. The

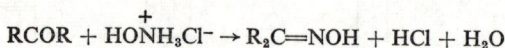

$$RCOR + HO\overset{+}{N}H_3Cl^- \rightarrow R_2C{=}NOH + HCl + H_2O$$

presence of pyridine is required to displace the equilibrium of the reaction and a potentiometric ending is necessary. Even so, all but the simplest aldehydes and ketones require long reaction times under reflux conditions. Details of this method can be found elsewhere.[6]

An alternative procedure involves reaction of the carbonyl compound with free hydroxyalmine (produced by adding 2-dimethylaminoethanol to hydroxylamine hydrochloride) and titration of the excess hydroxylamine with perchloric acid.[7]

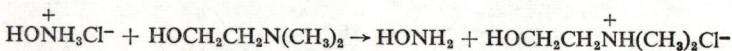

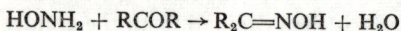

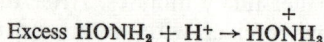

$$HO\overset{+}{N}H_3Cl^- + HOCH_2CH_2N(CH_3)_2 \rightarrow HONH_2 + HOCH_2CH_2\overset{+}{N}H(CH_3)_2Cl^-$$

$$HONH_2 + RCOR \rightarrow R_2C{=}NOH + H_2O$$

$$\text{Excess } HONH_2 + H^+ \rightarrow HO\overset{+}{N}H_3$$

Procedure. About 2 mM of monocarbonyl compound is placed in a 150 cm³ flask, exactly 20 cm³ of 0.25M 2-dimethylaminoethanol in 2-propanol is added, followed by 25 cm³ of 0.4M hydroxylamine hydrochloride reagent.* Most aldehydes and unhindered aliphatic ketones will react within about 15 minutes; hindered aliphatic ketones, diketones, and aryl ketones may require heating at 70°C for 45 minutes. That a sufficient reaction time has been allowed should be checked by reacting a further sample for a longer time.

The residual hydroxylamine is titrated potentiometrically with 0.2M perchloric acid† using glass and calomel electrodes. Alternatively five drops of indicator (70 mg martius yellow plus 4 mg methyl violet in 50 cm³ ethanol) are added and titration carried out to a colourless (or blue grey) end point. A blank determination is carried out alongside the sample determinations.

---

* 28 g in 300 cm³ methanol, made up to 1 dm³ with 2-propanol.
† 17 cm³ of 70% perchloric acid, made up to 1 dm³ with 2-methoxyethanol; standardised using tris(hydroxymethyl) aminomethane.
6 Mitchell in *Organic Analysis*, Vol. 1, Interscience, 1953, p. 243.
7 Fritz, Yamamura and Bradford, *Anal. Chem.* **31**, 260 (1959).

Determination of Unsaturation

Alkenyl Unsaturation

The preferred method for the determination of alkenes is that of room temperature catalytic hydrogenation, one mole of hydrogen being consumed by one mole of monoalkene,

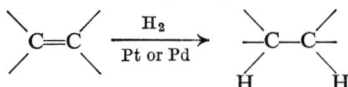

$$\diagdown \!\! \diagup \atop C=C \atop \diagup \!\! \diagdown \quad \xrightarrow[\text{Pt or Pd}]{H_2} \quad \begin{matrix} \diagdown \;\; \diagup \\ -C-C- \\ \diagup \;\; \diagdown \\ H \quad H \end{matrix}$$

although this procedure is inapplicable to the determination of alkenes which also contain other hydrogenatable functions such as alkynyl, aldehyde, nitro, nitrile groups.

The alternative method is that of halogen addition, which in general is more limited than catalytic hydrogenation in that the reactivity of alkenyl double bonds to halogen addition varies markedly with the structural environment of the double bond. Substituents which have a negative inductive effect and are close to the double bond greatly

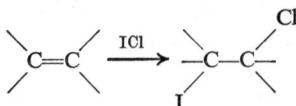

$$\diagdown \!\! \diagup \atop C=C \atop \diagup \!\! \diagdown \quad \xrightarrow{ICl} \quad \begin{matrix} \qquad Cl \\ \diagdown \;\; \diagup \\ -C-C- \\ \diagup \;\; \diagdown \\ I \end{matrix}$$

reduce the rate of halogen addition, and for this reason, conjugated dienes are reluctant to take up more than one mole of halogen at room temperature, while compounds such as crotonic, cinnamic, maleic and fumaric acids, and other compounds with electronegative groups close to the double bond, may undergo very little or no reaction with halogen under similar conditions. The use of higher temperatures or more active halogens e.g. bromine, chlorine, is prohibited, as undesirable substitution then occurs, as well as the desirable addition. (It is possible to assess the amount of substitution which takes place, by determination of the halogen acid produced, but such corrections often give low results.[8])

Catalytic hydrogenation. There are a number of different types of hydrogenation apparatus, the essentials of which are graphically represented in Fig. 16. The size of the sample to be hydrogenated will depend upon the degree of unsaturation of the compound and the type of burette B used, and may vary between about 20 and 2000 mg.

Procedure. The burette B is filled with water to the top of the tap TB and the tap is closed. The magnetic stirrer bar is placed in the hydrogenation flask F, together with the solvent (e.g. acetic acid or ethanol) and the platinum oxide (~10–20 mg) or palladium/charcoal catalyst (~50–80 mg). The top of the neck of the flask is lightly greased and the

8   Polgar and Jungnickel in *Organic Analysis*, Vol. 3, Interscience, 1956, p. 203.

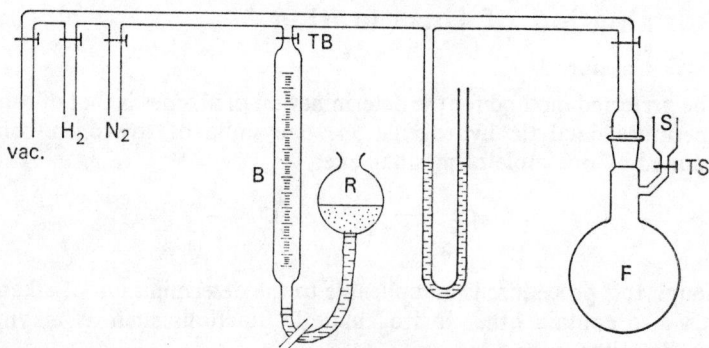

Fig. 16

flask connected to the apparatus which is evacuated (<25 torr), with stirring, and then filled with nitrogen to just below atmospheric pressure. The apparatus is again evacuated, filled with hydrogen (at this stage the catalyst will begin to take up hydrogen), evacuated, and filled with hydrogen again; this time the burette B is also filled with hydrogen. The reservoir R is adjusted to keep the apparatus under a slight vacuum, at which pressure it should be kept throughout the rest of the determination, by adjusting R periodically.

It is established that the catalyst has ceased to take up hydrogen by stopping the stirrer, reading the burette B at atmospheric pressure, stirring for five minutes, and then re-reading the burette. When the catalyst has been fully hydrogenated, a known quantity of the substance being examined is placed in the funnel S, either as a liquid or in solution (the volume of this must be known). Atmospheric temperature and pressure are noted and the gas in the apparatus is brought to atmospheric pressure using the reservoir R, and the burette reading taken. The sample is admitted to the flask F, using tap TS, taking care not to admit air, and the funnel S washed with three small successive portions of solvent (of known volume), admitting each to the flask. The stirrer is then restarted.

The diminution in gas volume is observed (R being adjusted accordingly) and, when no further change is noted, the stirrer is stopped and the burette read at atmospheric pressure. The stirrer is then restarted and the apparatus left for five minutes, after which time the stirrer is stopped and the burette reading again taken. This sequence is repeated until successive burette readings agree. The atmospheric pressure and temperature are again recorded and if these have changed since they were originally taken it will be necessary to know the total volume of gas in the apparatus. This can be calculated by obtaining burette readings at atmospheric pressure and at a pressure other than atmospheric by raising or lowering R and noting the pressure difference on the manometer. The error in the result obtained should not exceed $\pm 0.5\%$.

If the molecular weight of the substance is known, or can be approximated, the number of double bonds in the compound can then be obtained—or vice versa.

Halogenation. The quantity of halogen taken up by a substance is determined by back titration of excess reagent after the reaction is complete. If bromine is used for the determination of unsaturation then the amount of substitution which has taken place must be assessed by titrating the hydrogen bromide liberated, as well as the excess reagent, at the end of the reaction time.

Iodine monochloride, when used at room temperature and for moderate reaction times, does not have this tendency to substitute and therefore no acidity determination is necessary. The method can be used on a micro or macro scale using sample sizes in the range 3–3000 mg. Iodine monochloride is a fairly good oxidising agent so oxidisable groups will react; aromatic rings give iodinated substitution products.

The sample (100–200 mg) is weighed into a stoppered vessel and dissolved in carbon tetrachloride (10 cm³). Excess of iodine monochloride reagent* is added (25 cm³), the flask swirled and left in the dark for 15–60 minutes. The usual blank determinations are also carried out alongside the duplicate sample reactions. After the appropriate time the stopper is rinsed into the flask, potassium iodide is added (20 cm³ of a 15% aqueous solution) and the liberated iodine, which is equivalent to the excess reagent, determined with 0.1M sodium thiosulphate. The result obtained should be within 1.5% of theoretical for compounds which do not have substituent groups near to the unsaturated centre.

Alkynyl Unsaturation

Alkynyl unsaturation may be determined by catalytic hydrogenation in a similar manner to that used for alkenyl unsaturation but using a deactivated catalyst such as palladium on barium sulphate poisoned with quinoline (Lindlar catalyst). One mole of hydrogen is taken up by one mole of monoalkyne.

Compounds containing a hydrogen atom attached to a triply bonded carbon atom may be determined by treatment with aqueous alcoholic silver nitrate† and titration of the acid produced with 0.1M aqueous

$$RC\equiv CH + AgNO_3 \rightarrow RC\equiv CAg + HNO_3$$

sodium hydroxide, using methyl red–methylene blue indicator.

---

* Wijs reagent: 10 g iodine monochloride/300 cm³ chloroform/700 cm³ acetic acid.
† 50 g silver nitrate in 100 cm³ water, diluted to 1 dm³ with ethanol.

9 Determination of Some Physical Properties

Determination of Molecular Weight

Many of the methods discussed previously for the determination of elements and groups can be regarded as being methods of molecular weight determination, or at least as contributing towards the determination of molecular weight. For instance, if a compound is known to contain only one carboxylic acid group (and no other acidic or basic functions) titration of the pure compound with alkali will give the molecular weight of the acid. If the acid contains an unknown number of carboxylic acid groups, then the molecular weight cannot be determined absolutely by titration, but useful information can still be obtained. If the acid were shown to contain 10% of $-COOH$, then its molecular weight must be 450, 900, or some other multiple of 450, depending upon the number of carboxylic acid groups present. Similarly, a 10% hydroxyl group content would mean a molecular weight of 170, 340, etc., and 10% of elemental nitrogen would correspond to a molecular weight of 140, 180, etc. Similar information can be obtained for alkenes from a determination of unsaturation, for esters from the quantity of alkali required for hydrolysis, and so on.

When information of this type has been obtained a very accurate method of molecular weight determination is often not necessary; for example $\pm 10\%$ would be quite satisfactory to differentiate between possible molecular weights of 450 and 900.

Molecular Weight by Mass Spectrometry

The technique which gives the most accurate and precise value for molecular weights up to values in the 1000–2000 region (depending upon the type of instrument available) is undoubtedly mass spectrometry. The factor governing the applicability of this technique is the vapour pressure of the compound being examined, which should be such that microgram quantities will vapourise at temperatures up to about 300°C, under high vacuum conditions ($\sim 10^{-7}$ torr). This minute quantity of vapour is subjected to bombardment by an electron beam which ionises the molecule by ejecting an electron from it.

$$M \xrightarrow{+e} M^+ + 2e$$

The ion produced breaks down to give a host of other smaller species which are then separated according to their mass/charge ratio, detected, and recorded. The particles of highest mass/charge ratio are those of the original molecule which have lost one electron (called the parent ions) and the molecular weight of the parent ion can easily be obtained from the calibrated chart. Small quantities of impurities may not interfere with molecular weight determination by mass spectrometry.

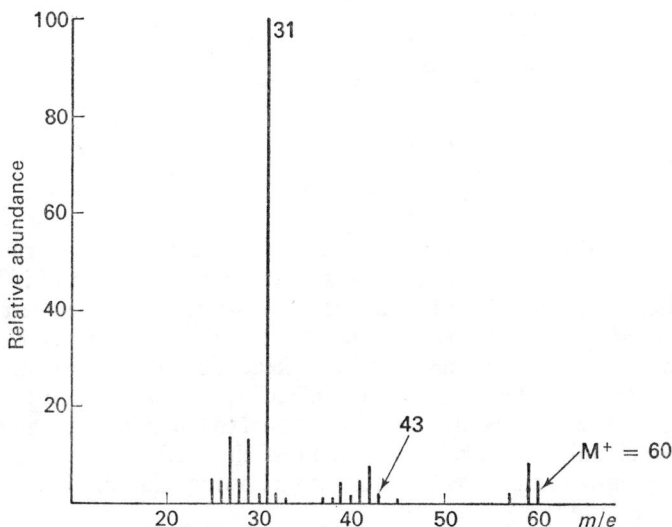

Fig. 17.   Mass spectrum of propan-1-ol.

This method is inapplicable to about 10% of all organic compounds which have the required volatility, due to their not giving detectable parent ions.

Vapour Pressure Methods

Vapour pressure methods of molecular weight determination depend upon the measurement of the difference between the vapour pressure of pure solvent and that of solvent containing the test material in solution, the vapour pressure of a solution being dependent upon its molarity.

The most convenient vapour pressure method for the determination of molecular weight is that of vapour pressure osmometry. Vapour pressure osmometers normally operate in the molecular weight range 60–25000, and the quantity of sample required can be as little as about 50 μg. The apparatus is diagrammatically represented in Fig. 18.

The atmosphere in the chamber C is allowed to become saturated with vapour of the chosen solvent at a suitable temperature. A drop of pure solvent (from V) is injected on to the reference thermistor T_R and a drop of solution, of equivalent volume (from U), on to the sample

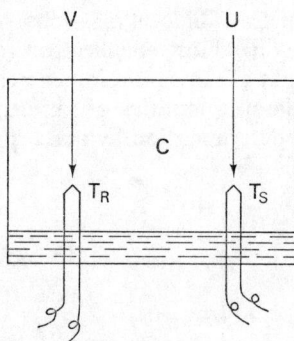

Fig. 18

thermistor T_S. Solvent vapour condenses on the sample thermistor (changing the concentration of the solution in the direction of infinite dilution) and causes this thermistor to be warmed. The temperature change brings about a change in the resistance of the thermistor (ΔR) which is recorded by means of a Wheatstone bridge circuit. The procedure is repeated with solutions of different concentrations.

The change in temperature is proportional to the number of moles of solute per unit weight of solvent, and as change in resistance varies directly with change in temperature the following relationship holds at infinite dilution:

$$R = K \cdot \frac{\text{wt. solute}}{\text{M.W. solute}} \cdot \frac{1}{\text{wt. solvent}}$$

$$= K \cdot \frac{C}{M}$$

$$\left[\frac{\Delta R}{C} \right]_{C \to 0} = \frac{K}{M}$$

where K = the molar constant for the thermistors and solvent used,
   C = concentration of solute in g/dm^3
   M = molecular weight of solute.

A wide range of solvents can be used for this procedure e.g. benzene, ethanol, chloroform, carbon tetrachloride, methylene chloride, water, ethyl acetate. The apparatus has first to be calibrated using a pure solute (e.g. benzil) of known molecular weight, at different concentration levels, and the molar constant (K) determined at zero concentration by plotting $\Delta R/C$ against concentration (Fig. 19). The molecular weight of the unknown compound is then determined in a similar manner.

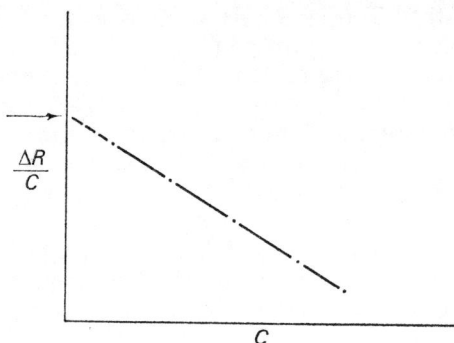

Fig. 19

Deviations of 2%, or better, from the true molecular weight are obtainable by this method, depending upon the instrument and the compound being examined.

Depression of the Melting/Freezing Point Method

The determination of molecular weight by observing the lowering of the melting point or freezing point of a solvent on the addition of the substance being examined, presumes that the lowering (Δt) is proportional to the number of moles of solute per unit weight of solvent.

$$\Delta t = K \cdot \frac{\text{wt. solute}}{\text{M.W. solute}} \cdot \frac{1}{\text{wt. solvent}}$$

Melting points can be obtained with greater reproducibility than freezing points.

In the Rast method of molecular weight determination, camphor is commonly employed as the solvent, though other solvents may be used (Table 3) depending upon the compound under examination. 0.2–2 mg of pure test material is required.

The value of the molecular lowering constant K (which is the depression equivalent to one mole of solute/g solvent), has to be determined for the chosen solvent using a pure compound of known molecular weight; naphthalene (m.p. 80°C), anthracene (216°), benzoic acid (121°) and azobenzene (68°) have been commonly used for this purpose.

Table 3.   Solvents suitable for use in the Rast molecular weight determination

Solvent	M.P. (°C)	Approx. K
Camphor	178	39×10^3
Camphene	49	31×10^3
Camphenilone	38	64×10^3
Cyclopentadecanone	66	21×10^3

Procedure. The tube T (Fig. 20) is weighed and the approximate quantity of sample required is forced from the filler tube F, using the rod R, into the bottom of the tube; the tube T is then reweighed. Care must be taken to ensure that all the sample is at the bottom of T and none is adhering to the walls of the tube. Solvent is then weighed into T (about ten times the weight of sample) by the same procedure. The tube T, which at this stage should be no more than a quarter full, is then sealed at a point remote from the contents and the substances mixed by rotating the tube at about 1–2°C above the melting point of the contents. A duplicate sample tube should be prepared. Solvent is then placed in a third tube and this tube sealed as before.

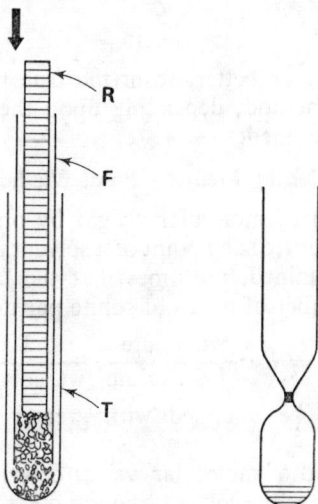

Fig. 20

All three tubes are rapidly heated in a suitable melting point apparatus until the solvent and solutions are liquid. The tubes are cooled until solidification occurs and then slowly reheated (0.5°C/minute), and the temperature noted when the last crystal vanishes in each tube. The tubes are cooled, and the sequence repeated until successive melting points agree to within ±0.2–0.3°C.

The molecular lowering constant (K) is obtained for the solvent used by carrying out the above procedure using a solute of known molecular weight.

Density of Liquids

The density of a liquid may be determined with a high degree of accuracy by means of small pyknometers of the type shown in Fig. 21.

Type (a) may be of about 1 cm³ capacity, type (b) about 0.1 cm³, and type (c) which is graduated, is suitable for the range 0.01–0.1 cm³.

The pyknometer has to be calibrated by weighing the apparatus empty and again when it contains a suitable pure liquid of known

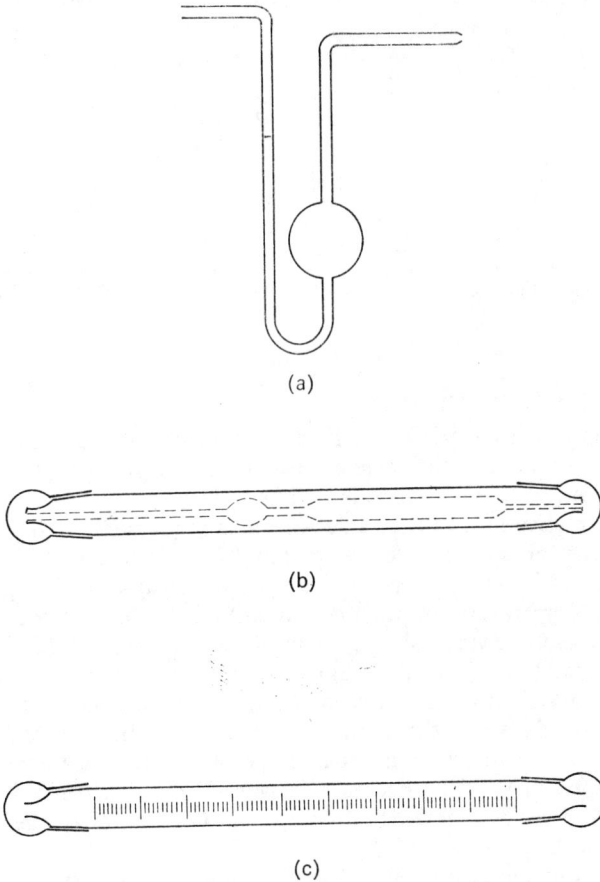

(a)

(b)

(c)

Fig. 21

density, commonly water. Type (a) is calibrated by drawing too much water into the pyknometer, which is then thermostatted, inclined in order that a piece of filter paper applied to the fine end absorbs a sufficient quantity of water to bring the liquid to the mark in the other limb.

The pyknometer is then removed from the thermostat bath, quickly and carefully dried, and weighed without delay, suspending the device by means of a platinum wire. The volume of the pyknometers of type (b) and (c) is determined in a similar manner.

The density of the substance under examination is determined in an analogous manner.

Table 4. The density of water (g/cm^3)[9]

°C								
0°C	0.999 868	8	0.999 877	16	0.998 972	24	0.997 327
1	927	9	809	17	804	25	075
2	968	10	728	18	625	26	0.996 814
3	992	11	634	19	435	27	544
4	1.000 000	12	526	20	234	28	264
5	0.999 992	13	406	21	022	29	0.995 976
6	968	14	273	22	0.997 801	30	678
7	930	15	129	23	569	31	372

9　*Handbook of Chemistry and Physics,* 49th edn., 1968–9, The Chemical Rubber Co., Ohio, p. F5.

Refractive Index

Liquids which have been purified should not only have the correct boiling point but also the correct refractive index. Refractive index is a better criterion of purity than is boiling point.

The refractive index (n) of a substance is fundamentally the ratio of the velocity of light in the compound to the velocity of light in a vacuum, and it can vary greatly with the wavelength of the light used and with temperature. In practice refractive index is usually measured relative to air, the value of $n_{(vac)}$ being about equal to $1.000\,27 \times n_{(air)}$. The sodium D line ($\lambda = 5893$ Å) is commonly used as the light source or alternatively white light is used and the result obtained is converted to a figure relative to the sodium D line. The radiation employed is recorded as a subscript and the temperature of measurement as a superscript e.g. n_D^{20}. The decrease in refractive index per 1°C rise in temperature varies slightly from compound to compound, the average value being about 4.5×10^{-4}.

Changes in refractive index from compound to compound can be related to changes in structure.[10] The molar refraction of a pure compound (R)

$$R = \frac{n^2 - 1}{n^2 + 2} \cdot \frac{M}{d}$$

is made up of incremental ΔR values associated with different parts of

10　Bauer and Fajans in *Techniques of Organic Chemistry, Vol. I, Physical Method of Organic Chemistry, Part II*, Ed. Weissberger, Interscience, 1949, p. 1141.

the structure of a molecule. Thus,

$$\Delta R_{D(C)} + 2\Delta R_{D(H)} = \Delta R_{D(CH_2)}$$

$$2.418 + 2 \times 1.100 = 4.618$$

Therefore the difference in molar refraction between adjacent members of a homologous series would be about 4.618 cm^3/mole. As there is a dependence of refractive index on structural features, this property can be of considerable assistance in the identification of unknown compounds.[10]

There are a number of methods available for the determination of refractive index. The procedure using the commonly available Abbé refractometer requires only a few drops of liquid and has an accuracy of about ±0.0002. The instrument consists essentially of two prisms (between which the liquid sample L is placed (Fig. 22)), attached to an adjustable arm A which in turn passes over a scale S, and a telescope system T. Daylight or artificial light is directed on to the prism assembly by means of a mirror, and the scale on the instrument gives the refractive index with respect to the sodium D line.

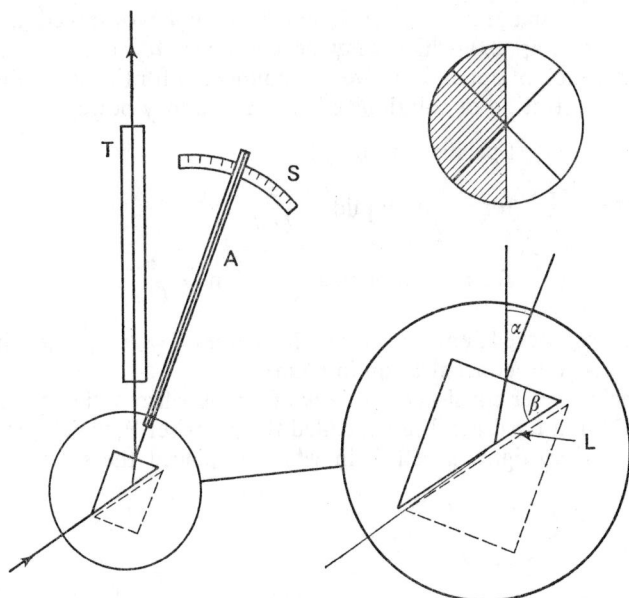

$$n = \sin \beta \, (N^2 - \sin^2\alpha)^{\frac{1}{2}} - \cos \beta \cdot \sin \alpha$$
$$(N = \text{refractive index of prism})$$

Fig. 22

If the determination is to be carried out at a temperature other than room temperature, liquid is circulated through the instrument from a thermostat bath for at least fifteen minutes prior to the determination being carried out. The prism assembly is opened and a few drops of the liquid are inserted between the prism faces, care being taken not to touch the prism surface with the dropper. The mirror is adjusted and the cross-wires focused by turning the eyepiece. The arm A is adjusted until the field of view is partly dark and the compensator knob (at the base of the telescope) adjusted to give a sharp edge to the dark field. The arm A is then adjusted until the edge of the dark field is coincident with intersection of the cross-wires (Fig. 22). The reading on the scale S is noted. The determination is repeated several times approaching the cross-wire intersection from alternate sides.

After the determination is complete, both prism faces should be cleaned with cotton wool soaked in a volatile solvent.

Determination of Optical Rotation

Most molecules which have no symmetry exhibit optical activity; that is, they rotate the plane of vibration of polarised light. The occurrence and magnitude of such effects may be observed using a polarimeter (Fig. 23) which may be employed for determining the specific rotation of optically active compounds, or for the determination of the concentration of solutions of known rotatory power.

Specific rotation, $[\alpha]$,

$$\text{for a pure liquid} = \frac{\alpha}{l \cdot d}$$

$$\text{for a compound in solution} = \frac{\alpha}{l \cdot c}$$

where α = observed angle of rotation in degrees, l = length tube in dm, d = density, c = concentration in g/cm^3.

The polarimeter consists essentially of a monochromatic light source (L; Fig. 23), a fixed nicol prism called the polariser P, which produces plane polarised light, a cell C in which is placed the substance, or

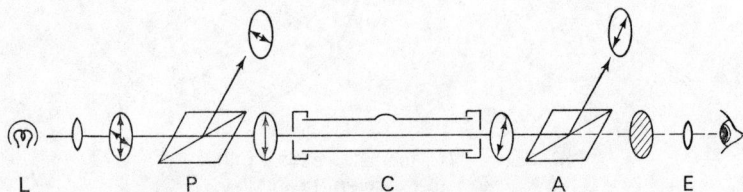

Fig. 23

solution of that substance, being examined, a second nicol prism called the analyser A which is connected to a 360° scale and an eyepiece E. The sodium D line is commonly used as the light source and the analyser and scale assembly is capable of rotation.

If the light is viewed through the eyepiece and the analyser slowly rotated, the field of view passes from an even brightness over the whole field, through a split image where one side of the field is bright and the other dark, to a state of minimum light transmittance with the field even again. This position of even darkness is 90° from the original even brightness position. Continuing to rotate the analyser, in the same direction as before, brings the other side of the split disc into darkness, and then eventually, at 180° from the original position of maximum brightness, even brightness is again obtained.

Procedure. If the specific rotation of a pure liquid is to be determined, then the empty cell is placed in the polarimeter and the zero position of the instrument determined. When the substance to be examined is in solution the cell is filled with pure solvent (commonly chloroform) prior to placing it in the instrument, taking care to exclude air from the field of view.

The cell (Fig. 24) is filled by unscrewing the cap X from one end of the cell, removing the glass disc D, filling the cell so that the surface of

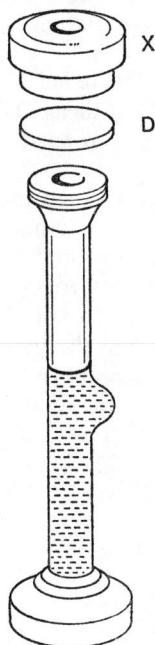

Fig. 24

the liquid forms a projecting meniscus, sliding the disc back into position and replacing the cap. Some polarimeter cells have a hump on the tube to accommodate residual bubbles of air.

The zero position of the polarimeter is obtained as follows. When the cell (containing solvent for determinations in solution) has attained room temperature the analyser is rotated in one direction until even darkness is obtained (this position is more easily discernible than that of maximum light transmittance) and the scale reading noted. The analyser is rotated slightly in the same direction as before, and the position of even darkness again obtained by approaching from the opposite direction. This series of operations is repeated several times until consistency is achieved whereupon the average of three pairs of readings is taken.

The cell is now filled with the pure liquid to be examined, or is washed out with, and then filled with, the solution under test. Care should be taken to fill the polarimeter cell rapidly when substances are being examined in volatile solvents, such as chloroform, as any delay may result in a change in the concentration of the solution being used.

When the cell has attained room temperature, the position of minimum light transmittance is obtained as before. The difference between the two values obtained gives the optical rotation (α) of the material in the cell, and the direction of rotation of the even darkness position indicates whether the material is dextro ($+$) or laevo ($-$) rotatory (clockwise and anticlockwise respectively).

The distance the light has travelled through the optically active medium (i.e. the internal length of the cell) is measured and the temperature of determination taken. The specific rotation is calculated as indicated previously (p. 266).

The whole procedure should be repeated with rather more dilute material if it is suspected that the compound has a sufficiently high specific rotation to make the magnitude or direction of rotation, as determined on the first sample, suspect.

Specific rotation is dependent to some extent on the solvent and wavelength of the light used, and the temperature and concentration of the solution. It is therefore necessary to specify these variables and results are therefore expressed in the following form, the concentration being given in g/dm^3.

$$[\alpha]_D^{20} = +66°, \text{ CHCl}_3 \text{ (1.06)}$$

Index